NEW! MyMathLab® Option Mathematical Ideas
by Miller/Heeren/Hornsby/Heeren

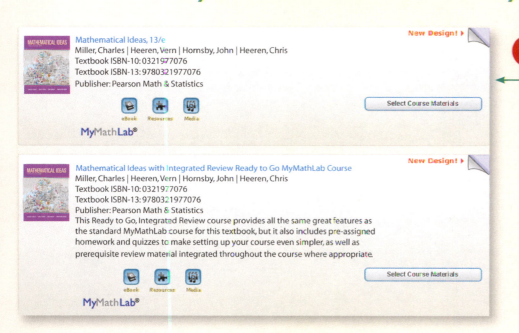

Mathematical Ideas, 13/e
Miller, Charles | Heeren, Vern | Hornsby, John | Heeren, Chris
Textbook ISBN-10: 0321977076
Textbook ISBN-13: 9780321977076
Publisher: Pearson Math & Statistics

New Design! ▸

eBook   Resources   Media

MyMath**Lab**®

Select Course Materials

Mathematical Ideas with Integrated Review Ready to Go MyMathLab Course
Miller, Charles | Heeren, Vern | Hornsby, John | Heeren, Chris
Textbook ISBN-10: 0321977076
Textbook ISBN-13: 9780321977076
Publisher: Pearson Math & Statistics
This Ready to Go, Integrated Review course provides all the same great features as the standard MyMathLab course for this textbook, but it also includes pre-assigned homework and quizzes to make setting up your course even simpler, as well as prerequisite review material integrated throughout the course where appropriate.

New Design! ▸

eBook   Resources   Media

MyMath**Lab**®

Select Course Materials

1 Standard MyMathLab
This option gives you maximum control over course organization and assignment creation.

2 MyMathLab with Integrated Review
This option offers a complete liberal arts math course along with integrated review of select topics from developmental math. Able to support a co-requisite course model, or any course where students will benefit from review of prerequisite skills, this Ready to Go **MyMathLab** solution includes prebuilt and pre-assigned assignments, making startup even easier!

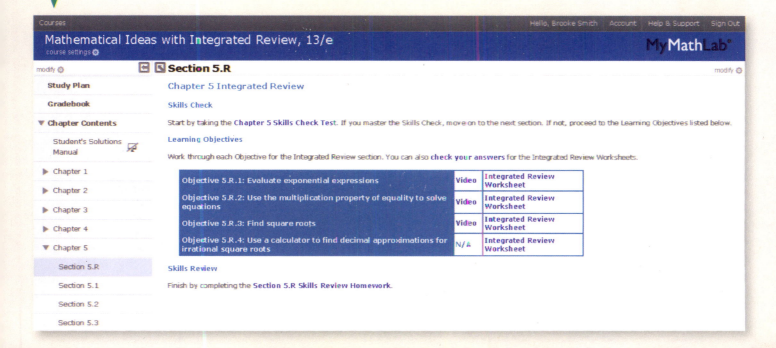

Courses   Hello, Brooke Smith | Account | Help & Support | Sign Out

Mathematical Ideas with Integrated Review, 13/e   **My**Math**Lab**®
course settings ⚙

modify ⚙   ◀ ▣ **Section 5.R**   modify ⚙

Study Plan	**Chapter 5 Integrated Review**
Gradebook	**Skills Check**
▼ Chapter Contents	Start by taking the **Chapter 5 Skills Check Test**. If you master the Skills Check, move on to the next section. If not, proceed to the Learning Objectives listed below.
Student's Solutions Manual	**Learning Objectives**
▶ Chapter 1	Work through each Objective for the Integrated Review section. You can also **check your answers** for the Integrated Review Worksheets.

Objective 5.R.1: Evaluate exponential expressions	Video	Integrated Review Worksheet
Objective 5.R.2: Use the multiplication property of equality to solve equations	Video	Integrated Review Worksheet
Objective 5.R.3: Find square roots	Video	Integrated Review Worksheet
Objective 5.R.4: Use a calculator to find decimal approximations for irrational square roots	N/A	Integrated Review Worksheet

▶ Chapter 2
▶ Chapter 3
▶ Chapter 4
▼ Chapter 5
  Section 5.R
  Section 5.1
  Section 5.2
  Section 5.3

Skills Review

Finish by completing the **Section 5.R Skills Review Homework**.

Mathematical Ideas captures the interest of non-majors by inspiring them to see mathematics as something interesting, relevant, and utterly practical. With a new focus on jobs, professions and careers as the context in which to frame the math, *Mathematical Ideas* demonstrates the importance math can play in everyday life while drawing students into the content.

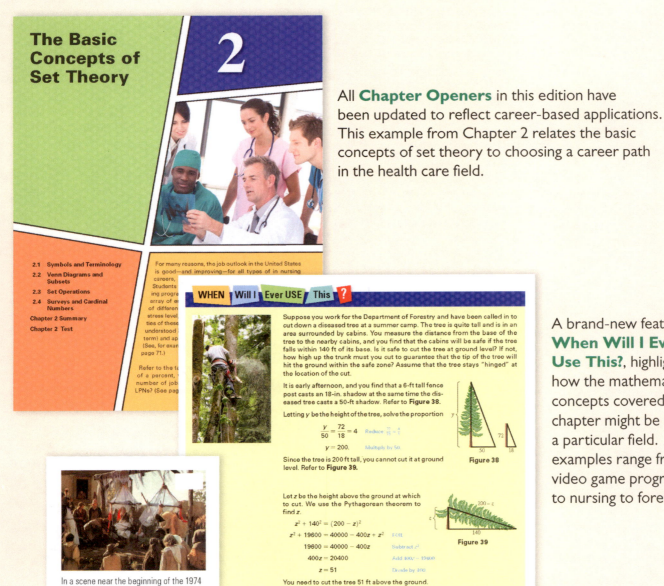

All **Chapter Openers** in this edition have been updated to reflect career-based applications. This example from Chapter 2 relates the basic concepts of set theory to choosing a career path in the health care field.

A brand-new feature, **When Will I Ever Use This?**, highlights how the mathematical concepts covered in the chapter might be used in a particular field. Career examples range from video game programming to nursing to forestry.

In a scene near the beginning of the 1974 film *Monty Python and the Holy Grail*, an amazing application of **poor logic** leads to the apparent demise of a supposed witch. Some peasants have forced a young woman to wear a nose made of wood. The convoluted argument they make is this: Witches and wood are both burned, and because witches are made of wood, and wood floats, and ducks also float, if she weighs the same as a duck, then she is made of wood and, therefore, is a witch!

Margin notes appear frequently throughout the text. These provide relevant examples from media and literature in addition to historical anecdotes and current research.

THIRTEENTH EDITION

MATHEMATICAL
IDEAS

Charles D. Miller

Vern E. Heeren
American River College

John Hornsby
University of New Orleans

Christopher Heeren
American River College

AND

Margaret L. Morrow
Pittsburgh State University of New York
for the chapter on Graph Theory

Jill Van Newenhizen
Lake Forest College
for the chapter on Voting and Apportionment

PEARSON

Boston   Columbus   Hoboken   Indianapolis   New York   San Francisco
Amsterdam   Cape Town   Dubai   London   Madrid   Milan   Munich   Paris   Montréal   Toronto
Delhi   Mexico City   São Paulo   Sydney   Hong Kong   Seoul   Singapore   Taipei   Tokyo

Editorial Director: Chris Hoag
Editor in Chief: Anne Kelly
Senior Acquisitions Editor: Marnie Greenhut
Editorial Assistant: Lucia Kim
Program Manager: Patty Bergin
Project Manager: Sherry Berg
Program Management Team Lead: Karen Wernholm
Project Management Team Lead: Peter Silvia
Media Producer: Nicholas Sweeny
TestGen Content Manager: John Flanagan
MathXL Content Developer: Bob Carroll
Marketing Manager: Alicia Frankel
Marketing Assistant: Brooke Smith
Senior Author Support/Technology Specialist: Joe Vetere
Rights and Permissions Project Manager: Diahanne Lucas
Senior Procurement Specialist: Carol Melville
Associate Director of Design: Andrea Nix
Program Design Lead: Beth Paquin
Text Design, Production Coordination, Composition, and Illustrations:
   Cenveo Publisher Services
Cover Design: Infiniti
Cover Image: Sergey Nivens/Shutterstock

Library of Congress Cataloging-in-Publication Data
Miller, Charles D. (Charles David), 1942-1986.
  Mathematical ideas. – 13th edition / Charles D. Miller, Vern E. Heeren, American River College, John Hornsby, University of New Orleans, Christopher Heeren, American River College.
    pages cm
  ISBN 0-321-97707-6 (student edition)
  1. Mathematics–Textbooks. I. Heeren, Vern E., author. II. Hornsby, John, 1949-III. Heeren, Christopher. IV. Title.
QA39.3.M55 2015
510–dc23   2014032895

www.pearsonhighered.com

ISBN 13: 978-0-321-97707-6
ISBN 10: 0-321-97707-6

Phor my phriend Phlash Phelps, whose Phunny Pharm helps me get through my mornings—JOHNNY (1119)

To my beloved wife, Carole, for decades of inspiration and support—VERN

To Heather, for your undying love and encouragement—CHRIS

CONTENTS

v

6 The Real Numbers and Their Representations 219

7 The Basic Concepts of Algebra 291

8 | Graphs, Functions, and Systems of Equations and Inequalities 365

11 Probability 583

12 Statistics 643

NOTE: Trigonometry module and Metrics module available in MyMathLab or online
at www.pearsonhighered.com/mathstatsresources.

PREFACE

After twelve editions and over four decades, *Mathematical Ideas* continues to be one of the most popular textbooks in liberal arts mathematics education. We are proud to present the thirteenth edition of a text that offers non-physical science students a practical coverage that connects mathematics to the world around them. It is a flexible text that has evolved alongside changing trends but remains steadfast to its original objectives.

Mathematical Ideas is written with a variety of students in mind. It is well suited for several courses, including those geared toward the aforementioned liberal arts audience and survey courses in mathematics or finite mathematics. Students taking these courses will pursue careers in nursing and healthcare, the construction trades, communications, hospitality, information technology, criminal justice, retail management and sales, computer programming, political science, school administration, and a myriad of other careers. Accordingly, we have chosen to increase our focus on showcasing how the math in this course will be relevant in this wide array of career options.

- Chapter openers now address how the chapter topics can be applied within the context of work and future careers.

- We made sure to retain the hundreds of examples and exercises from the previous edition that pertain to these interests.

- Every chapter also contains the brand new **When Will I Ever Use This?** features that help students connect mathematics to the workplace.

Interesting and mathematically pertinent movie and television applications and references are still interspersed throughout the chapters.

Ample topics are included for a two-term course, yet the variety of topics and flexibility of sequence makes the text suitable for shorter courses as well. Our main objectives continue to be comprehensive coverage, appropriate organization, clear exposition, an abundance of examples, and well-planned exercise sets with numerous applications.

New to This Edition

- New chapter openers connect the mathematics of the chapter to a particular career area, or in some cases, to an everyday life situation that will be important to people in virtually any career.

- *When Will I Ever Use This*? features in each chapter also connect chapter topics to career or workplace situations and answer that age-old question.

- Career applications have taken on greater prominence.

- Every section of every chapter now begins with a list of clear learning objectives for the student.

- An extensive summary at the end of each chapter includes the following components.

  ○ A list of **Key Terms** for each section of the chapter

  ○ **New Symbols,** with definitions, to clarify newly introduced symbols

  ○ **Test Your Word Power** questions that allow students to test their knowledge of new vocabulary

  ○ A **Quick Review** that gives a brief summary of concepts covered in the chapter, along with examples illustrating those concepts

- All exercise sets have once again been updated, with over 1000 new or modified exercises, many with a new emphasis on career applications.

- Since Intermediate Algebra is often a prerequisite for the liberal arts course, the algebra chapters have been streamlined to focus in on key concepts, many of which will aid in comprehension of other chapters' content.

- The presentation has been made more uniform whenever clarity for the reader could be served.

- The general style has been freshened, with more pedagogical use of color, new photos and art, and opening of the exposition.

- NEW! An Integrated Review MyMathLab course option provides embedded review of select developmental topics in a Ready to Go format with assignments pre-assigned. This course solution can be used in a co-requisite course model, or simply to help under-prepared students master prerequisite skills and concepts.

- Expanded online resources

  - **NEW! Interactive, conceptual videos** with assignable MML questions walk students through a concept and then ask them to answer a question within the video. If students answer correctly, the concept is summarized. If students select one of the two incorrect answers, the video continues focusing on why students probably selected that answer and works to correct that line of thinking and explain the concept. Then students get another chance to answer a question to prove mastery.

  - **NEW! Learning Catalytics**   This student engagement, assessment and classroom intelligence system gives instructors real-time feedback on student learning.

  - **NEW! "When Will I Ever Use This?" videos** bring the ideas in the feature to life in a fun, memorable way.

  - **NEW! An Integrated Review MyMathLab** course option provides an embedded review of selected developmental topics. Assignments are pre-assigned in this course, which includes a Skills Check quiz on skills that students will need in order to learn effectively at the chapter level. Students who demonstrate mastery can move on to the *Mathematical Ideas* content, while students who need additional review can polish up their skills by using the videos supplied and can benefit from the practice they gain from the Integrated Review Worksheets. This course solution can be used either in a co-requisite course model, or simply to help underprepared students master prerequisite skills and concepts.

  - The Trigonometry and Metrics content that was previously in the text is now found in the MyMathLab course, including the assignable MML questions.

  - Extensions previously in the text are now found in the MyMathLab course, along with any assignable MML questions.

Overview of Chapters

- **Chapter 1 (The Art of Problem Solving)** introduces the student to inductive reasoning, pattern recognition, and problem-solving techniques. We continue to provide exercises based on the monthly Calendar from *Mathematics Teacher* and have added new ones throughout this edition. The new chapter opener recounts the solving of the Rubik's cube by a college professor. The *When Will I Ever Use This?* feature (p. 31) shows how estimation techniques may be used by a group home employee charged with holiday grocery shopping.

- **Chapter 2 (The Basic Concepts of Set Theory)** includes updated examples and exercises on surveys. The chapter opener and the *When Will I Ever Use This?* feature (p. 73) address the future job outlook for the nursing profession and the allocation of work crews in the building trade, respectively.

- **Chapter 3 (Introduction to Logic)** introduces the fundamental concepts of inductive and deductive logic. The chapter opener connects logic with fantasy literature, and new exercises further illustrate this relationship. A new *For Further Thought* (p. 99) and new exercises address logic gates in computers. One *When Will I Ever Use This?* feature (p. 108) connects circuit logic to the design and installation of home monitoring systems. Another (p. 128) shows a pediatric nurse applying a logical flowchart and truth tables to a child's vaccination protocol.

- **Chapter 4 (Numeration Systems)** covers historical numeration systems, including Egyptian, Roman, Chinese, Babylonian, Mayan, Greek, and Hindu-Arabic systems. A connection between base conversions in positional numeration systems and computer network design is suggested in the new chapter opener and illustrated in the *When Will I Ever Use This?* feature (p. 168), a new example, and new exercises.

- **Chapter 5 (Number Theory)** presents an introduction to the prime and composite numbers, the Fibonacci sequence, and a cross section of related historical developments, including the fairly new topic of "prime number splicing." The largest currently known prime numbers of various categories are identified, and recent progress on Goldbach's conjecture and the twin prime conjecture are noted. The chapter opener and one *When Will I Ever Use This?* feature (p. 189) apply cryptography and modular arithmetic to criminal justice, relating to cyber security. Another *When Will I Ever Use This?* feature (p. 205) shows how a nurse may use the concept of least common denominator in determining proper drug dosage.

- **Chapter 6 (The Real Numbers and Their Representations)** introduces some of the basic concepts of real numbers, their various forms of representation, and operations of arithmetic with them. The chapter opener and *When Will I Ever Use This?* feature (p. 273) connect percents and basic algebraic procedures to pricing, markup and discount, student grading, and market share analysis, as needed by a retail manager, a teacher, a salesperson, a fashion merchandiser, and a business owner.

- **Chapter 7 (The Basic Concepts of Algebra)** can be used to present the basics of algebra (linear and quadratic equations, applications, exponents, polynomials, and factoring) to students for the first time, or as a review of previous courses. The chapter opener connects proportions to an automobile owner's determination of fuel mileage, and the *When Will I Ever Use This?* feature (p. 330) relates inequalities to a test-taker's computation of the score needed to maintain a certain grade point average.

- **Chapter 8 (Graphs, Functions, and Systems of Equations and Inequalities)** is the second of our two algebra chapters. It continues with graphs, equations, and applications of linear, quadratic, exponential, and logarithmic functions and models, along with systems of equations. The chapter opener shows how an automobile owner can use a linear graph to relate price per gallon, amount purchased, and total cost. The *When Will I Ever Use This?* feature (p. 416) connects logarithms with the interpretation of earthquake reporting in the news.

- **Chapter 9 (Geometry)** covers elementary plane geometry, transformational geometry, basic geometric constructions, non-Euclidean geometry, and chaos and fractals. Section 9.7 now includes projective geometry. At reviewer request, the discussion of networks (the Königsberg Bridge problem) has been moved to Chapter 14 (Graph Theory). The chapter opener and one *When Will I Ever Use This?* feature (p. 497) connect geometric volume formulas to a video game programmer's job of designing the visual field of a game screen. A second *When Will I Ever Use This?* feature (p. 470) relates right triangle geometry to a forester's determining of safe tree-felling parameters.

- **Chapter 10 (Counting Methods)** focuses on elementary counting techniques, in preparation for the probability chapter. The chapter opener relates how a restaurateur used counting methods to help design the sales counter signage in a new restaurant. The *When Will I Ever Use This?* feature (p. 534) describes an entrepreneur's use of probability and sports statistics in designing a game and in building a successful company based on it.

- **Chapter 11 (Probability)** covers the basics of probability, odds, and expected value. The chapter opener relates to the professions of weather forecaster, actuary, baseball manager, and corporate manager, applying probability, statistics, and expected value to interpreting forecasts, determining insurance rates, selecting optimum strategies, and making business decisions. One *When Will I Ever Use This?* feature (p. 586) shows how a tree diagram helps a decision maker provide equal chances of winning to three players in a game of chance. A second such feature (p. 606) shows how knowledge of probability can help a television game show contestant determine the best winning strategy.

- **Chapter 12 (Statistics)** is an introduction to statistics that focuses on the measures of central tendency, dispersion, and position and discusses the normal distribution and its applications. The chapter opener and two *When Will I Ever Use This?* features (pp. 656, 661) connect probability and graph construction and interpretation to how a psychological therapist may motivate and carry out treatment for alcohol and tobacco addiction.

- **Chapter 13 (Personal Financial Management)** provides the student with the basics of the mathematics of finance as applied to inflation, consumer debt, and house buying. We also include a section on investing, with emphasis on stocks, bonds, and mutual funds. Tables, examples, and exercises have been updated to reflect current interest rates and investment returns. New margin notes feature smart apps for financial calculations. Additions in response to reviewer requests include a *When Will I Ever Use This?* feature (p. 741) connecting several topics of the chapter to how a financial planner can provide comparisons between renting and buying a house, and exercises comparing different mortgage options. Another *When Will I Ever Use This?* feature (p. 732) explores the cost-effectiveness of solar energy, using chapter topics essential for a solar energy salesperson. The chapter opener connects the time value of money to how a financial planner can help clients make wise financial decisions.

- **Chapter 14 (Graph Theory)** covers the basic concepts of graph theory and its applications. The chapter opener shows how a writer can apply graph theory to the analysis of poetic rhyme. One *When Will I Ever Use This?* feature (p. 800) connects graph theory to how a postal or delivery service manager could determine the most efficient delivery routes. Another (p. 818) tells of a unique use by an entrepreneur who developed a business based on finding time-efficient ways to navigate theme parks.

- **Chapter 15 (Voting and Apportionment)** deals with issues in voting methods and apportionment of representation, topics that have become increasingly popular in liberal arts mathematics courses. The Adams method of apportionment, as well as the Huntington-Hill method (currently used in United States presidential elections) are now included in the main body of the text. To illustrate the important work of a political consultant, the chapter opener connects different methods of analyzing votes. One *When Will I Ever Use This?* feature (p. 859) relates voting methods to the functioning of governing boards. Another (p. 874) gives an example of how understanding apportionment methods can help in the work of a school administrator.

Course Outline Considerations

Chapters in the text are, in most cases, independent and may be covered in the order chosen by the instructor. The few exceptions are as follows:

- Chapter 6 contains some material dependent on the ideas found in Chapter 5.
- Chapter 6 should be covered before Chapter 7 if student background so dictates.
- Chapters 7 and 8 form an algebraic "package" and should be covered in sequential order.
- A thorough coverage of Chapter 11 depends on knowledge of Chapter 10 material, although probability can be covered without teaching extensive counting methods by avoiding the more difficult exercises.

Features of the Thirteenth Edition

NEW! Chapter Openers In keeping with the career theme, chapter openers address a situation related to a particular career. All are new to this edition. Some openers include a problem that the reader is asked to solve. We hope that you find these chapter openers useful and practical.

ENHANCED! Varied Exercise Sets We continue to present a variety of exercises that integrate drill, conceptual, and applied problems, and there are over 1000 new or modified exercises in this edition. The text contains a wealth of exercises to provide students with opportunities to practice, apply, connect, and extend the mathematical skills they are learning. We have updated the exercises that focus on real-life data and have retained their titles for easy identification. Several chapters are enriched with new applications, particularly Chapters 6, 7, 8, 11, 12, and 13. We continue to use graphs, tables, and charts when appropriate. Many of the graphs use a style similar to that seen by students in today's print and electronic media.

UPDATED! Emphasis on Real Data in the Form of Graphs, Charts, and Tables We continue to use up-to-date information from magazines, newspapers, and the Internet to create real applications that are relevant and meaningful.

Problem-Solving Strategies Special paragraphs labeled "Problem-Solving Strategy" relate the discussion of problem-solving strategies to techniques that have been presented earlier.

For Further Thought These entries encourage students to share their reasoning processes among themselves to gain a deeper understanding of key mathematical concepts.

ENHANCED! Margin Notes This popular feature is a hallmark of this text and has been retained and updated where appropriate. These notes are interspersed throughout the text and are drawn from various sources, such as lives of mathematicians, historical vignettes, anecdotes on mathematics textbooks of the past, newspaper and magazine articles, and current research in mathematics.

Optional Graphing Technology We continue to provide sample graphing calculator screens to show how technology can be used to support results found analytically. It is not essential, however, that a student have a graphing calculator to study from this text. *The technology component is optional.*

NEW! Chapter Summaries Extensive summaries at the end of each chapter include Key Terms, New Symbols with definitions, Test Your Word Power vocabulary checks, and a Quick Review that provides a brief summary of concepts (with examples) covered in the chapter.

Chapter Tests Each chapter concludes with a chapter test so that students can check their mastery of the material.

Resources For Success

MyMathLab® Online Course (access code required)

MyMathLab delivers **proven results** in helping individual students succeed. It provides **engaging experiences** that personalize, stimulate, and measure learning for each student. And it comes from an **experienced partner** with educational expertise and an eye on the future. **MyMathLab** helps prepare students and gets them thinking more conceptually and visually through the following features:

Personalized Homework ▶

Attaching a personalized homework assignment to a quiz or test enables students to focus on the topics they did not master and those with which they need additional practice, providing an individualized experience.

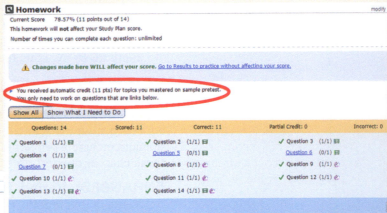

◀ Adaptive Study Plan

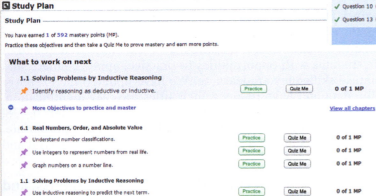

The Study Plan makes studying more efficient and effective for every student. Performance and activity are assessed continually in real time. Then data and analytics are used to provide personalized content-reinforcing concepts that target each student's strengths and weaknesses.

◀ Interactive Concept Videos

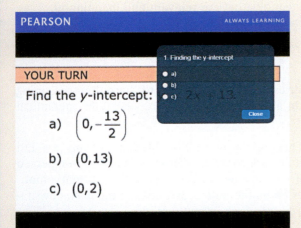

NEW! Conceptual videos require students' input as they walk students through a concept and pause to ask them questions. Correct answers are followed by reinforcement of the concept. Incorrect answers are followed by a video that explains the correct answer, as well as addressing the misconception that may have led to that particular mistake. Instructors can also assign trackable exercises that correspond to the videos to check student understanding.

Learning Catalytics ▶

NEW! Integrated into **MyMathLab**, the Learning Catalytics feature uses students' devices in the classroom for an engagement, assessment, and classroom intelligence system that gives instructors real-time feedback on student learning.

Skills for Success Modules are integrated within the **MyMathLab** course to help students succeed in college courses and prepare for future professions.

Instructor Resources

Additional resources can be downloaded from www.pearsonhighered.com or hardcopy resources can be ordered from your sales representative.

Integrated Review Ready to Go MyMathLab® Course

The Ready to Go **MyMathLab** course option makes it even easier to get started, including author-chosen preassigned homework, integrated review of prerequisite topics, and more.

TestGen®

TestGen® (www.pearsoned.com/testgen) enables instructors to build, edit, print, and administer tests using a computerized bank of questions developed to cover all the objectives of the text.

PowerPoint® Lecture Slides

Fully editable slides correlated with the textbook are available.

Annotated Instructor's Edition

When possible, answers are on the page with the exercises. Longer answers are in the back of the book.

Instructor's Resource and Solutions Manual

This manual includes fully worked solutions to all text exercises, as well as the Collaborative Investigations that were formerly in the text.

Instructor's Testing Manual

This manual includes tests with answer keys for each chapter of the text.

Student Resources

Additional resources are available to support student success.

Updated Video Program

Available in **MyMathLab**, video lectures cover every section in the text and have been updated for this edition where necessary. New interactive concept videos and new "*When Will I Ever Use This?*" videos complete the video package, reinforcing students' conceptual understanding, while also engaging them with the math in context.

Student Solutions Manual

This manual provides detailed worked-out solutions to odd-numbered exercises.

Integrated Review Worksheets

Intended to be used with the Integrated Review **MyMathLab** course, these worksheets give students an opportunity to review and practice prerequisite topics from developmental math that are needed for each chapter in *Mathematical Ideas*.

ACKNOWLEDGMENTS

We wish to thank the following reviewers for their helpful comments and suggestions for this and previous editions of the text. (Reviewers of the thirteenth edition are noted with an asterisk.)

H. Achepohl, *College of DuPage*

Shahrokh Ahmadi, *Northern Virginia Community College*

Richard Andrews, *Florida A&M University*

Cindy Anfinson, *Palomar College*

Erika Asano, *University of South Florida, St. Petersburg*

Elaine Barber, *Germanna Community College*

Anna Baumgartner, *Carthage College*

James E. Beamer, *Northeastern State University*

*Brad Beauchamp, *Vernon College*

Elliot Benjamin, *Unity College*

Jaime Bestard, *Barry University*

Joyce Blair, *Belmont University*

Gus Brar, *Delaware County Community College*

Roger L. Brown, *Davenport College*

Douglas Burke, *Malcolm X College*

John Busovicki, *Indiana University of Pennsylvania*

Ann Cascarelle, *St. Petersburg Junior College*

Kenneth Chapman, *St. Petersburg Junior College*

Gordon M. Clarke, *University of the Incarnate Word*

M. Marsha Cupitt, *Durham Technical Community College*

James Curry, *American River College*

Rosemary Danaher, *Sacred Heart University*

Ken Davis, *Mesa State College*

Nancy Davis, *Brunswick Community College*

George DeRise, *Thomas Nelson Community College*

Catherine Dermott, *Hudson Valley Community College*

Greg Dietrich, *Florida Community College at Jacksonville*

Vincent Dimiceli, *Oral Roberts University*

*Qiang Dotzel, *University of Missouri, St. Louis*

Diana C. Dwan, *Yavapai College*

Laura Dyer, *Belleville Area College*

Jan Eardley, *Barat College*

Joe Eitel, *Folsom College*

Azin Enshai, *American River College*

Gayle Farmer, *Northeastern State University*

Michael Farndale, *Waldorf College*

Gordon Feathers, *Passaic County Community College*

Thomas Flohr, *New River Community College*

Bill Fulton, *Black Hawk College—East*

Anne Gardner, *Wenatchee Valley College*

Justin M. Gash, *Franklin College*

Donald Goral, *Northern Virginia Community College*

Glen Granzow, *Idaho State University*

Larry Green, *Lake Tahoe Community College*

Arthur D. Grissinger, *Lock Haven University*

Don Hancock, *Pepperdine University*

Denis Hanson, *University of Regina*

Marilyn Hasty, *Southern Illinois University*

Shelby L. Hawthorne, *Thomas Nelson Community College*

Jeff Heiking, *St. Petersburg Junior College*

Laura Hillerbrand, *Broward Community College*

Corinne Irwin, *University of Texas at Austin*

*Neha Jain, *Northern Virginia Community College*

Jacqueline Jensen, *Sam Houston State University*

Emanuel Jinich, *Endicott College*

Frank Juric, *Brevard Community College–Palm Bay*

Karla Karstens, *University of Vermont*

Najam Khaja, *Centennial College*

Hilary Kight, *Wesleyan College*

Barbara J. Kniepkamp, *Southern Illinois University at Edwardsville*

Suda Kunyosying, *Shepherd College*

Yu-Ju Kuo, *Indiana University of Pennsylvania*

Stephane Lafortune, *College of Charleston*

Pam Lamb, *J. Sargeant Reynolds Community College*

John Lattanzio, *Indiana University of Pennsylvania*

John W. Legge, *Pikeville College*

Dawn Locklear, *Crown College*

Bin Lu, *California State University, Sacramento*

Leo Lusk, *Gulf Coast Community College*

Sherrie Lutsch, *Northwest Indian College*

Rhonda Macleod, *Florida State University*

Andrew Markoe, *Rider University*

Darlene Marnich, *Point Park College*

Victoria Martinez, *Okaloosa Walton Community College*

Chris Mason, *Community College of Vermont*

Mark Maxwell, *Maryville University*

Carol McCarron, *Harrisburg Area Community College*

Delois McCormick, *Germanna Community College*

Daisy McCoy, *Lyndon State College*

Cynthia McGinnis, *Okaloosa Walton Community College*

Vena McGrath, *Davenport College*

Robert Moyer, *Fort Valley State University*

Shai Neumann, *Brevard Community College*

Barbara Nienstedt, *Gloucester County College*

Chaitanya Nigam, *Gateway Community-Technical College*

Vladimir Nikiforov, *University of Memphis*

Vicky Ohlson, *Trenholm State Technical College*

Jean Okumura, *Windward Community College*

Stan Perrine, *Charleston Southern University*

*Mary Anne Petruska, *Pensacola State College*

Bob Phillips, *Mesabi Range Community College*

Kathy Pinchback, *University of Memphis*

*Fatima Prioleau, *Borough of Manhattan Community College*

Priscilla Putman, *New Jersey City University*

Scott C. Radtke, *Davenport College*

Doraiswamy Ramachandran, *California State University, Sacramento*

John Reily, *Montclair State University*

Beth Reynolds, *Mater Dei College*

Shirley I. Robertson, *High Point University*

Andrew M. Rockett, *CW Post Campus of Long Island University*

Kathleen Rodak, *St. Mary's College of Ave Maria University*

Cynthia Roemer, *Union County College*

Lisa Rombes, *Washtenaw Community College*

Abby Roscum, *Marshalltown Community College*

*Catherine A. Sausville, *George Mason University*

D. Schraeder, *McLennan Community College*

Wilfred Schulte, *Cosumnes River College*

Melinda Schulteis, *Concordia University*

Gary D. Shaffer, *Allegany College of Maryland*

Doug Shaw, *University of North Iowa*

Jane Sinibaldi, *York College of Pennsylvania*

Nancy Skocik, *California University of Pennsylvania*

Larry Smith, *Peninsula College*

Marguerite Smith, *Merced College*

Charlene D. Snow, *Lower Columbia College*

H. Jeannette Stephens, *Whatcom Community College*

Suzanne J. Stock, *Oakton Community College*

Dawn M. Strickland, *Winthrop University*

Dian Thom, *McKendree College*

Claude C. Thompson, *Hollins University*

Mark Tom, *College of the Sequoias*

Ida Umphers, *University of Arkansas at Little Rock*

Karen Villarreal, *University of New Orleans*

Dr. Karen Walters, *Northern Virginia Community College*

Wayne Wanamaker, *Central Florida Community College*

David Wasilewski, *Luzerne County Community College*

William Watkins, *California State University, Northridge*

Alice Williamson, *Sussex County Community College*

Susan Williford, *Columbia State Community College*

Tom Witten, *Southwest Virginia Community College*

Fred Worth, *Henderson State University*

Rob Wylie, *Carl Albert State College*

Henry Wyzinski, *Indiana University Northwest*

A project of this magnitude cannot be accomplished without the help of many other dedicated individuals. Marnie Greenhut served as acquisitions editor for this edition. Carol Merrigan provided excellent production supervision. Anne Kelly, Greg Tobin, Sherry Berg, Alicia Frankel, and Patty Bergin of Pearson gave us their unwavering support.

Terry McGinnis gave her usual excellent behind-the-scenes guidance. Thanks go to Dr. Margaret L. Morrow of Plattsburgh State University and Dr. Jill Van Newenhizen of Lake Forest College, who wrote the material on graph theory and voting/apportionment, respectively. Paul Lorczak, Renato Mirollo, Beverly Fusfield, Jack Hornsby, and Perian Herring did an outstanding job of accuracy- and answer-checking. And finally, we thank our loyal users over these many editions for making this book one of the most successful in its market.

Vern E. Heeren
John Hornsby
Christopher Heeren

ABOUT THE AUTHORS

Vern Heeren grew up in the Sacramento Valley of California. After earning a Bachelor of Arts degree in mathematics, with a minor in physics, at Occidental College, and completing his Master of Arts degree in mathematics at the University of California, Davis, he began a 38-year teaching career at American River College, teaching math and a little physics. He coauthored *Mathematical Ideas* in 1968 with office mate Charles Miller, and he has enjoyed researching and revising it over the years. It has been a joy for him to complete the thirteenth edition, along with long-time coauthor John Hornsby, and now also with son Christopher.

These days, besides pursuing his mathematical interests, Vern enjoys spending time with his wife Carole and their family, exploring the wonders of nature near their home in central Oregon.

John Hornsby joined the author team of Margaret Lial, Charles Miller, and Vern Heeren in 1988. In 1990, the sixth edition of *Mathematical Ideas* became the first of nearly 150 titles he has coauthored for Scott Foresman, HarperCollins, Addison-Wesley, and Pearson in the years that have followed. His books cover the areas of developmental and college algebra, precalculus, trigonometry, and mathematics for the liberal arts. He is a native and resident of New Roads, Louisiana.

Christopher Heeren is a native of Sacramento, California. While studying engineering in college, he had an opportunity to teach a math class at a local high school, and this sparked both a passion for teaching and a change of major. He received a Bachelor of Arts degree and a Master of Arts degree, both in mathematics, from California State University, Sacramento. Chris has taught mathematics at the middle school, high school, and college levels, and he currently teaches at American River College in Sacramento. He has a continuing interest in using technology to bring mathematics to life. When not writing, teaching, or preparing to teach, Chris enjoys spending time with his lovely wife Heather and their three children (and two dogs and a guinea pig).

The Art of Problem Solving

1

Professor Terry Krieger, of Rochester (Minnesota) Community College, shares his thoughts about why he decided to become a mathematics teacher. He is an expert at the Rubik's Cube. Here, he explains how he mastered this classic problem.

*From a very young age I always enjoyed solving problems, especially problems involving numbers and patterns. There is something inherently beautiful in the process of discovering mathematical truth. Mathematics may be the only discipline in which different people, using wildly varied but logically sound methods, will arrive at the **same** correct result—not just once, but every time! It is this aspect of mathematics that led me to my career as an educator. As a mathematics instructor, I get to be part of, and sometimes guide, the discovery process.*

I received a Rubik's Cube as a gift my junior year of high school. I was fascinated by it. I devoted the better part of three months to solving it for the first time, sometimes working 3 or 4 hours per day on it.

There was a lot of trial and error involved. I devised a process that allowed me to move only a small number of pieces at a time while keeping other pieces in their places. Most of my moves affect only three or four of the 26 unique pieces of the puzzle. What sets my solution apart from those found in many books is that I hold the cube in a consistent position and work from the top to the bottom. Most book solutions work upward from the bottom.

My first breakthrough came when I realized that getting a single color on one face of the cube was not helpful if the colors along the edges of that face were placed improperly. In other words, it does no good to make the top of the cube all white if one of the edges along the white top shows green, yellow, and blue. It needs to be all green, for example.

I worked on the solution so much that I started seeing cube moves in my sleep. In fact, I figured out the moves for one of my most frustrating sticking points while sleeping. I just woke up knowing how to do it.

The eight corners of the cube represented a particularly difficult challenge for me. Finding a consistent method for placing the corners appropriately took many, many hours. To this day, the amount of time that it takes for me to solve a scrambled cube depends largely on the amount of time that it takes for me to place the corners.

When I first honed my technique, I was able to consistently solve the cube in 2 to 3 minutes. My average time is now about 65 seconds. My fastest time is 42 seconds.

Since figuring out how to solve the cube, I have experimented with other possible color patterns that can be formed. The most complicated one I have created leaves the cube with three different color stripes on all six faces. I have never met another person who can accomplish this arrangement.

1.1   SOLVING PROBLEMS BY INDUCTIVE REASONING

OBJECTIVES

1   Be able to distinguish between inductive and deductive reasoning.

2   Understand that in some cases, inductive reasoning may not lead to valid conclusions.

Characteristics of Inductive and Deductive Reasoning

The development of mathematics can be traced to the Egyptian and Babylonian cultures (3000 B.C.–A.D. 260) as a necessity for counting and problem solving. To solve a problem, a cookbook-like recipe was given, and it was followed repeatedly to solve similar problems. By observing that a specific method worked for a certain type of problem, the Babylonians and the Egyptians concluded that the same method would work for any similar type of problem. Such a conclusion is called a *conjecture*. **A conjecture** is an educated guess based on repeated observations of a particular process or pattern.

The method of reasoning just described is called *inductive reasoning*.

INDUCTIVE REASONING

Inductive reasoning is characterized by drawing a general conclusion (making a conjecture) from repeated observations of specific examples. The conjecture may or may not be true.

In testing a conjecture obtained by inductive reasoning, it takes only one example that does not work to prove the conjecture false. Such an example is called a **counterexample.**

Inductive reasoning provides a powerful method of drawing conclusions, but there is no assurance that the observed conjecture will always be true. For this reason, mathematicians are reluctant to accept a conjecture as an absolute truth until it is formally proved using methods of *deductive reasoning*. Deductive reasoning characterized the development and approach of Greek mathematics, as seen in the works of Euclid, Pythagoras, Archimedes, and others. During the classical Greek period (600 B.C.–A.D. 450), general concepts were applied to specific problems, resulting in a structured, logical development of mathematics.

DEDUCTIVE REASONING

Deductive reasoning is characterized by applying general principles to specific examples.

We now look at examples of these two types of reasoning. In this chapter, we often refer to the **natural, or counting, numbers.**

$$1, 2, 3, \ldots \quad \text{Natural (counting) numbers}$$

$$\uparrow$$

Ellipsis points

The three dots (*ellipsis points*) indicate that the numbers continue indefinitely in the pattern that has been established. The most probable rule for continuing this pattern is "Add 1 to the previous number," and this is indeed the rule that we follow.

Now consider the following list of natural numbers:

$$2, 9, 16, 23, 30.$$

What is the next number of this list? What is the pattern? After studying the numbers, we might see that $2 + 7 = 9$, and $9 + 7 = 16$. Do we add 16 and 7 to get 23? Do we add 23 and 7 to get 30? Yes. It seems that any number in the given list can be found by adding 7 to the preceding number, so the next number in the list would be $30 + 7 = 37$.

We set out to find the "next number" by reasoning from observation of the numbers in the list. We may have jumped from these observations to the general statement that any number in the list is 7 more than the preceding number. This is an example of inductive reasoning.

By using inductive reasoning, we concluded that 37 was the next number. Suppose the person making up the list has another answer in mind. The list of numbers

$$2, 9, 16, 23, 30$$

actually gives the dates of Mondays in June if June 1 falls on a Sunday. The next Monday after June 30 is July 7. With this pattern, the list continues as

$$2, 9, 16, 23, 30, 7, 14, 21, 28, \ldots.$$

See the calendar in **Figure 1.** The correct answer would then be 7. The process used to obtain the rule "add 7" in the preceding list reveals a main flaw of inductive reasoning. *We can never be sure that what is true in a specific case will be true in general. Inductive reasoning does not guarantee a true result, but it does provide a means of making a conjecture.*

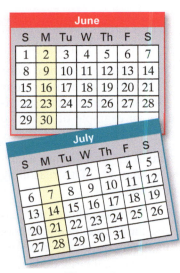

Figure 1

We now review some basic notation. Throughout this book, we use *exponents* to represent repeated multiplication.

$$\text{Base} \rightarrow 4^3 = 4 \cdot 4 \cdot 4 = 64 \qquad \text{4 is used as a factor 3 times.}$$
$$\uparrow$$
$$\text{Exponent}$$

EXPONENTIAL EXPRESSION

If a is a number and n is a counting number $(1, 2, 3, \dots)$, then the exponential expression a^n is defined as follows.

$$a^n = \underbrace{a \cdot a \cdot a \cdot \dots \cdot a}_{n \text{ factors of } a}$$

The number a is the **base** and n is the **exponent.**

With deductive reasoning, we use general statements and apply them to specific situations. For example, a basic rule for converting feet to inches is to multiply the number of feet by 12 in order to obtain the equivalent number of inches. This can be expressed as a formula.

$$\text{Number of inches} = 12 \times \text{number of feet}$$

This general rule can be applied to any specific case. For example, the number of inches in 3 feet is $12 \times 3 = 36$ inches.

Reasoning through a problem usually requires certain *premises*. A **premise** can be an assumption, law, rule, widely held idea, or observation. Then reason inductively or deductively from the premises to obtain a **conclusion.** The premises and conclusion make up a **logical argument.**

EXAMPLE 1 Identifying Premises and Conclusions

Identify each premise and the conclusion in each of the following arguments. Then tell whether each argument is an example of inductive or deductive reasoning.

(a) Our house is made of brick. Both of my next-door neighbors have brick houses. Therefore, all houses in our neighborhood are made of brick.

(b) All keyboards have the symbol @. I have a keyboard. My keyboard has the symbol @.

(c) Today is Tuesday. Tomorrow will be Wednesday.

Solution

(a) The premises are "Our house is made of brick" and "Both of my next-door neighbors have brick houses." The conclusion is "Therefore, all houses in our neighborhood are made of brick." Because the reasoning goes from specific examples to a general statement, the argument is an example of inductive reasoning (although it may very well be faulty).

(b) Here the premises are "All keyboards have the symbol @" and "I have a keyboard." The conclusion is "My keyboard has the symbol @." This reasoning goes from general to specific, so deductive reasoning was used.

(c) There is only one premise here, "Today is Tuesday." The conclusion is "Tomorrow will be Wednesday." The fact that Wednesday immediately follows Tuesday is being used, even though this fact is not explicitly stated. Because the conclusion comes from general facts that apply to this special case, deductive reasoning was used. ∎

While inductive reasoning may, at times, lead to false conclusions, in many cases it does provide correct results if we look for the most *probable* answer.

The Fibonacci Sequence

$$1, 1, 2, 3, 5, 8, 13, 21, \ldots$$

In the 2003 movie *A Wrinkle in Time*, young Charles Wallace, played by David Dorfman, is challenged to identify a particular sequence of numbers. He correctly identifies it as the **Fibonacci sequence.**

EXAMPLE 2   Predicting the Next Number in a Sequence

Use inductive reasoning to determine the *probable* next number in each list below.

(a) 5, 9, 13, 17, 21, 25, 29   **(b)** 1, 1, 2, 3, 5, 8, 13, 21   **(c)** 2, 4, 8, 16, 32

Solution

(a) Each number in the list is obtained by adding 4 to the previous number. The probable next number is $29 + 4 = 33$. (This is an example of an *arithmetic sequence*.)

(b) Beginning with the third number in the list, 2, each number is obtained by adding the two previous numbers in the list. That is,

$$1 + 1 = 2, \quad 1 + 2 = 3, \quad 2 + 3 = 5,$$

and so on. The probable next number in the list is $13 + 21 = 34$. (These are the first few terms of the *Fibonacci sequence*.)

(c) It appears here that to obtain each number after the first, we must double the previous number. Therefore, the probable next number is $32 \times 2 = 64$. (This is an example of a *geometric sequence*.) ■

EXAMPLE 3   Predicting the Product of Two Numbers

Consider the list of equations. Predict the next multiplication fact in the list.

$$37 \times \ \ 3 = 111$$
$$37 \times \ \ 6 = 222$$
$$37 \times \ \ 9 = 333$$
$$37 \times 12 = 444$$

Solution

The left side of each equation has two factors, the first 37 and the second a multiple of 3, beginning with 3. Each product (answer) consists of three digits, all the same, beginning with 111 for 37×3. Thus, the next multiplication fact would be

$$37 \times 15 = 555, \quad \text{which is indeed true.}$$ ■

Pitfalls of Inductive Reasoning

There are pitfalls associated with inductive reasoning. A classic example involves the maximum number of regions formed when chords are constructed in a circle. When two points on a circle are joined with a line segment, a *chord* is formed.

Locate a single point on a circle. Because no chords are formed, a single interior region is formed. See **Figure 2(a)** on the next page. Locate two points and draw a chord. Two interior regions are formed, as shown in **Figure 2(b)**. Continue this pattern. Locate three points, and draw all possible chords. Four interior regions are formed, as shown in **Figure 2(c)**. Four points yield 8 regions and five points yield 16 regions. See **Figures 2(d) and 2(e).**

The results of the preceding observations are summarized in **Table 1.** The pattern formed in the column headed "Number of Regions" is the same one we saw in **Example 2(c),** where we predicted that the next number would be 64. It seems here that for each additional point on the circle, the number of regions doubles.

Table 1

Number of Points	Number of Regions
1	1
2	2
3	4
4	8
5	16

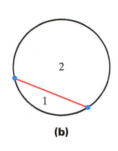 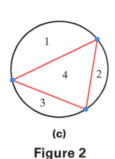 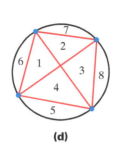 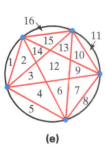

(a)   (b)   (c)   (d)   (e)

Figure 2

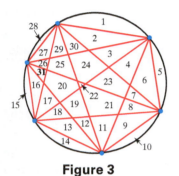

Figure 3

A reasonable inductive conjecture would be that for six points, 32 regions would be formed. But as **Figure 3** indicates, there are *only 31 regions*. The pattern of doubling ends when the sixth point is considered. Adding a seventh point would yield 57 regions. The numbers obtained here are

$$1, 2, 4, 8, 16, 31, 57.$$

For *n* points on the circle, the number of regions is given by the formula

$$\frac{n^4 - 6n^3 + 23n^2 - 18n + 24}{24}\,{}^* .$$

1.1 EXERCISES

In Exercises 1–16, determine whether the reasoning is an example of deductive or inductive reasoning.

1. The next number in the pattern 2, 4, 6, 8, 10 is 12.

2. My dog barked and woke me up at 1:02 a.m., 2:03 a.m., and 3:04 a.m. So he will bark again and wake me up at 4:05 a.m.

3. To find the perimeter *P* of a square with side of length *s*, I can use the formula $P = 4s$. So the perimeter of a square with side of length 7 inches is $4 \times 7 = 28$ inches.

4. A company charges a 10% re-stocking fee for returning an item. So when I return a radio that cost $150, I will only get $135 back.

5. If the mechanic says that it will take seven days to repair your SUV, then it will actually take ten days. The mechanic says, "I figure it'll take exactly one week to fix it, ma'am." Then you can expect it to be ready ten days from now.

6. If you take your medicine, you'll feel a lot better. You take your medicine. Therefore, you'll feel a lot better.

7. It has rained every day for the past six days, and it is raining today as well. So it will also rain tomorrow.

8. Carrie's first five children were boys. If she has another baby, it will be a boy.

9. The 2000 movie *Cast Away* stars Tom Hanks as the only human survivor of a plane crash, stranded on a tropical island. He approximates his distance from where the plane lost radio contact to be 400 miles (a radius), and uses the formula for the area of a circle,

$$\text{Area} = \pi \, (\text{radius})^2$$

to determine that a search party would have to cover an area of over 500,000 square miles to look for him and his "pal" Wilson.

---

*For more information on this and other similar patterns, see "Counting Pizza Pieces and Other Combinatorial Problems," by Eugene Maier, in the January 1988 issue of *Mathematics Teacher,* pp. 22–26.

10. If the same number is subtracted from both sides of a true equation, the new equation is also true. I know that $9 + 18 = 27$. Therefore, $(9 + 18) - 13 = 27 - 13$.

11. If you build it, they will come. You build it. Therefore, they will come.

12. All men are mortal. Socrates is a man. Therefore, Socrates is mortal.

13. It is a fact that every student who ever attended Delgado University was accepted into graduate school. Because I am attending Delgado, I can expect to be accepted to graduate school, too.

14. For the past 126 years, a rare plant has bloomed in Columbia each summer, alternating between yellow and green flowers. Last summer, it bloomed with green flowers, so this summer it will bloom with yellow flowers.

15. In the sequence $5, 10, 15, 20, 25, \ldots$, the most probable next number is 30.

16. (This anecdote is adapted from a story by Howard Eves in *In Mathematical Circles*.) A scientist had a group of 100 fleas, and one by one he would tell each flea "Jump," and the flea would jump. Then with the same fleas, he yanked off their hind legs and repeated "Jump," but the fleas would not jump. He concluded that when a flea has its hind legs yanked off, it cannot hear.

17. Discuss the differences between inductive and deductive reasoning. Give an example of each.

18. Give an example of faulty inductive reasoning.

Determine the most probable next term in each of the following lists of numbers.

19. $6, 9, 12, 15, 18$

20. $13, 18, 23, 28, 33$

21. $3, 12, 48, 192, 768$

22. $32, 16, 8, 4, 2$

23. $3, 6, 9, 15, 24, 39$

24. $\dfrac{1}{3}, \dfrac{3}{5}, \dfrac{5}{7}, \dfrac{7}{9}, \dfrac{9}{11}$

25. $\dfrac{1}{2}, \dfrac{3}{4}, \dfrac{5}{6}, \dfrac{7}{8}, \dfrac{9}{10}$

26. $1, 4, 9, 16, 25$

27. $1, 8, 27, 64, 125$

28. $2, 6, 12, 20, 30, 42$

29. $4, 7, 12, 19, 28, 39$

30. $27, 21, 16, 12, 9$

31. $5, 3, 5, 5, 3, 5, 5, 5, 3, 5, 5, 5, 5, 3, 5, 5, 5, 5$

32. $8, 2, 8, 2, 2, 8, 2, 2, 2, 8, 2, 2, 2, 2, 8, 2, 2, 2, 2$

33. Construct a list of numbers similar to those in **Exercise 19** such that the most probable next number in the list is 60.

34. Construct a list of numbers similar to those in **Exercise 30** such that the most probable next number in the list is 8.

Use the list of equations and inductive reasoning to predict the next equation, and then verify your conjecture.

35.
$(9 \times 9) + 7 = 88$
$(98 \times 9) + 6 = 888$
$(987 \times 9) + 5 = 8888$
$(9876 \times 9) + 4 = 88{,}888$

36.
$(1 \times 9) + 2 = 11$
$(12 \times 9) + 3 = 111$
$(123 \times 9) + 4 = 1111$
$(1234 \times 9) + 5 = 11{,}111$

37.
$3367 \times 3 = 10{,}101$
$3367 \times 6 = 20{,}202$
$3367 \times 9 = 30{,}303$
$3367 \times 12 = 40{,}404$

38.
$15873 \times 7 = 111{,}111$
$15873 \times 14 = 222{,}222$
$15873 \times 21 = 333{,}333$
$15873 \times 28 = 444{,}444$

39.
$34 \times 34 = 1156$
$334 \times 334 = 111{,}556$
$3334 \times 3334 = 11{,}115{,}556$

40.
$11 \times 11 = 121$
$111 \times 111 = 12{,}321$
$1111 \times 1111 = 1{,}234{,}321$

41.
$$3 = \frac{3(2)}{2}$$
$$3 + 6 = \frac{6(3)}{2}$$
$$3 + 6 + 9 = \frac{9(4)}{2}$$
$$3 + 6 + 9 + 12 = \frac{12(5)}{2}$$

42.
$2 = 4 - 2$
$2 + 4 = 8 - 2$
$2 + 4 + 8 = 16 - 2$
$2 + 4 + 8 + 16 = 32 - 2$

43.
$5(6) = 6(6 - 1)$
$5(6) + 5(36) = 6(36 - 1)$
$5(6) + 5(36) + 5(216) = 6(216 - 1)$
$5(6) + 5(36) + 5(216) + 5(1296) = 6(1296 - 1)$

44.

$$3 = \frac{3(3-1)}{2}$$

$$3 + 9 = \frac{3(9-1)}{2}$$

$$3 + 9 + 27 = \frac{3(27-1)}{2}$$

$$3 + 9 + 27 + 81 = \frac{3(81-1)}{2}$$

45.

$$\frac{1}{2} = 1 - \frac{1}{2}$$

$$\frac{1}{2} + \frac{1}{4} = 1 - \frac{1}{4}$$

$$\frac{1}{2} + \frac{1}{4} + \frac{1}{8} = 1 - \frac{1}{8}$$

$$\frac{1}{2} + \frac{1}{4} + \frac{1}{8} + \frac{1}{16} = 1 - \frac{1}{16}$$

46.

$$\frac{1}{1 \cdot 2} = \frac{1}{2}$$

$$\frac{1}{1 \cdot 2} + \frac{1}{2 \cdot 3} = \frac{2}{3}$$

$$\frac{1}{1 \cdot 2} + \frac{1}{2 \cdot 3} + \frac{1}{3 \cdot 4} = \frac{3}{4}$$

$$\frac{1}{1 \cdot 2} + \frac{1}{2 \cdot 3} + \frac{1}{3 \cdot 4} + \frac{1}{4 \cdot 5} = \frac{4}{5}$$

Legend has it that the great mathematician Carl Friedrich Gauss (1777–1855) at a very young age was told by his teacher to find the sum of the first 100 counting numbers. While his classmates toiled at the problem, Carl simply wrote down a single number and handed the correct answer in to his teacher. The young Carl explained that he observed that there were 50 pairs of numbers that each added up to 101. (See below.) So the sum of all the numbers must be $50 \times 101 = 5050$.

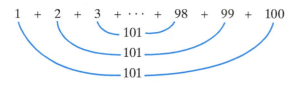

50 sums of 101 = 50 × 101 = 5050

Use the method of Gauss to find each sum.

47. $1 + 2 + 3 + \cdots + 200$ **48.** $1 + 2 + 3 + \cdots + 400$

49. $1 + 2 + 3 + \cdots + 800$ **50.** $1 + 2 + 3 + \cdots + 2000$

51. Modify the procedure of Gauss to find the sum $1 + 2 + 3 + \cdots + 175$.

52. Explain in your own words how the procedure of Gauss can be modified to find the sum $1 + 2 + 3 + \cdots + n$, where n is an odd natural number. (When an odd natural number is divided by 2, it leaves a remainder of 1.)

53. Modify the procedure of Gauss to find the sum $2 + 4 + 6 + \cdots + 100$.

54. Use the result of **Exercise 53** to find the sum $4 + 8 + 12 + \cdots + 200$.

55. What is the most probable next number in this list?

$$12, 1, 1, 1, 2, 1, 3$$

(*Hint:* Think about a clock with chimes.)

56. What is the next term in this list?

$$O, T, T, F, F, S, S, E, N, T$$

(*Hint:* Think about words and their relationship to numbers.)

57. Choose any three-digit number with all different digits, and follow these steps.

  (a) Reverse the digits, and subtract the smaller from the larger. Record your result.

  (b) Choose another three-digit number and repeat this process. Do this as many times as it takes for you to see a pattern in the different results you obtain. (*Hint:* What is the middle digit? What is the sum of the first and third digits?)

  (c) Write an explanation of this pattern.

58. Choose any number, and follow these steps.

  (a) Multiply by 2.   **(b)** Add 6.

  (c) Divide by 2.   **(d)** Subtract the number you

  (e) Record your result.   started with.

  Repeat the process, except in Step (b), add 8. Record your final result. Repeat the process once more, except in Step (b), add 10. Record your final result.

  (f) Observe what you have done. Then use inductive reasoning to explain how to predict the final result.

59. Complete the following.

142,857 × 1 = _____	142,857 × 2 = _____
142,857 × 3 = _____	142,857 × 4 = _____
142,857 × 5 = _____	142,857 × 6 = _____

What pattern exists in the successive answers? Now multiply 142,857 by 7 to obtain an interesting result.

60. Refer to **Figures 2(b)–(e)** and **Figure 3**. Instead of counting interior regions of the circle, count the chords formed. Use inductive reasoning to predict the number of chords that would be formed if seven points were used.

1.2   AN APPLICATION OF INDUCTIVE REASONING: NUMBER PATTERNS

OBJECTIVES

1 Be able to recognize arithmetic and geometric sequences.

2 Be able to apply the method of successive differences to predict the next term in a sequence.

3 Be able to recognize number patterns.

4 Be able to use sum formulas.

5 Be able to recognize triangular, square, and pentagonal numbers.

Number Sequences

An ordered list of numbers such as

$$3, 9, 15, 21, 27, \ldots$$

is called a *sequence*. A **number sequence** is a list of numbers having a first number, a second number, a third number, and so on, called the **terms** of the sequence.

The sequence that begins

$$5, 9, 13, 17, 21, \ldots$$

is an *arithmetic sequence,* or *arithmetic progression*. In an **arithmetic sequence,** each term after the first is obtained by adding the same number, called the **common difference,** to the preceding term. To find the common difference, choose any term after the first and subtract from it the preceding term. If we choose $9 - 5$ (the second term minus the first term), for example, we see that the common difference is 4. To find the term following 21, we add 4 to get $21 + 4 = 25$.

The sequence that begins

$$2, 4, 8, 16, 32, \ldots$$

is a *geometric sequence,* or *geometric progression*. In a **geometric sequence,** each term after the first is obtained by multiplying the preceding term by the same number, called the **common ratio.** To find the common ratio, choose any term after the first and divide it by the preceding term. If we choose $\frac{4}{2}$ (the second term divided by the first term), for example, we see that the common ratio is 2. To find the term following 32, we multiply by 2 to get $32 \cdot 2 = 64$.

EXAMPLE 1 **Identifying Arithmetic and Geometric Sequences**

For each sequence, determine if it is an *arithmetic sequence,* a *geometric sequence,* or *neither*. If it is either arithmetic or geometric, give the next term in the sequence.

(a) $5, 10, 15, 20, 25, \ldots$ **(b)** $3, 12, 48, 192, 768, \ldots$ **(c)** $1, 4, 9, 16, 25, \ldots$

Solution

(a) If we choose *any* term after the first term, and subtract the preceding term, we find that the common difference is 5.

$$10 - 5 = 5 \qquad 15 - 10 = 5 \qquad 20 - 15 = 5 \qquad 25 - 20 = 5$$

Therefore, this is an arithmetic sequence. The next term in the sequence is

$$25 + 5 = 30.$$

(b) If any term after the first is multiplied by 4, the following term is obtained.

$$\frac{12}{3} = 4 \qquad \frac{48}{12} = 4 \qquad \frac{192}{48} = 4 \qquad \frac{768}{192} = 4$$

Therefore, this is a geometric sequence. The next term in the sequence is

$$768 \cdot 4 = 3072.$$

(c) Although there is a pattern here (the terms are the squares of the first five counting numbers), there is neither a common difference nor a common ratio. This is neither an arithmetic nor a geometric sequence. ∎

Successive Differences

Some sequences may present more difficulty than our earlier examples when making a conjecture about the next term. Often the **method of successive differences** may be applied in such cases. Consider the sequence

$$2, 6, 22, 56, 114, \ldots.$$

Because the next term is not obvious, subtract the first term from the second term, the second from the third, the third from the fourth, and so on.

Now repeat the process with the sequence 4, 16, 34, 58 and continue repeating until the difference is a constant value, as shown in line (4).

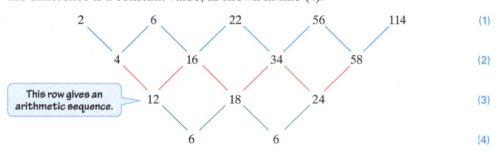

Once a line of constant values is obtained, simply work "backward" by adding until the desired term of the given sequence is obtained. Thus, for this pattern to continue, another 6 should appear in line (4), meaning that the next term in line (3) would have to be $24 + 6 = 30$. The next term in line (2) would be $58 + 30 = 88$. Finally, the next term in the given sequence would be $114 + 88 = \mathbf{202}$.

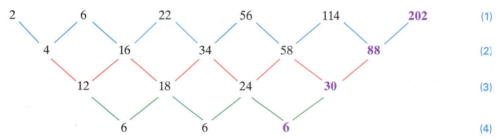

EXAMPLE 2 Using Successive Differences

Use successive differences to determine the next number in each sequence.

(a) $14, 22, 32, 44, \ldots$ **(b)** $5, 15, 37, 77, 141, \ldots$

Solution

(a) Subtract a term from the one that follows it, and continue until a pattern is observed.

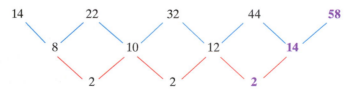

Once the row of 2s was obtained and extended, we were able to obtain

$$12 + 2 = 14, \quad \text{and} \quad 44 + 14 = 58 \quad \text{as shown above.}$$

The next number in the sequence is **58**.

(b) Proceed as before to obtain the following diagram.

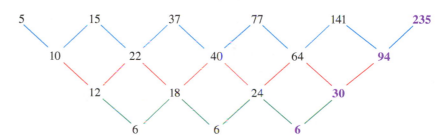

The numbers in the "diagonal" at the far right were obtained by adding: $24 + 6 = 30$, $64 + 30 = 94$, and $141 + 94 = 235$. The next number in the sequence is **235**. ∎

The method of successive differences does not always work. For example, try it on the Fibonacci sequence in **Example 2(b)** of **Section 1.1** and see what happens.

Number Patterns and Sum Formulas

Observe the following number pattern.

$$1 = 1^2$$
$$1 + 3 = 2^2$$
$$1 + 3 + 5 = 3^2$$
$$1 + 3 + 5 + 7 = 4^2$$
$$1 + 3 + 5 + 7 + 9 = 5^2$$

In each case, the left side of the equation is the indicated sum of consecutive odd counting numbers beginning with 1, and the right side is the square of the number of terms on the left side. Inductive reasoning would suggest that the next line in this pattern is as follows.

$$1 + 3 + 5 + 7 + 9 + 11 = 6^2$$

Evaluating each side shows that each side simplifies to 36.

We cannot conclude that this pattern will continue indefinitely, because observation of a finite number of examples does *not* guarantee that the pattern will continue. However, mathematicians have proved that this pattern does indeed continue indefinitely, using a method of proof called **mathematical induction.** (See any standard college algebra text.)

Any even counting number may be written in the form $2k$, where k is a counting number. It follows that the kth odd counting number is written $2k - 1$. For example, the **third** odd counting number, 5, can be written

$$2(3) - 1.$$

Using these ideas, we can write the result obtained above as follows.

SUM OF THE FIRST n ODD COUNTING NUMBERS

If n is any counting number, then the following is true.

$$1 + 3 + 5 + \cdots + (2n - 1) = n^2$$

EXAMPLE 3 Predicting the Next Equation in a List

In each of the following, several equations are given illustrating a suspected number pattern. Determine what the next equation would be, and verify that it is indeed a true statement.

(a)
$$1^2 = 1^3$$
$$(1+2)^2 = 1^3 + 2^3$$
$$(1+2+3)^2 = 1^3 + 2^3 + 3^3$$
$$(1+2+3+4)^2 = 1^3 + 2^3 + 3^3 + 4^3$$

(b)
$$1 = 1^3$$
$$3 + 5 = 2^3$$
$$7 + 9 + 11 = 3^3$$
$$13 + 15 + 17 + 19 = 4^3$$

(c)
$$1 = \frac{1 \cdot 2}{2}$$
$$1 + 2 = \frac{2 \cdot 3}{2}$$
$$1 + 2 + 3 = \frac{3 \cdot 4}{2}$$
$$1 + 2 + 3 + 4 = \frac{4 \cdot 5}{2}$$

(d)
$$12{,}345{,}679 \times 9 = 111{,}111{,}111$$
$$12{,}345{,}679 \times 18 = 222{,}222{,}222$$
$$12{,}345{,}679 \times 27 = 333{,}333{,}333$$
$$12{,}345{,}679 \times 36 = 444{,}444{,}444$$
$$12{,}345{,}679 \times 45 = 555{,}555{,}555$$

Notice that there is no 8 here.

Solution

(a) The left side of each equation is the square of the sum of the first n counting numbers, and the right side is the sum of their cubes. The next equation in the pattern would be

$$(1 + 2 + 3 + 4 + 5)^2 = 1^3 + 2^3 + 3^3 + 4^3 + 5^3.$$

Each side simplifies to 225, so the pattern is true for this equation.

(b) The left sides of the equations contain the sum of odd counting numbers, starting with the first (1) in the first equation, the second and third (3 and 5) in the second equation, the fourth, fifth, and sixth (7, 9, and 11) in the third equation, and so on. Each right side contains the cube (third power) of the number of terms on the left side. Following this pattern, the next equation would be

$$21 + 23 + 25 + 27 + 29 = 5^3,$$

which can be verified by computation.

(c) The left side of each equation gives the indicated sum of the first n counting numbers, and the right side is always of the form

$$\frac{n(n + 1)}{2}.$$

For the pattern to continue, the next equation would be

$$1 + 2 + 3 + 4 + 5 = \frac{5 \cdot 6}{2}.$$

Because each side simplifies to 15, the pattern is true for this equation.

(d) In each case, the first factor on the left is 12,345,679 and the second factor is a multiple of 9 (that is, 9, 18, 27, 36, 45). The right side consists of a nine-digit number, all digits of which are the same (that is, 1, 2, 3, 4, 5). For the pattern to continue, the next equation would be as follows.

$$12{,}345{,}679 \times 54 = 666{,}666{,}666$$

Verify that this is a true statement.

The patterns established in **Examples 3(a) and 3(c)** can be written as follows.

SPECIAL SUM FORMULAS

For any counting number n, the following are true.

$$(1 + 2 + 3 + \cdots + n)^2 = 1^3 + 2^3 + 3^3 + \cdots + n^3$$

$$1 + 2 + 3 + \cdots + n = \frac{n(n + 1)}{2}$$

We can provide a general deductive argument showing how the second equation is obtained.

Let S represent the sum $1 + 2 + 3 + \cdots + n$. This sum can also be written as $S = n + (n - 1) + (n - 2) + \cdots + 1$. Write these two equations as follows.

$$S = 1 \qquad + 2 \qquad + 3 \qquad + \cdots + n$$

$$\underline{S = n \qquad + (n - 1) + (n - 2) + \cdots + 1}$$

$$2S = (n + 1) + (n + 1) + (n + 1) + \cdots + (n + 1) \qquad \text{Add the corresponding sides.}$$

$$2S = n(n + 1) \qquad \text{There are } n \text{ terms of } n + 1.$$

$$S = \frac{n(n + 1)}{2} \qquad \text{Divide both sides by 2.}$$

Figurate Numbers

Pythagoras and his Pythagorean brotherhood studied numbers of geometric arrangements of points, such as **triangular numbers, square numbers,** and **pentagonal numbers. Figure 4** illustrates the first few of each of these types of numbers.

The **figurate numbers** possess numerous interesting patterns. For example, every square number greater than 1 is the sum of two consecutive triangular numbers. ($9 = 3 + 6, 25 = 10 + 15$, and so on.)

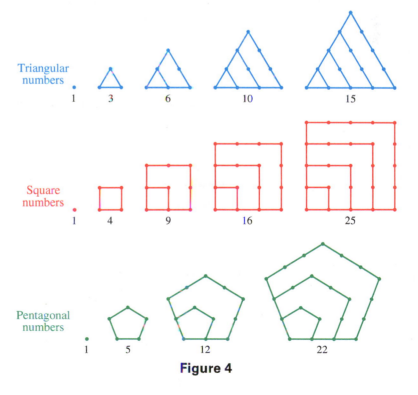

Figure 4

Every pentagonal number can be represented as the sum of a square number and a triangular number. (For example, $5 = 4 + 1$ and $12 = 9 + 3$.) Many other such relationships exist.

In the expression T_n, n is called a **subscript**. T_n is read **"T sub n,"** and it represents the triangular number in the nth position in the sequence. For example,

$$T_1 = 1, \quad T_2 = 3, \quad T_3 = 6, \quad \text{and} \quad T_4 = 10.$$

S_n and P_n represent the nth square and pentagonal numbers, respectively.

FORMULAS FOR TRIANGULAR, SQUARE, AND PENTAGONAL NUMBERS

For any natural number n, the following are true.

The nth triangular number is given by $\quad T_n = \dfrac{n(n + 1)}{2}$.

The nth square number is given by $\quad\quad S_n = n^2$.

The nth pentagonal number is given by $\quad P_n = \dfrac{n(3n - 1)}{2}$.

EXAMPLE 4 Using the Formulas for Figurate Numbers

Use the formulas to find each of the following.

(a) seventh triangular number   **(b)** twelfth square number

(c) sixth pentagonal number

Solution

(a) $T_7 = \dfrac{n(n + 1)}{2} = \dfrac{7(7 + 1)}{2} = \dfrac{7(8)}{2} = \dfrac{56}{2} = 28$ Formula for a triangular number, with $n = 7$

(b) $S_{12} = n^2 = 12^2 = 144$ Formula for a square number, with $n = 12$

$\quad\quad\quad\quad\quad\quad$ $12^2 = 12 \cdot 12$

Inside the brackets, multiply first and then subtract.

(c) $P_6 = \dfrac{n(3n - 1)}{2} = \dfrac{6[3(6) - 1]}{2} = \dfrac{6(18 - 1)}{2} = \dfrac{6(17)}{2} = 51$ ∎

EXAMPLE 5 Illustrating a Figurate Number Relationship

Show that the sixth pentagonal number is equal to the sum of 6 and 3 times the fifth triangular number.

Solution

From **Example 4(c)**, $P_6 = 51$. The fifth triangular number is **15**. Thus,

$$51 = 6 + 3(15) = 6 + 45 = 51.$$ ∎

The general relationship examined in **Example 5** can be written as follows.

$$P_n = n + 3 \cdot T_{n-1} \quad (n \geq 2)$$

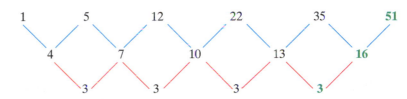

EXAMPLE 6 **Predicting the Value of a Pentagonal Number**

The first five pentagonal numbers are 1, 5, 12, 22, 35. Use the method of successive differences to predict the sixth pentagonal number.

Solution

```
1       5       12      22      35      51
    4       7       10      13      16
        3       3       3       3
```

After the second line of successive differences, we work backward to find that the sixth pentagonal number is **51**, which was also found in **Example 4(c).** ∎

FOR FURTHER THOUGHT

Kaprekar Constants

Take any four-digit number whose digits are all different. Arrange the digits in decreasing order, and then arrange them in increasing order. Now subtract. Repeat the process, called the *Kaprekar routine,* until the same result appears.

For example, suppose that we choose a number whose digits are 1, 5, 7, and 9, such as 1579.

Note that we have obtained the number 6174, and the process will lead to 6174 again. The number 6174 is called a **Kaprekar constant.** This number 6174 will always be generated eventually if this process is applied to such a four-digit number.

For Group or Individual Investigation

1. Apply the process of Kaprekar to a four-digit number of your choice, in which the digits are all different. How many steps did it take for you to arrive at 6174?

2. See **Exercises 77 and 78** in the exercise set that follows.

9751	8721	7443	9963	6642	7641
−1579	−1278	−3447	−3699	−2466	−1467
8172	7443	3996	6264	4176	**6174**

1.2 EXERCISES

For each sequence, determine if it is an arithmetic sequence, a geometric sequence, or neither. If it is either arithmetic or geometric, give the next term in the sequence.

1. 6, 16, 26, 36, 46, . . .

2. 8, 16, 24, 32, 40, . . .

3. 5, 15, 45, 135, 405, . . .

4. 2, 12, 72, 432, 2592, . . .

5. 1, 8, 27, 81, 243, . . .

6. 2, 8, 18, 32, 50, . . .

7. 256, 128, 64, 32, 16, . . .

8. 4096, 1024, 256, 64, 16, . . .

9. 1, 3, 4, 7, 11, . . .

10. 0, 1, 1, 2, 3, . . .

11. 12, 14, 16, 18, 20, . . .

12. 10, 50, 90, 130, 170, . . .

Use the method of successive differences to determine the next number in each sequence.

13. 1, 4, 11, 22, 37, 56, . . .

14. 3, 14, 31, 54, 83, 118, . . .

15. 6, 20, 50, 102, 182, 296, . . .

16. 1, 11, 35, 79, 149, 251, . . .

17. 0, 12, 72, 240, 600, 1260, 2352, . . .

18. 2, 57, 220, 575, 1230, 2317, . . .

19. 5, 34, 243, 1022, 3121, 7770, 16799, . . .

20. 3, 19, 165, 771, 2503, 6483, 14409, . . .

21. Refer to **Figures 2 and 3** in **Section 1.1.** The method of successive differences can be applied to the sequence of interior regions,

$$1, 2, 4, 8, 16, 31,$$

to find the number of regions determined by seven points on the circle. What is the next term in this sequence? How many regions would be determined by eight points? Verify this using the formula given at the end of that section.

22. The 1952 film *Hans Christian Andersen* stars Danny Kaye as the Danish writer of fairy tales. In a scene outside a schoolhouse window, Kaye sings a song to an inchworm. *Inchworm* was written for the film by the composer Frank Loesser and has been recorded by many artists, including Paul McCartney and Kenny Loggins. It was once featured on an episode of *The Muppets* and sung by Charles Aznavour.

As Kaye sings the song, the children in the school room are heard chanting addition facts: $2 + 2 = 4$, $4 + 4 = 8$, $8 + 8 = 16$, and so on.

(a) Use patterns to state the next addition fact (as heard in the movie).

(b) If the children were to extend their facts to the next four in the pattern, what would those facts be?

In Exercises 23–32, several equations are given illustrating a suspected number pattern. Determine what the next equation would be, and verify that it is indeed a true statement.

23. $(1 \times 9) - 1 = 8$
$(21 \times 9) - 1 = 188$
$(321 \times 9) - 1 = 2888$

24. $(1 \times 8) + 1 = 9$
$(12 \times 8) + 2 = 98$
$(123 \times 8) + 3 = 987$

25. $999,999 \times 2 = 1,999,998$
$999,999 \times 3 = 2,999,997$

26. $101 \times 101 = 10,201$
$10,101 \times 10,101 = 102,030,201$

27. $3^2 - 1^2 = 2^3$
$6^2 - 3^2 = 3^3$
$10^2 - 6^2 = 4^3$
$15^2 - 10^2 = 5^3$

28. $1 = 1^2$
$1 + 2 + 1 = 2^2$
$1 + 2 + 3 + 2 + 1 = 3^2$
$1 + 2 + 3 + 4 + 3 + 2 + 1 = 4^2$

29. $2^2 - 1^2 = 2 + 1$
$3^2 - 2^2 = 3 + 2$
$4^2 - 3^2 = 4 + 3$

30. $1^2 + 1 = 2^2 - 2$
$2^2 + 2 = 3^2 - 3$
$3^2 + 3 = 4^2 - 4$

31. $1 = 1 \times 1$
$1 + 5 = 2 \times 3$
$1 + 5 + 9 = 3 \times 5$

32. $1 + 2 = 3$
$4 + 5 + 6 = 7 + 8$
$9 + 10 + 11 + 12 = 13 + 14 + 15$

Use the formula $S = \dfrac{n(n + 1)}{2}$ to find each sum.

33. $1 + 2 + 3 + \cdots + 300$

34. $1 + 2 + 3 + \cdots + 500$

35. $1 + 2 + 3 + \cdots + 675$

36. $1 + 2 + 3 + \cdots + 825$

Use the formula $S = n^2$ to find each sum. (Hint: To find n, add 1 to the last term and divide by 2.)

37. $1 + 3 + 5 + \cdots + 101$

38. $1 + 3 + 5 + \cdots + 49$

39. $1 + 3 + 5 + \cdots + 999$

40. $1 + 3 + 5 + \cdots + 301$

41. Use the formula for finding the sum

$$1 + 2 + 3 + \cdots + n$$

to discover a formula for finding the sum

$$2 + 4 + 6 + \cdots + 2n.$$

42. State in your own words the following formula discussed in this section.

$$(1 + 2 + 3 + \cdots + n)^2 = 1^3 + 2^3 + 3^3 + \cdots + n^3$$

43. Explain how the following diagram geometrically illustrates the formula $1 + 3 + 5 + 7 + 9 = 5^2$.

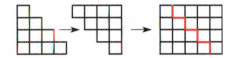

44. Explain how the following diagram geometrically illustrates the formula $1 + 2 + 3 + 4 = \frac{4 \times 5}{2}$.

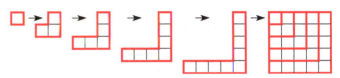

45. Use patterns to complete the table below.

Figurate Number	1st	2nd	3rd	4th	5th	6th	7th	8th
Triangular	1	3	6	10	15	21		
Square	1	4	9	16	25			
Pentagonal	1	5	12	22				
Hexagonal	1	6	15					
Heptagonal	1	7						
Octagonal	1							

46. The first five triangular, square, and pentagonal numbers can be obtained using sums of terms of sequences as shown below.

Triangular	Square	Pentagonal
$1 = 1$	$1 = 1$	$1 = 1$
$3 = 1 + 2$	$4 = 1 + 3$	$5 = 1 + 4$
$6 = 1 + 2 + 3$	$9 = 1 + 3 + 5$	$12 = 1 + 4 + 7$
$10 = 1 + 2 + 3 + 4$	$16 = 1 + 3 + 5 + 7$	$22 = 1 + 4 + 7 + 10$
$15 = 1 + 2 + 3 + 4 + 5$	$25 = 1 + 3 + 5 + 7 + 9$	$35 = 1 + 4 + 7 + 10 + 13$

Notice the successive differences of the added terms on the right sides of the equations. The next type of figurate number is the **hexagonal** number. (A hexagon has six sides.) Use the patterns above to predict the first five hexagonal numbers.

47. Eight times any triangular number, plus 1, is a square number. Show that this is true for the first four triangular numbers.

48. Divide the first triangular number by 3 and record the remainder. Divide the second triangular number by 3 and record the remainder. Repeat this procedure several more times. Do you notice a pattern?

49. Repeat **Exercise 48,** but instead use square numbers and divide by 4. What pattern is determined?

50. Exercises 48 and 49 are specific cases of the following: When the numbers in the sequence of n-agonal numbers are divided by n, the sequence of remainders obtained is a repeating sequence. Verify this for $n = 5$ and $n = 6$.

51. Every square number can be written as the sum of two triangular numbers. For example, $16 = 6 + 10$. This can be represented geometrically by dividing a square array of dots with a line as shown.

The triangular arrangement above the line represents 6, the one below the line represents 10, and the whole arrangement represents 16. Show how the square numbers 25 and 36 may likewise be geometrically represented as the sum of two triangular numbers.

52. A fraction is in *lowest terms* if the greatest common factor of its numerator and its denominator is 1. For example, $\frac{3}{8}$ is in lowest terms, but $\frac{4}{12}$ is not.

  (a) For $n = 2$ to $n = 8$, form the fractions

$$\frac{n\text{th square number}}{(n + 1)\text{st square number}}.$$

  (b) Repeat part (a) with triangular numbers.

  (c) Use inductive reasoning to make a conjecture based on your results from parts (a) and (b), observing whether the fractions are in lowest terms.

*In addition to the formulas for T_n, S_n, and P_n, the following formulas are true for **hexagonal** numbers (H), **heptagonal** numbers (Hp), and **octagonal** numbers (O):*

$$H_n = \frac{n(4n - 2)}{2}, \quad Hp_n = \frac{n(5n - 3)}{2}, \quad O_n = \frac{n(6n - 4)}{2}.$$

Use these formulas to find each of the following.

53. the sixteenth square number

54. the eleventh triangular number

55. the ninth pentagonal number

56. the seventh hexagonal number

57. the tenth heptagonal number

58. the twelfth octagonal number

59. Observe the formulas given for H_n, Hp_n, and O_n, and use patterns and inductive reasoning to predict the formula for N_n, the nth **nonagonal** number. (A nonagon has nine sides.) Then use the fact that the sixth nonagonal number is 111 to further confirm your conjecture.

60. Use the result of **Exercise 59** to find the tenth nonagonal number.

Use inductive reasoning to answer each question.

61. If you add two consecutive triangular numbers, what kind of figurate number do you get?

62. If you add the squares of two consecutive triangular numbers, what kind of figurate number do you get?

63. Square a triangular number. Square the next triangular number. Subtract the smaller result from the larger. What kind of number do you get?

64. Choose a value of n greater than or equal to 2. Find T_{n-1}, multiply it by 3, and add n. What kind of figurate number do you get?

In an arithmetic sequence, the nth term a_n is given by the formula

$$a_n = a_1 + (n - 1)d,$$

where a_1 is the first term and d is the common difference. Similarly, in a geometric sequence, the nth term is given by

$$a_n = a_1 \cdot r^{n-1}.$$

Here r is the common ratio. In Exercises 65–76, use these formulas to determine the indicated term in the given sequence.

65. The eleventh term of 2, 6, 10, 14, . . .

66. The sixteenth term of 5, 15, 25, 35, . . .

67. The 21st term of 19, 39, 59, 79, . . .

68. The 36th term of 8, 38, 68, 98, . . .

69. The 101st term of $\frac{1}{2}$, 1, $\frac{3}{2}$, 2, . . .

70. The 151st term of 0.75, 1.50, 2.25, 3.00, . . .

71. The eleventh term of 2, 4, 8, 16, . . .

72. The ninth term of 1, 4, 16, 64, . . .

73. The 12th term of 1, $\frac{1}{2}$, $\frac{1}{4}$, $\frac{1}{8}$, . . .

74. The 10th term of 1, $\frac{1}{3}$, $\frac{1}{9}$, $\frac{1}{27}$, . . .

75. The 8th term of 40, 10, $\frac{5}{2}$, $\frac{5}{8}$, . . .

76. The 9th term of 10, 2, $\frac{2}{5}$, $\frac{2}{25}$, . . .

77. In the *For Further Thought* investigation of this section, we saw that the number 6174 is a Kaprekar constant. Use the procedure described there, starting with a three-digit number of your choice whose digits are all different. You should arrive at a particular three-digit number that has the same property described for 6174. What is this three-digit number?

78. Applying the Kaprekar routine to a five-digit number does not reach a single repeating result but instead reaches one of the following ten numbers and then cycles repeatedly through a subset of these ten numbers.

53955, 59994, 61974, 62964, 63954,

71973, 74943, 75933, 82962, 83952

  (a) Start with the number 45986 and determine which one of the ten numbers above is reached first.

  (b) Start with a five-digit number of your own, and determine which one of the ten numbers is eventually reached first.

The mathematical array of numbers known as **Pascal's triangle** consists of rows of numbers, each of which contains one more entry than the previous row. The first six rows are shown here.

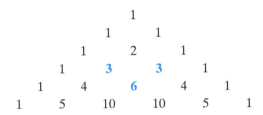

Refer to this array to answer Exercises 79–82.

79. Each row begins and ends with a 1. Discover a method whereby the other entries in a row can be determined from the entries in the row immediately above it. (*Hint:* See the entries in color above.) Find the next three rows of the triangle, and prepare a copy of the first nine rows for later reference.

80. Find the sum of the entries in each of the first eight rows. What is the pattern that emerges? Predict the sum of the entries in the ninth row, and confirm your prediction.

81. The first six rows of the triangle are arranged "flush left" here.

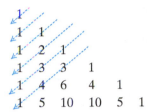

Add along the blue diagonal lines. Write these sums in order from left to right. What sequence is this?

82. Find the values of the first four powers of the number 11, starting with 11^0, which by definition is equal to 1. Predict what the next power of 11 will equal by observing the rows of Pascal's triangle. Confirm your prediction by actual computation.

1.3   STRATEGIES FOR PROBLEM SOLVING

OBJECTIVES

1 Know George Polya's four-step method of problem solving.

2 Be able to apply various strategies for solving problems.

A General Problem-Solving Method

In the first two sections of this chapter we stressed the importance of pattern recognition and the use of inductive reasoning in solving problems. Probably the most famous study of problem-solving techniques was developed by George Polya (1888–1985), among whose many publications was the modern classic *How to Solve It*. In this book, Polya proposed a four-step method for problem solving.

POLYA'S FOUR-STEP METHOD FOR PROBLEM SOLVING

Step 1   **Understand the problem.** You cannot solve a problem if you do not understand what you are asked to find. The problem must be read and analyzed carefully. You may need to read it several times. After you have done so, ask yourself, *"What must I find?"*

Step 2   **Devise a plan.** There are many ways to attack a problem. Decide what plan is appropriate for the particular problem you are solving.

Step 3   **Carry out the plan.** Once you know how to approach the problem, carry out your plan. You may run into "dead ends" and unforeseen roadblocks, but be persistent.

Step 4   **Look back and check.** Check your answer to see that it is reasonable. *Does it satisfy the conditions of the problem? Have you answered all the questions the problem asks? Can you solve the problem a different way and come up with the same answer?*

In Step 2 of Polya's problem-solving method, we are told to devise a plan. Here are some strategies that may prove useful.

George Polya, author of the classic *How to Solve It*, died at the age of 97 on September 7, 1985. A native of Budapest, Hungary, he was once asked why there were so many good mathematicians to come out of Hungary at the turn of the century. He theorized that it was because mathematics is the cheapest science. It does not require any expensive equipment, only pencil and paper.

Polya authored or coauthored more than 250 papers in many languages, wrote a number of books, and was a brilliant lecturer and teacher. Yet, interestingly enough, he never learned to drive a car.

Problem-Solving Strategies

- Make a table or a chart.
- Look for a pattern.
- Solve a similar, simpler problem.
- Draw a sketch.
- Use inductive reasoning.
- Write an equation and solve it.
- If a formula applies, use it.
- Work backward.
- Guess and check.
- Use trial and error.
- Use common sense.
- Look for a "catch" if an answer seems too obvious or impossible.

Using a Table or Chart

EXAMPLE 1 Solving Fibonacci's Rabbit Problem

A man put a pair of rabbits in a cage. During the first month the rabbits produced no offspring but each month thereafter produced one new pair of rabbits. If each new pair thus produced reproduces in the same manner, how many pairs of rabbits will there be at the end of 1 year? (This problem is a famous one in the history of mathematics and first appeared in *Liber Abaci*, a book written by the Italian mathematician Leonardo Pisano (also known as Fibonacci) in the year 1202.)

Solution

Step 1　**Understand the problem.** We can reword the problem as follows:

> *How many pairs of rabbits will the man have at the end of one year if he starts with one pair, and they reproduce this way: During the first month of life, each pair produces no new rabbits, but each month thereafter each pair produces one new pair?*

Step 2　**Devise a plan.** Because there is a definite pattern to how the rabbits will reproduce, we can construct **Table 2.**

Table 2

Month	Number of Pairs at Start	Number of New Pairs Produced	Number of Pairs at End of Month
1st			
2nd			
3rd			
4th			
5th			
6th			
7th			
8th			
9th			
10th			
11th			
12th			

Fibonacci (1170–1250) discovered the sequence named after him in a problem on rabbits. Fibonacci ("son of Bonaccio") is one of several names for Leonardo of Pisa. His father managed a warehouse in present-day Bougie (or Bejaia), in Algeria. Thus it was that Leonardo Pisano studied with a Moorish teacher and learned the "Indian" numbers that the Moors and other Moslems brought with them in their westward drive.

Fibonacci wrote books on algebra, geometry, and trigonometry.

The answer will go here.

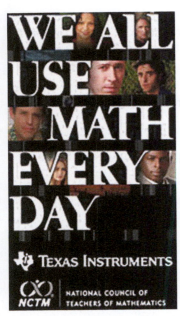

On January 23, 2005, the CBS television network presented the first episode of *NUMB3RS*, a show focusing on how mathematics is used in solving crimes. David Krumholtz plays Charlie Eppes, a brilliant mathematician who assists his FBI agent brother (Rob Morrow).

In the first-season episode "Sabotage" (2/25/2005), one of the agents admits that she was not a good math student, and Charlie uses the **Fibonacci sequence** and its relationship to nature to enlighten her.

The sequence shown in color in **Table 3** is the Fibonacci sequence, mentioned in **Example 2(b)** of **Section 1.1.**

Step 3 **Carry out the plan.** At the start of the first month, there is only one pair of rabbits. No new pairs are produced during the first month, so there is $1 + 0 = 1$ pair present at the end of the first month. This pattern continues. In **Table 3,** we add the number in the first column of numbers to the number in the second column to get the number in the third.

Table 3

Month	Number of Pairs at Start	+	Number of New Pairs Produced	=	Number of Pairs at End of Month	
1st	1		0		1	$1 + 0 = 1$
2nd	1		1		2	$1 + 1 = 2$
3rd	2		1		3	$2 + 1 = 3$
4th	3		2		5	.
5th	5		3		8	.
6th	8		5		13	.
7th	13		8		21	.
8th	21		13		34	.
9th	34		21		55	.
10th	55		34		89	.
11th	89		55		144	.
12th	144		89		**233**	$144 + 89 = 233$

The answer is the final entry.

There will be **233** pairs of rabbits at the end of one year.

Step 4 **Look back and check.** Go back and make sure that we have interpreted the problem correctly. Double-check the arithmetic. We have answered the question posed by the problem, so the problem is solved. ■

Working Backward

EXAMPLE 2 **Determining a Wager at the Track**

Ronnie goes to the racetrack with his buddies on a weekly basis. One week he tripled his money, but then lost $12. He took his money back the next week, doubled it, but then lost $40. The following week he tried again, taking his money back with him. He quadrupled it, and then played well enough to take that much home, a total of $224. How much did he start with the first week?

Solution

This problem asks us to find Ronnie's starting amount. Since we know his final amount, the method of working backward can be applied.

Because his final amount was $224 and this represents four times the amount he started with on the third week, we *divide* $224 by 4 to find that he started the third week with $56. Before he lost $40 the second week, he had this $56 plus the $40 he lost, giving him $96.

Augustus De Morgan was an English mathematician and philosopher, who served as professor at the University of London. He wrote numerous books, one of which was *A Budget of Paradoxes*. His work in set theory and logic led to laws that bear his name and are covered in other chapters.

The $96 represented double what he started with, so he started with $96 *divided by* 2, or $48, the second week. Repeating this process once more for the first week, before his $12 loss he had

$$\$48 + \$12 = \$60,$$

which represents triple what he started with. Therefore, he started with

$$\$60 \div 3 = \$20. \quad \text{Answer}$$

To check, observe the following equations that depict winnings and losses.

> ***First week:*** $(3 \times \$20) - \$12 = \$60 - \$12 = \$48$
> ***Second week:*** $(2 \times \$48) - \$40 = \$96 - \$40 = \$56$
> ***Third week:*** $(4 \times \$56) = \$224$ His final amount ∎

Using Trial and Error

Recall that $5^2 = 5 \cdot 5 = 25$. That is, 5 squared is 25. Thus, 25 is called a **perfect square,** a term that we use in **Example 3.**

$$1, \quad 4, \quad 9, \quad 16, \quad 25, \quad 36, \quad \text{and so on} \quad \text{Perfect squares}$$

EXAMPLE 3 Finding Augustus De Morgan's Birth Year

The mathematician Augustus De Morgan lived in the nineteenth century. He made the following statement: "I was x years old in the year x^2." In what year was he born?

Solution

We must find the year of De Morgan's birth. The problem tells us that he lived in the nineteenth century, which is another way of saying that he lived during the 1800s. One year of his life was a perfect square, so we must find a number between 1800 and 1900 that is a perfect square. Use trial and error.

$$42^2 = 42 \cdot 42 = 1764$$
$$43^2 = 43 \cdot 43 = 1849 \quad \text{←} \quad \boxed{\text{1849 is between 1800 and 1900.}}$$
$$44^2 = 44 \cdot 44 = 1936$$

The only natural number whose square is between 1800 and 1900 is 43, because $43^2 = 1849$. Therefore, De Morgan was 43 years old in 1849. The final step in solving the problem is to subtract 43 from 1849 to find the year of his birth.

$$1849 - 43 = 1806 \quad \text{←} \quad \boxed{\text{He was born in 1806.}}$$

To check this answer, look up De Morgan's birth date in a book dealing with mathematics history, such as *An Introduction to the History of Mathematics,* Sixth Edition, by Howard W. Eves. ∎

Guessing and Checking

As mentioned above, $5^2 = 25$. The inverse procedure for squaring a number is called taking the **square root.** We indicate the positive square root using a **radical symbol** $\sqrt{}$. Thus, $\sqrt{25} = 5$. Also,

$$\sqrt{4} = 2, \quad \sqrt{9} = 3, \quad \sqrt{16} = 4, \quad \text{and so on.} \quad \text{Square roots}$$

The next problem deals with a square root and dates back to Hindu mathematics, circa 850.

EXAMPLE 4 **Finding the Number of Camels**

One-fourth of a herd of camels was seen in the forest. Twice the square root of that herd had gone to the mountain slopes, and 3 times 5 camels remained on the river-bank. What is the numerical measure of that herd of camels?

Solution

The numerical measure of a herd of camels must be a counting number. Because the problem mentions "one-fourth of a herd" and "the square root of that herd," the number of camels must be both a multiple of 4 and a perfect square, so that only whole numbers are used. The least counting number that satisfies both conditions is 4.

We write an equation where x represents the numerical measure of the herd, and then substitute 4 for x to see if it is a solution.

One-fourth of the herd	+	Twice the square root of that herd	+	3 times 5 camels	=	The numerical measure of the herd.
$\frac{1}{4}x$	+	$2\sqrt{x}$	+	$3 \cdot 5$	=	x

$$\frac{1}{4}(4) + 2\sqrt{4} + 3 \cdot 5 \stackrel{?}{=} 4 \quad \text{Let } x = 4.$$

$$1 + 4 + 15 \stackrel{?}{=} 4 \quad \sqrt{4} = 2$$

$$20 \neq 4$$

Because 4 is not the solution, try **16**, the next perfect square that is a multiple of 4.

$$\frac{1}{4}(16) + 2\sqrt{16} + 3 \cdot 5 \stackrel{?}{=} 16 \quad \text{Let } x = 16.$$

$$4 + 8 + 15 \stackrel{?}{=} 16 \quad \sqrt{16} = 4$$

$$27 \neq 16$$

Because 16 is not a solution, try **36**.

$$\frac{1}{4}(36) + 2\sqrt{36} + 3 \cdot 5 \stackrel{?}{=} 36 \quad \text{Let } x = 36.$$

$$9 + 12 + 15 \stackrel{?}{=} 36 \quad \sqrt{36} = 6$$

$$36 = 36$$

Thus, 36 is the numerical measure of the herd.

Check: "One-fourth of 36, plus twice the square root of 36, plus 3 times 5" gives 9 plus 12 plus 15, which equals 36. ∎

Considering a Similar, Simpler Problem

EXAMPLE 5 **Finding the Units Digit of a Power**

The digit farthest to the right in a counting number is called the *ones* or *units* digit, because it tells how many ones are contained in the number when grouping by tens is considered. What is the ones (or units) digit in 2^{4000}?

Solution

Recall that 2^{4000} means that 2 is used as a factor 4000 times.

$$2^{4000} = \underbrace{2 \times 2 \times 2 \times \cdots \times 2}_{4000 \text{ factors}}$$

Mathematics to Die For In the 1995 movie *Die Hard: With a Vengeance*, detective John McClane (Bruce Willis) is tormented by the villain Simon Gruber (Jeremy Irons), who has planted a bomb in a park. Gruber presents McClane and his partner Zeus Carver (Samuel L. Jackson) with a riddle for disarming it, but there is a time limit. The riddle requires that they get exactly 4 gallons of water using 3-gallon and 5-gallon jugs having no markers. They are able to solve it and defuse the bomb. Can you do it? (See the next page for the answer.)

Solution to the Jugs-of-Water Riddle
This is one way to do it: With both jugs empty, fill the 3-gallon jug and pour its contents into the 5-gallon jug. Then fill the 3-gallon jug again, and pour it into the 5-gallon jug until the latter is filled. There is now $(3 + 3) - 5 = 1$ gallon in the 3-gallon jug. Empty the 5-gallon jug, and pour the 1 gallon of water from the 3-gallon jug into the 5-gallon jug. Finally, fill the 3-gallon jug and pour all of it into the 5-gallon jug, resulting in $1 + 3 = 4$ gallons in the 5-gallon jug.

(*Note:* There is another way to solve this problem. See if you can discover the alternative solution.)

To answer the question, we examine some smaller powers of 2 and then look for a pattern. We start with the exponent 1 and look at the first twelve powers of 2.

$$2^1 = 2 \qquad 2^5 = 32 \qquad 2^9 = 512$$
$$2^2 = 4 \qquad 2^6 = 64 \qquad 2^{10} = 1024$$
$$2^3 = 8 \qquad 2^7 = 128 \qquad 2^{11} = 2048$$
$$2^4 = 16 \qquad 2^8 = 256 \qquad 2^{12} = 4096$$

Notice that in any one of the four rows above, the ones digit is the same all the way across the row. The final row, which contains the exponents 4, 8, and 12, has the ones digit 6. Each of these exponents is divisible by 4, and because 4000 is divisible by 4, we can use inductive reasoning to predict that the units digit in 2^{4000} is 6.

(*Note:* The units digit for any other power can be found if we divide the exponent by 4 and consider the remainder. Then compare the result to the list of powers above. For example, to find the units digit of 2^{543}, divide 543 by 4 to get a quotient of 135 and a remainder of 3. The units digit is the same as that of 2^3, which is 8.) ∎

Drawing a Sketch

EXAMPLE 6 **Connecting the Dots**

An array of nine dots is arranged in a 3×3 square, as shown in **Figure 5.** Is it possible to join the dots with exactly four straight line segments if you are not allowed to pick up your pencil from the paper and may not trace over a segment that has already been drawn? If so, show how.

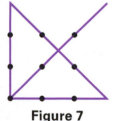

Figure 5

Solution

Figure 6 shows three attempts. In each case, something is wrong. In the first sketch, one dot is not joined. In the second, the figure cannot be drawn without picking up your pencil from the paper or tracing over a line that has already been drawn. In the third figure, all dots have been joined, but you have used five line segments as well as retraced over the figure.

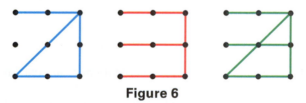

Figure 6

The conditions of the problem can be satisfied, as shown in **Figure 7.** We "went outside of the box," which was not prohibited by the conditions of the problem. This is an example of creative thinking—we used a strategy that often is not considered at first. ∎

Figure 7

Using Common Sense

> **Problem-Solving Strategies**
>
> Some problems involve a "catch." They seem too easy or perhaps impossible at first because we tend to overlook an obvious situation. Look carefully at the use of language in such problems. And, of course, never forget to use common sense.

EXAMPLE 7 **Determining Coin Denominations**

Two currently minted United States coins together have a total value of $1.05. One is not a dollar. What are the two coins?

Solution

Our initial reaction might be, "The only way to have two such coins with a total of $1.05 is to have a nickel and a dollar, but the problem says that one of them is not a dollar." This statement is indeed true. What we must realize here is that the one that is not a dollar is the nickel, and the *other* coin is a dollar! So the two coins are a dollar and a nickel. ∎

1.3   EXERCISES

One of the most popular features in the journal Mathematics Teacher, *published by the National Council of Teachers of Mathematics, is the monthly calendar. It provides an interesting, unusual, or challenging problem for each day of the month. Some of these exercises, and others later in this text, are chosen from these calendars (the day, month, and year of publication of each problem are indicated). The authors want to thank the many contributors for permission to use these problems.*

Use the various problem-solving strategies to solve each problem. In many cases there is more than one possible approach, so be creative.

1. **Broken Elevator**   A man enters a building on the first floor and runs up to the third floor in 20 seconds. At this rate, how many seconds would it take for the man to run from the first floor up to the sixth floor? (October 3, 2010)

2. **Saving Her Dollars**   Every day Sally saved a penny, a dime, and a quarter. What is the least number of days required for her to save an amount equal to an integral (counting) number of dollars? (January 11, 2012)

3. **Do You Have a Match?**   Move 4 of the matches in the figure to create exactly 3 equilateral triangles. (An *equilateral triangle* has all three sides the same length.) (February 20, 2011)

4. **Sudoku**   Sudoku is an $n \times n$ puzzle that requires the solver to fill in all the squares using the integers 1 through n. Each row, column, and subrectangle contains exactly one of each number. Complete the $n \times n$ puzzle. (June 25, 2012)

			4
2			
	1		
		1	

5. **Break This Code**   Each letter of the alphabet is assigned an integer, starting with A = 0, B = 1, and so on. The numbers repeat after every seven letters, so that G = 6, H = 0, and I = 1, continuing on to Z. What two-letter word is represented by the digits 16? (October 3, 2011)

6. **A Real Problem**   We are given the following sequence:

    PROBLEMSOLVINGPROBLEMSOLVINGPROB . . .

    If the pattern continues, what letter will be in the 2012th position? (November 1, 2012)

7. **How Old Is Mommy?**   A mother has two children whose ages differ by 5 years. The sum of the squares of their ages is 97. The square of the mother's age can be found by writing the squares of the children's ages one after the other as a four-digit number. How old is the mother? (March 9, 2012)

8. **An Alarming Situation**   You have three alarms in your room. Your cell phone alarm is set to ring every 30 minutes, your computer alarm is set to ring every 20 minutes, and your clock alarm is set to ring every 45 minutes. If all three alarms go off simultaneously at 12:34 p.m., when is the next time that they will go off simultaneously? (November 2, 2012)

9. *Laundry Day* Every Monday evening, a mathematics teacher stops by the dry cleaners, drops off the shirts that he wore for the week, and picks up his previous week's load. If he wears a clean shirt every day, including Saturday and Sunday, what is the minimum number of shirts that he can own? (April 1, 2012)

10. *Pick an Envelope* Three envelopes contain a total of six bills. One envelope contains two $10 bills, one contains two $20 bills, and the third contains one $10 and one $20 bill. A label on each envelope indicates the sum of money in one of the other envelopes. It is possible to select one envelope, see one bill in that envelope, and then state the contents of all of the envelopes. Which envelope should you choose? (May 1, 2012)

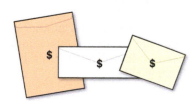

11. *Class Members* A classroom contains an equal number of boys and girls. If 8 girls leave, twice as many boys as girls remain. What was the original number of students present? (May 24, 2008)

12. *Give Me a Digit* Given a two-digit number, make a three-digit number by putting a 6 as the rightmost digit. Then add 6 to the resulting three-digit number and remove the rightmost digit to obtain another two-digit number. If the result is 76, what is the original two-digit number? (October 18, 2009)

13. *Missing Digit* Look for a pattern and find the missing digit x.

3	2	4	8
7	2	1	3
8	4	x	5
4	3	6	9

(February 14, 2009)

14. *Abundancy* An integer $n > 1$ is **abundant** if the sum of its proper divisors (positive integer divisors smaller than n) is greater than n. Find the smallest abundant integer. (November 27, 2009)

15. *Cross-Country Competition* The schools in an athletic conference compete in a cross-country meet to which each school sends three participants. Erin, Katelyn, and Iliana are the three representatives from one school. Erin finished the race in the middle position; Katelyn finished after Erin, in the 19th position; and Iliana finished 28th. How many schools took part in the race? (May 27, 2008)

16. *Gone Fishing* Four friends go fishing one day and bring home a total of 11 fish. If each person caught at least 1 fish, then which of the following *must* be true?

   A. One person caught exactly 2 fish.

   B. One person caught exactly 3 fish.

   C. One person caught fewer than 3 fish.

   D. One person caught more than 3 fish.

   E. Two people each caught more than 1 fish.

   (May 24, 2008)

17. *Cutting a Square in Half* In how many ways can a single straight line cut a square in half? (October 2, 2008)

18. *You Lie!* Max, Sam, and Brett were playing basketball. One of them broke a window, and the other two saw him break it. Max said, "I am innocent." Sam said, "Max and I are both innocent." Brett said, "Max and Sam are both innocent." If only one of them is telling the truth, who broke the window? (September 21, 2008)

19. *Bookworm Snack* A 26-volume encyclopedia (one for each letter) is placed on a bookshelf in alphabetical order from left to right. Each volume is 2 inches thick, including the front and back covers. Each cover is $\frac{1}{4}$ inch thick. A bookworm eats straight through the encyclopedia, beginning inside the front cover of volume A and ending after eating through the back cover of volume Z. How many inches of book did the bookworm eat? (November 12, 2008)

20. *Pick a Card, Any Card* Three face cards from an ordinary deck of playing cards lie facedown in a horizontal row and are arranged such that immediately to the right of a king is a queen or two queens, immediately to the left of a queen is a queen or two queens, immediately to the left of a heart is a spade or two spades, and immediately to the right of a spade is a spade or two spades. Name the three cards in order. (April 23, 2008)

21. *Catwoman's Cats* If you ask Batman's nemesis, Catwoman, how many cats she has, she answers with a riddle: "Five-sixths of my cats plus seven." How many cats does Catwoman have? (April 20, 2003)

22. **Pencil Collection** Bob gave four-fifths of his pencils to Barbara, then he gave two-thirds of the remaining pencils to Bonnie. If he ended up with ten pencils for himself, with how many did he start? (October 12, 2003)

23. **Adding Gasoline** The gasoline gauge on a van initially read $\frac{1}{8}$ full. When 15 gallons were added to the tank, the gauge read $\frac{3}{4}$ full. How many more gallons are needed to fill the tank? (November 25, 2004)

24. **Gasoline Tank Capacity** When 6 gallons of gasoline are put into a car's tank, the indicator goes from $\frac{1}{4}$ of a tank to $\frac{5}{8}$. What is the total capacity of the gasoline tank? (February 21, 2004)

25. **Number Pattern** What is the relationship between the rows of numbers?

18,	38,	24,	46,	42
8,	24,	8,	24,	8

(May 26, 2005)

26. **Locking Boxes** You and I each have one lock and a corresponding key. I want to mail you a box with a ring in it, but any box that is not locked will be emptied before it reaches its recipient. How can I safely send you the ring? (Note that you and I each have keys to our own lock but not to the other lock.) (May 4, 2004)

27. **Number in a Sequence** In the sequence 16, 80, 48, 64, A, B, C, D, each term beyond the second term is the arithmetic mean (average) of the two previous terms. What is the value of D? (April 26, 2004)

28. **Unknown Number** Cindy was asked by her teacher to subtract 3 from a certain number and then divide the result by 9. Instead, she subtracted 9 and then divided the result by 3, giving an answer of 43. What would her answer have been if she had worked the problem correctly? (September 3, 2004)

29. **Unfolding and Folding a Box** An unfolded box is shown below.

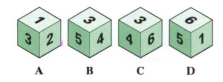

Which figure shows the box folded up? (November 7, 2001)

30. **Vertical Symmetry in States' Names** (If a vertical line is drawn through the center of a figure, and the left and right sides are reflections of each other across this line, the figure is said to have vertical symmetry.) When spelled with all capital letters, each letter in HAWAII has vertical symmetry. Find the name of a state whose letters all have vertical and horizontal symmetry. (September 11, 2001)

31. **Labeling Boxes** You are working in a store that has been very careless with the stock. Three boxes of socks are each incorrectly labeled. The labels say *red socks*, *green socks*, and *red and green socks*. How can you relabel the boxes correctly by taking only one sock out of one box, without looking inside the boxes? (October 22, 2001)

32. **Mr. Green's Age** At his birthday party, Mr. Green would not directly tell how old he was. He said, "If you add the year of my birth to this year, subtract the year of my tenth birthday and the year of my fiftieth birthday, and then add my present age, the result is eighty." How old was Mr. Green? (December 14, 1997)

33. **Sum of Hidden Dots on Dice** Three dice with faces numbered 1 through 6 are stacked as shown. Seven of the eighteen faces are visible, leaving eleven faces hidden on the back, on the bottom, and between dice. The total number of dots not visible in this view is _____.

A. 21

B. 22

C. 31

D. 41

E. 53

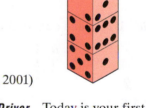

(September 17, 2001)

34. **Age of the Bus Driver** Today is your first day driving a city bus. When you leave downtown, you have twenty-three passengers. At the first stop, three people exit and five people get on the bus. At the second stop, eleven people exit and eight people get on the bus. At the third stop, five people exit and ten people get on. How old is the bus driver? (April 1, 2002)

35. **Matching Triangles and Squares** How can you connect each square with the triangle that has the same number? Lines cannot cross, enter a square or triangle, or go outside the diagram. (October 15, 1999)

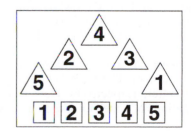

36. *Forming Perfect Square Sums* How must one place the integers from 1 to 15 in each of the spaces below in such a way that no number is repeated and the sum of the numbers in any two consecutive spaces is a perfect square? (November 11, 2001)

37. *Difference Triangle* Balls numbered 1 through 6 are arranged in a **difference triangle.** Note that in any row, the difference between the larger and the smaller of two successive balls is the number of the ball that appears below them. Arrange balls numbered 1 through 10 in a difference triangle. (May 6, 1998)

38. *Clock Face* By drawing two straight lines, divide the face of a clock into three regions such that the numbers in the regions have the same total. (October 28, 1998)

39. *Alphametric* If a, b, and c are digits for which

$$\begin{array}{r} 7\ a\ 2 \\ -4\ 8\ b \\ \hline c\ 7\ 3, \end{array}$$

then $a + b + c =$ _____.

A. 14   **B.** 15   **C.** 16   **D.** 17   **E.** 18

(September 22, 1999)

40. *Perfect Square* Only one of these numbers is a perfect square. Which one is it? (October 8, 1997)

  329476   389372   964328
    326047   724203

41. *Sleeping on the Way to Grandma's House* While traveling to his grandmother's for Christmas, George fell asleep halfway through the journey. When he awoke, he still had to travel half the distance that he had traveled while sleeping. For what part of the entire journey had he been asleep? (December 25, 1998)

42. *Buckets of Water* You have brought two unmarked buckets to a stream. The buckets hold 7 gallons and 3 gallons of water, respectively. How can you obtain exactly 5 gallons of water to take home? (October 19, 1997)

43. *Counting Puzzle (Rectangles)* How many rectangles are in the figure? (March 27, 1997)

44. *Digit Puzzle* Place each of the digits 1, 2, 3, 4, 5, 6, 7, and 8 in separate boxes so that boxes that share common corners do not contain successive digits. (November 29, 1997)

45. *Palindromic Number* (*Note*: A **palindromic number** is a number whose digits read the same left to right as right to left. For example, 383, 12321, and 9876789 are palindromic.) The odometer of the family car read 15951 when the driver noticed that the number was palindromic. "Curious," said the driver to herself. "It will be a long time before that happens again." But 2 hours later, the odometer showed a new palindromic number. (*Author's note:* Assume it was the next possible one.) How fast was the car driving in those 2 hours? (December 26, 1998)

46. *How Much Is That Doggie in the Window?* A man wishes to sell a puppy for $11. A customer who wants to buy it has only foreign currency. The exchange rate for the foreign currency is as follows: 11 round coins = $15, 11 square coins = $16, 11 triangular coins = $17. How many of each coin should the customer pay? (April 20, 2008)

47. *Final Digits of a Power of 7* What are the final two digits of 7^{1997}? (November 29, 1997)

48. *Units Digit of a Power of 3* If you raise 3 to the 324th power, what is the units digit of the result?

49. *Summing the Digits* When $10^{50} - 50$ is expressed as a single whole number, what is the sum of its digits? (April 7, 2008)

50. **Frog Climbing up a Well** A frog is at the bottom of a 20-foot well. Each day it crawls up 4 feet, but each night it slips back 3 feet. After how many days will the frog reach the top of the well?

51. **Units Digit of a Power of 7** What is the units digit in 7^{491}?

52. **Money Spent at a Bazaar** Christine bought a book for $10 and then spent half her remaining money on a train ticket. She then bought lunch for $4 and spent half her remaining money at a bazaar. She left the bazaar with $8. How much money did she start with?

53. **Going Postal** Joanie wants to mail a package that requires $1.53 in postage. If she has only 5-cent and 8-cent stamps, what is the smallest number of stamps she could use that would total exactly $1.53? (August 20, 2008)

54. **Counting Puzzle (Squares)** How many squares are in the figure?

55. **Matching Socks** A drawer contains 20 black socks and 20 white socks. If the light is off and you reach into the drawer to get your socks, what is the minimum number of socks you must pull out in order to be sure that you have a matching pair?

56. **Counting Puzzle (Triangles)** How many triangles are in the figure?

57. **Perfect Number** A **perfect number** is a counting number that is equal to the sum of all its counting number divisors except itself. For example, 28 is a perfect number because its divisors other than itself are 1, 2, 4, 7, and 14, and $1 + 2 + 4 + 7 + 14 = 28$. What is the least perfect number?

58. **Naming Children** Becky's mother has three daughters. She named her first daughter Penny and her second daughter Nichole. What did she name her third daughter?

59. **Growth of a Lily Pad** A lily pad grows so that each day it doubles its size. On the twentieth day of its life, it completely covers a pond. On what day was the pond half covered?

60. **Interesting Property of a Sentence** Comment on an interesting property of this sentence: "A man, a plan, a canal, Panama." (*Hint:* See **Exercise 45.**)

61. **High School Graduation Year of Author** One of the authors of this book graduated from high school in the year that satisfies these conditions: (1) The sum of the digits is 23; (2) The hundreds digit is 3 more than the tens digit; (3) No digit is an 8. In what year did he graduate?

62. **Where in the World Is Matt Lauer?** Matt Lauer is one of the hosts of the *Today* show on the NBC television network. From time to time he travels the world and is in a new location each day of the week, which is unknown even to his co-hosts back in the studio in New York. On one day, he decided to give them a riddle as a hint to where he would be on the following day. Here's the riddle:

    This country is an ANAGRAM of a SYNONYM of a HOMOPHONE of an EVEN PRIME NUMBER.

    In what country was Matt going to be the following day? (If you are unfamiliar with some of the terms in capital letters, look up their definitions.)

63. **Adam and Eve's Assets** Eve said to Adam, "If you give me one dollar, then we will have the same amount of money." Adam then replied, "Eve, if you give me one dollar, I will have double the amount of money you are left with." How much does each have?

64. **Missing Digits Puzzle** In the addition problem below, some digits are missing, as indicated by the blanks. If the problem is done correctly, what is the sum of the missing digits?

$$
\begin{array}{r}
\_\ 3\ 5 \\
8\ \_\ 6 \\
+\ 1\ 4\ \_ \\
\hline
\_\ 4\ 0\ 8
\end{array}
$$

65. **Missing Digits Puzzle** Fill in the blanks so that the multiplication problem below uses all digits $0, 1, 2, 3, \ldots, 9$ exactly once and is worked correctly.

$$
\begin{array}{r}
\_\ 0\ 2 \\
\times\quad\ \ 3\ \_ \\
\hline
5,\ \_\ \_\ \_
\end{array}
$$

66. *Magic Square* A **magic square** is a square array of numbers that has the property that the sum of the numbers in any row, column, or diagonal is the same. Fill in the square below so that it becomes a magic square, and all digits 1, 2, 3, . . . , 9 are used exactly once.

6		8
	5	
		4

67. *Magic Square* Refer to **Exercise 66.** Complete the magic square below so that all counting numbers 1, 2, 3, . . . , 16 are used exactly once, and the sum in each row, column, or diagonal is 34.

6		9	
	15	14	
11	10		
16	13		

68. *Decimal Digit* What is the 100th digit in the decimal representation for $\frac{1}{7}$?

69. *Pitches in a Baseball Game* What is the minimum number of pitches that a baseball player who pitches a complete game can make in a regulation 9-inning baseball game?

70. *Weighing Coins* You have eight coins. Seven are genuine and one is a fake, which weighs a little less than the other seven. You have a balance scale, which you may use only three times. Tell how to locate the bad coin in three weighings. (Then show how to detect the bad coin in only *two* weighings.)

71. *Geometry Puzzle* When the diagram shown is folded to form a cube, what letter is opposite the face marked Z?

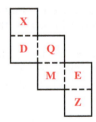

72. *Geometry Puzzle* Draw the following figure without picking up your pencil from the paper and without tracing over a line you have already drawn.

73. *Geometry Puzzle* Repeat **Exercise 72** for this figure.

74. *Books on a Shelf* Volumes 1 and 2 of *The Complete Works of Wally Smart* are standing in numerical order from left to right on your bookshelf. Volume 1 has 450 pages and Volume 2 has 475 pages. Excluding the covers, how many pages are between page 1 of Volume 1 and page 475 of Volume 2?

75. *Paying for a Mint* Brian has an unlimited number of cents (pennies), nickels, and dimes. In how many different ways can he pay 15¢ for a chocolate mint? (For example, one way is 1 dime and 5 pennies.)

76. *Teenager's Age* A teenager's age increased by 2 gives a perfect square. Her age decreased by 10 gives the square root of that perfect square. She is 5 years older than her brother. How old is her brother?

77. *Area and Perimeter* Triangle *ABC* has sides 10, 24, and 26 cm long. A rectangle that has an area equal to that of the triangle is 3 cm wide. Find the perimeter of the rectangle. (November 13, 2008)

78. *Making Change* In how many different ways can you make change for a half dollar using currently minted U.S. coins, if cents (pennies) are not allowed?

79. *Ages* James, Dan, Jessica, and Cathy form a pair of married couples. Their ages are 36, 31, 30, and 29. Jessica is married to the oldest person in the group. James is older than Jessica but younger than Cathy. Who is married to whom, and what are their ages?

80. *Final Digit* What is the last digit of $49{,}327^{1783}$? (April 11, 2009)

81. *Geometry Puzzle* What is the maximum number of small squares in which we may place crosses (✕) and not have any row, column, or diagonal completely filled with crosses? Illustrate your answer.

82. *Making Change* Webster has some pennies, dimes, and quarters in his pocket. When Josefa asks him for change for a dollar, Webster discovers that he cannot make the change exactly. What is the largest possible total value of the coins in his pocket? (October 5, 2009)

WHEN Will I Ever USE This ?

Suppose that you are an employee of a group home, and you have been assigned to provide the necessary items for a Thanksgiving meal for ten of the residents of the home, who will be preparing the meal for themselves. You decide that the dinner will be a traditional one: turkey, stuffing, cranberry sauce, yams, green bean casserole, rolls, and iced tea. You must go to the supermarket for these items and have decided to use cash to pay for them. Of course, you do not want to approach the checkout counter and not have enough for the purchase. You also know that a sales tax of 8.75% will be added to the total. You take your allotment of $80 in cash with you.

In a situation like this, it is not difficult to get a fairly accurate idea of what the total will be by mentally rounding each item (up or down) to the nearest dollar and keeping a running total along the way as you place items in the shopping cart. Here is an example of this procedure.

Item	Actual Cost	Estimate
18-lb turkey	$26.82	$30
12-pack of dinner rolls	2.29	2
15-oz container of margarine	2.79	3
40-oz can of yams	3.34	3
28-oz can of cranberry sauce	3.97	4
26-oz can of cream of mushroom soup	2.99	3
14-oz bag of herb stuffing	3.59	4
1-lb bag of pecans	8.79	9
50-oz can of green beans	2.59	3
28-oz can of onion flakes	2.19	2
22-bag pack of tea bags	4.29	4

Before reaching the checkout counter, look at the items in the basket (move from the top to the bottom in the third column) and add them, rounding off whenever doing so helps to simplify the computation. Here is one of many ways this can be done. (Remember, this is not an *exact* computation, so thought processes will vary.)

"30 plus 2 gives 32; I notice that the next three items total 10, so 32 plus 10 equals 42, which I will round down to 40; plus 3 gives 43; plus 4 gives 47, which I will round up to 50; I will round 9 to 10 and add 50 to 10 to give 60; the final three items total 9, which I will round down to 8 (because I rounded up just before) and add to 60 to get 68; 68 is about 70."

Before tax is added, the items total about $70. The sales tax is 8.75%, so round this to 10%. Taking 10% of a number is simple: Just move the decimal point one unit to the left. In this case, the tax will be about $7. So add 7 to 70 to get an approximate total of $77. It looks like $80 will cover the total cost.

Now, as the cashier rings up the purchase, the screen shows that the actual total cost is $63.65, the sales tax is $5.57, and the grand total is $69.22. The estimate is a bit high in this example (because most of the roundoffs were "upward"), but this is what "being in the ballpark" means: The estimate is close enough for our purpose.

Happy Thanksgiving.

| 1.4 | **NUMERACY IN TODAY'S WORLD** |

OBJECTIVES

1 Be able to use a calculator for routine mathematical operations.

2 Be able to use estimation techniques.

3 Be able to interpret information by reading circle, bar, and line graphs.

4 Be able to use writing skills to convey information about mathematics.

The familiar term *literacy* applies to language in the same way that the term **numeracy** applies to mathematics. It is virtually impossible to function in the world today without understanding fundamental number concepts. The basic ideas of calculating, estimating, interpreting data from graphs, and conveying mathematics via language and writing are among the skills required to be "numerate."

Calculation

The search for easier ways to calculate and compute has culminated in the development of hand-held calculators and computers. For the general population, a calculator that performs the operations of arithmetic and a few other functions is sufficient. These are known as **four-function calculators.** Students who take higher mathematics courses (engineers, for example) usually need the added power of **scientific calculators. Graphing calculators,** which actually plot graphs on small screens, are also available. *Always refer to your owner's manual if you need assistance in performing an operation with your calculator. If you need further help, ask your instructor or another student who is using the same model.*

Today's smartphones routinely include a calculator application (app). For example, Apple's iPhone has an app that serves as a four-function calculator when the phone is held vertically, but it becomes a scientific calculator when held horizontally. Furthermore, graphing calculator apps are available at little or no cost. Although it is not necessary to have a graphing calculator to study the material presented in this text, we occasionally include graphing calculator screens to support results obtained or to provide supplemental information.[*]

The screens that follow illustrate some common entries and operations.

3+9		
		12
7−2		
		5
4*5		
		20

$3 + 9 = 12$
$7 - 2 = 5$
$4 \times 5 = 20$

A

24/20	
	1.2
Ans▶Frac	
	$\frac{6}{5}$
5−(8−7)	
	4

$\frac{24}{20} = 1.2$
$1.2 = \frac{6}{5}$
$5 - (8 - 7) = 4$

B

7^2	
	49
5^3	
	125
$\sqrt{81}$	
	9

$7^2 = 49$
$5^3 = 125$
$\sqrt{81} = 9$

C

Screen A illustrates how two numbers can be added, subtracted, or multiplied. Screen B shows how two numbers can be divided, how the decimal quotient (stored in the memory cell Ans) can be converted into a fraction, and how parentheses can be used in a computation. Screen C shows how a number can be squared, how it can be cubed, and how its square root can be taken.

Shown here is an example of a **calculator app** from a smartphone. Since the introduction of hand-held calculators in the early 1970s, the methods of everyday arithmetic have been drastically altered. One of the first consumer models available was the Texas Instruments SR-10, which sold for nearly $150 in 1973. It could perform the four operations of arithmetic and take square roots, but it could do very little more.

$\sqrt[3]{27}$	
	3
$\sqrt[4]{16}$	
	2
5^{-1}	
	.2

$\sqrt[3]{27} = 3$
$\sqrt[4]{16} = 2$
$5^{-1} \left(\text{or} \frac{1}{5} \right) = .2$

D

π	
	3.141592654
5!	
	120
6265804*8980591	
	5.627062301E13

$\pi \approx 3.141592654$
$5!$ (or $1 \times 2 \times 3 \times 4 \times 5$) $= 120$
$6,265,804 \times 8,980,591 \approx 5.627062301 \times 10^{13}$
$\approx$ indicates "is approximately equal to"

E

---

[*]Because it is one of the most popular graphing calculators, we include screens similar to those generated by the TI-83 Plus and TI-84 Plus from Texas Instruments.

The popular **TI-84 Plus** graphing calculator is shown here.

Screen D shows how other roots (cube root and fourth root) can be found, and how the reciprocal of a number can be found using -1 as an exponent. Screen E shows how π can be accessed with its own special key, how a *factorial* (as indicated by !) can be found, and how a result might be displayed in *scientific notation*. (The "E13" following 5.627062301 means that this number is multiplied by 10^{13}. This answer is still only an approximation, because the product $6{,}265{,}804 \times 8{,}980{,}591$ contains more digits than the calculator can display.)

In **Section 1.3** we presented a list of problem-solving strategies. As Terry Krieger points out in the chapter opener, "mathematics may be the only discipline in which different people, using wildly varied but logically sound methods, will arrive at the same correct result" with respect to solving problems. Sometimes more than one strategy can be used in a particular situation. In **Example 1,** we present a type of problem that has been around for thousands of years in various forms, and is solved by observing a pattern within a table. One ancient form of this problem deals with doubling a kernel of corn for each square on a checkerbord. (See http://mathforum.org/sanders/geometry/GP11Fable.html.) It illustrates an example of exponential growth, covered in more detail in a later chapter.

EXAMPLE 1 Calculating a Sum That Involves a Pattern

Following her success in *I Love Lucy,* Lucille Ball starred in *The Lucy Show,* which aired for six seasons on CBS in the 1960s. She worked for Mr. Mooney (Gale Gordon), who was very careful with his money.

In the September 26, 1966, show "Lucy, the Bean Queen," Lucy learned a lesson about saving money. Mr. Mooney had refused to lend her $1500 to buy furniture, because he claimed she did not know the value of money. He explained to her that if she were to save so that she would have one penny on Day 1, two pennies on Day 2, four pennies on Day 3, and so on, she would have more than enough money to buy her furniture after only nineteen days. Use a calculator to verify this fact.

Solution

A calculator will help in constructing **Table 4.**

Table 4

Day Number	Accumulated Savings on That Day	Day Number	Accumulated Savings on That Day
1	$0.01	11	$10.24
2	0.02	12	20.48
3	0.04	13	40.96
4	0.08	14	81.92
5	0.16	15	163.84
6	0.32	16	327.68
7	0.64	17	655.36
8	1.28	18	1310.72
9	2.56	19	**2621.44**
10	5.12		

So, indeed, Lucy will have accumulated $2621.44 on day 19, which is enough to buy the furniture.) ∎

Estimation

Although calculators can make life easier when it comes to computations, many times we need only estimate an answer to a problem, and in these cases, using a calculator may not be necessary or appropriate.

EXAMPLE 2 Estimating an Appropriate Number of Birdhouses

A birdhouse for swallows can accommodate up to 8 nests. How many birdhouses would be necessary to accommodate 58 nests?

Solution

If we divide 58 by 8 either by hand or with a calculator, we get 7.25. Can this possibly be the desired number? Of course not, because we cannot consider fractions of birdhouses. Do we need 7 or 8 birdhouses?

To provide nesting space for the nests left over after the 7 birdhouses (as indicated by the decimal fraction), we should plan to use 8 birdhouses. In this problem, we must round our answer *up* to the next counting number. ∎

EXAMPLE 3 Approximating Average Number of Yards per Carry

In 2013, Fred Jackson of the Buffalo Bills carried the football a total of 206 times for 890 yards (*Source:* www.nfl.com). Approximate his average number of yards per carry that year.

Solution

Because we are told only to find Jackson's approximate average, we can say that he carried about 200 times for about 900 yards, and his average was therefore about $\frac{900}{200} = 4.5$ yards per carry. (A calculator shows that his average to the nearest tenth was 4.3 yards per carry. Verify this.) ∎

Interpretation of Graphs

In a **circle graph,** or **pie chart,** a circle is used to indicate the total of all the data categories represented. The circle is divided into sectors, or wedges (like pieces of a pie), whose sizes show the relative magnitudes of the categories. The sum of all the fractional parts must be 1 (for one whole circle).

In the introduction to his book *Innumeracy: Mathematical Illiteracy and Its Consequences,* Temple University professor **John Allen Paulos** writes

> ***Innumeracy,*** *an inability to deal comfortably with the fundamental notions of number and chance, plagues far too many otherwise knowledgeable citizens.*
> *. . .*
> *(W)e were watching the news and the TV weathercaster announced that there was a 50 percent chance of rain for Saturday and a 50 percent chance for Sunday, and concluded that there was therefore a 100 percent chance of rain that weekend. . . . (U)nlike other failings which are hidden, mathematical illiteracy is often flaunted. "I can't even balance my checkbook." "I'm a people person, not a numbers person." Or "I always hated math."*

EXAMPLE 4 Interpreting Information in a Circle Graph

In a recent month there were about 2100 million (2.1 billion) Internet users worldwide. The circle graph in **Figure 8** shows the approximate shares of these users living in various regions of the world.

(a) Which region had the largest share of Internet users? What was that share?

(b) Estimate the number of Internet users in North America.

(c) How many actual Internet users were there in North America?

Worldwide Internet Users by Region

- North America 13%
- Other 20.5%
- Europe 22.5%
- Asia 44%

Source: www.internetworldstats.com

Figure 8

Solution

(a) In the circle graph, the sector for Asia is the largest, so Asia had the largest share of Internet users, 44%.

(b) A share of 13% can be rounded down to 10%. Then find 10% of 2100 million by finding $\frac{1}{10}$ of 2100, or 210. There were *about* 210 million users in North America. (This estimate is low, because we rounded *down*.)

(c) To find the actual number of users, find 13% of 2100 million. We do this by multiplying 0.13×2100 million.

$$\underbrace{0.13}_{13\%} \quad \times \quad \underbrace{2100 \text{ million}}_{\text{total}} \quad = \quad \underbrace{273 \text{ million}}_{\substack{\text{Actual number} \\ \text{of users in} \\ \text{North America}}}$$

■

A **bar graph** is used to show comparisons. It consists of a series of bars (or simulations of bars) arranged either vertically or horizontally. In a bar graph, values from two categories are paired with each other (for example, years with dollar amounts).

EXAMPLE 5 Interpreting Information in a Bar Graph

The bar graph in **Figure 9** shows annual per-capita spending on health care in the United States for the years 2004 through 2009.

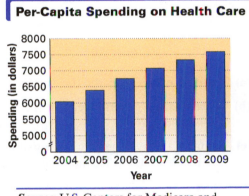

Per-Capita Spending on Health Care

Source: U.S. Centers for Medicare and Medicaid Services.

Figure 9

(a) In what years was per-capita health care spending greater than $7000?

(b) Estimate per-capita health care spending in 2004 and 2008.

(c) Describe the change in per-capita spending as the years progressed.

Solution

(a) Locate 7000 on the vertical axis and follow the line across to the right. Three years—2007, 2008, and 2009—have bars that extend above the line for 7000, so per-capita health care spending was greater than $7000 in those years.

(b) Locate the top of the bar for 2004 and move horizontally across to the vertical scale to see that it is about 6000. Per-capita health care spending for 2004 was about $6000.

Similarly, follow the top of the bar for 2008 across to the vertical scale to see that it lies a little more than halfway between 7000 and 7500, so per-capita health care spending in 2008 was about $7300.

(c) As the years progressed, per-capita spending on health care increased steadily, from about $6000 in 2004 to about $7500 in 2009. ∎

A **line graph** is used to show changes or trends in data over time. To form a line graph, we connect a series of points representing data with line segments.

EXAMPLE 6   Interpreting Information in a Line Graph

The line graph in **Figure 10** shows average prices of a gallon of regular unleaded gasoline in the United States for the years 2004 through 2011.

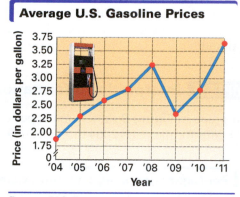

Source: U.S. Department of Energy.

Figure 10

(a) Between which years did the average price of a gallon of gasoline decrease?

(b) What was the general trend in the average price of a gallon of gasoline from 2004 through 2008?

(c) Estimate the average price of a gallon of gasoline in 2004 and in 2008. About how much did the price increase between 2004 and 2008?

Solution

(a) The line between 2008 and 2009 falls, so the average price of a gallon of gasoline decreased from 2008 to 2009.

(b) The line graph rises from 2004 to 2008, so the average price of a gallon of gasoline increased over those years.

(c) Move up from 2004 on the horizontal scale to the point plotted for 2004. This point is about halfway between the lines on the vertical scale for $1.75 and $2.00. Halfway between $1.75 and $2.00 would be about $1.88. Therefore, a gallon of gasoline cost about $1.88 in 2004.

Similarly, locate the point plotted for 2008. Moving across to the vertical scale, the graph indicates that the price for a gallon of gasoline in 2008 was about $3.25.

Between 2004 and 2008, the average price of a gallon of gasoline increased by about

$$\$3.25 - \$1.88 = \$1.37.$$ ∎

Mathematical writing takes many forms. One of the most famous author/ mathematicians was **Charles Dodgson** (1832–1898), who used the pen name **Lewis Carroll.**

Dodgson was a mathematics lecturer at Oxford University in England. Queen Victoria told Dodgson how much she enjoyed *Alice's Adventures in Wonderland* and how much she wanted to read his next book; he is said to have sent her *Symbolic Logic,* his most famous mathematical work.

The *Alice* books made Carroll famous. Late in life, however, Dodgson shunned attention and denied that he and Carroll were the same person, even though he gave away hundreds of signed copies to children and children's hospitals.

Communicating Mathematics through Language Skills

Research has indicated that the ability to express mathematical observations in writing can serve as a positive force in one's continued development as a mathematics student. The implementation of writing in the mathematics class can use several approaches.

One way of using writing in mathematics is to keep a **journal** in which you spend a few minutes explaining what happened in class that day. The journal entries may be general or specific, depending on the topic covered, the degree to which you understand the topic, your interest level at the time, and so on. Journal entries are usually written in informal language.

Although journal entries are for the most part informal writings in which the student's thoughts are allowed to roam freely, entries in **learning logs** are typically more formal. An instructor may pose a specific question for a student to answer in a learning log. In this text, we intersperse in each exercise set exercises that require written answers that are appropriate for answering in a learning log.

EXAMPLE 7 Writing an Answer to a Conceptual Exercise

Exercise 17 of **Section 1.1** reads as follows.

> *Discuss the differences between inductive and deductive reasoning. Give an example of each.*

Write a short paragraph to answer this exercise.

Solution
Here is one possible response.

> Deductive reasoning occurs when you go from general ideas to specific ones. For example, I know that I can multiply both sides of $\frac{1}{2}x = 6$ by 2 to get $x = 12$, because I can multiply both sides of any equation by whatever I want (except 0). Inductive reasoning goes the other way. If I have a general conclusion from specific observations, that's inductive reasoning. Example – in the numbers 4, 8, 12, 16, and so on, I can conclude that the next number is 20, since I always add 4 to get the next number.

The motto "Publish or perish" has long been around, implying that a scholar in pursuit of an academic position must publish in a journal in his or her field. There are numerous such journals in mathematics research and/or mathematics education. The National Council of Teachers of Mathematics publishes *Teaching Children Mathematics, Mathematics Teaching in the Middle School, Mathematics Teacher, Journal for Research in Mathematics Education, Mathematics Teacher Educator,* and *Student Explorations in Mathematics.* Refer to the Web site www.nctm.org to access these journals, or refer to print copies in your local library.

Writing a report on a journal article can help you understand what mathematicians do and what ideas mathematics teachers use to convey concepts to their students. Many professors in mathematics survey courses require short term papers of their students. In doing such research, students can become aware of the plethora of books and articles on mathematics and mathematicians, many written specifically for the layperson.

A list of important mathematicians, philosophers, and scientists follows.

Abel, N.	Cardano, G.	Gauss, C.	Noether, E.
Agnesi, M. G.	Copernicus, N.	Hilbert, D.	Pascal, B.
Agnesi, M. T.	De Morgan, A.	Kepler, J.	Plato
Al-Khowârizmi	Descartes, R.	Kronecker, L.	Polya, G.
Apollonius	Euler, L.	Lagrange, J.	Pythagoras
Archimedes	Fermat, P.	Leibniz, G.	Ramanujan, S.
Aristotle	Fibonacci	L'Hôspital, G.	Riemann, G.
Babbage, C.	(Leonardo	Lobachevsky, N.	Russell, B.
Bernoulli, Jakob	of Pisa)	Mandelbrot, B.	Somerville, M.
Bernoulli,	Galileo (Galileo	Napier, J.	Tartaglia, N.
Johann	Galilei)	Nash, J.	Whitehead, A.
Cantor, G.	Galois, E.	Newton, I.	Wiles, A.

The following topics in the history and development of mathematics can also be used for term papers.

Babylonian mathematics	Pascal's triangle
Egyptian mathematics	The origins of probability theory
The origin of zero	Women in mathematics
Plimpton 322	Mathematical paradoxes
The Rhind papyrus	Unsolved problems in mathematics
Origins of the Pythagorean theorem	The four-color theorem
The regular (Platonic) solids	The proof of Fermat's Last Theorem
The Pythagorean brotherhood	The search for large primes
The Golden Ratio (Golden Section)	Fractal geometry
The three famous construction problems of the Greeks	The co-inventors of calculus
The history of the approximations of π	The role of the computer in the study of mathematics
Euclid and his *Elements*	Mathematics and music
Early Chinese mathematics	Police mathematics
Early Hindu mathematics	The origins of complex numbers
Origin of the word *algebra*	Goldbach's conjecture
Magic squares	The use of the Internet in mathematics education
Figurate numbers	The development of graphing calculators
The Fibonacci sequence	
The Cardano/Tartaglia controversy	Mathematics education reform movement
Historical methods of computation (logarithms, the abacus, Napier's rods, the slide rule, etc.)	Multicultural mathematics
	The Riemann Hypothesis

1.4 | EXERCISES

Perform the indicated operations, and give as many digits in your answer as shown on your calculator display. (The number of displayed digits may vary depending on the model used.)

1. $39.7 + (8.2 - 4.1)$

2. $2.8 \times (3.2 - 1.1)$

3. $\sqrt{5.56440921}$

4. $\sqrt{37.38711025}$

5. $\sqrt[3]{418.508992}$

6. $\sqrt[3]{700.227072}$

7. 2.67^2

8. 3.49^3

9. 5.76^5

10. 1.48^6

11. $\dfrac{14.32 - 8.1}{2 \times 3.11}$

12. $\dfrac{12.3 + 18.276}{3 \times 1.04}$

13. $\sqrt[5]{1.35}$

14. $\sqrt[6]{3.21}$

15. $\dfrac{\pi}{\sqrt{2}}$

16. $\dfrac{2\pi}{\sqrt{3}}$

17. $\sqrt[4]{\dfrac{2143}{22}}$

18. $\dfrac{12{,}345{,}679 \times 72}{\sqrt[3]{27}}$

19. $\dfrac{\sqrt{2}}{\sqrt[3]{6}}$

20. $\dfrac{\sqrt[3]{12}}{\sqrt{3}}$

21. Choose any number consisting of five digits. Multiply it by 9 on your calculator. Now add the digits in the answer. If the sum is more than 9, add the digits of this sum, and repeat until the sum is less than 10. Your answer will always be 9. Repeat the exercise with a number consisting of six digits. Does the same result hold?

22. Use your calculator to *square* the following two-digit numbers ending in 5: 15, 25, 35, 45, 55, 65, 75, 85. Write down your results, and examine the pattern that develops. Then use inductive reasoning to predict the value of 95^2. Write an explanation of how you can mentally square a two-digit number ending in 5.

Perform each calculation and observe the answers. Then fill in the blank with the appropriate response.

23. $\left(\dfrac{-3}{-8}\right)$; $\left(\dfrac{-5}{-4}\right)$; $\left(\dfrac{-2.7}{-4.3}\right)$

Dividing a negative number by another negative number gives a _____ product.
<u>(negative/positive)</u>

24.

Multiplying a negative number by a positive number gives a _____ product.
<u>(negative/positive)</u>

25. $\boxed{5 \cdot 6^0}$; $\boxed{\pi^0}$; $\boxed{2^0}$; $\boxed{120^0}$

Raising a nonzero number to the power 0 gives a result of _____.

26. $\boxed{1^2}$; $\boxed{1^3}$; $\boxed{1^{-3}}$; $\boxed{1^0}$

Raising 1 to any power gives a result of _____.

27. $\boxed{\dfrac{1}{7}}$; $\boxed{\dfrac{1}{-9}}$; $\boxed{\dfrac{1}{3}}$; $\boxed{\dfrac{1}{-8}}$

The sign of the reciprocal of a number is _____ the sign of the number.
<u>(the same as/different from)</u>

28. $\boxed{5 \div 0}$; $\boxed{9 \div 0}$; $\boxed{0 \div 0}$

Dividing a number by 0 gives a(n) _____ on a calculator.

29. $\boxed{0 \div 8}$; $\boxed{0 \div -2}$; $\boxed{0 \div \pi}$

Zero divided by a nonzero number gives a quotient of _____.

30. $\boxed{\sqrt{-3}}$; $\boxed{\sqrt{-4}}$; $\boxed{\sqrt{-10}}$

Taking the square root of a negative number gives a(n) _____ on a calculator.

31. $\boxed{-3 \ast -4 \ast -5}$; $\boxed{-3 \ast -4 \ast -5 \ast -6 \ast -7}$;

$\boxed{-3 \ast -4 \ast -5 \ast -6 \ast -7 \ast -8 \ast -9}$

Multiplying an *odd* number of negative numbers gives a _____ product.
<u>(positive/negative)</u>

32. $\boxed{-3 \ast -4}$; $\boxed{-3 \ast -4 \ast -5 \ast -6}$;

$\boxed{-3 \ast -4 \ast -5 \ast -6 \ast -7 \ast -8}$

Multiplying an *even* number of negative numbers gives a _____ product.
<u>(positive/negative)</u>

33. Find the decimal representation of $\frac{1}{6}$ on your calculator. Following the decimal point will be a 1 and a string of 6s. The final digit will be a 7 if your calculator *rounds off* or a 6 if it *truncates*. Which kind of calculator do you have?

34. Choose any three-digit number and enter the digits into a calculator. Then enter them again to get a six-digit number. Divide this six-digit number by 7. Divide the result by 13. Divide the result by 11. What is interesting about your answer? Explain why this happens.

35. Choose any digit except 0. Multiply it by 429. Now multiply the result by 259. What is interesting about your answer? Explain why this happens.

36. Refer to **Example 1.** If Lucy continues to double her savings amount each day, on what day will she become a millionaire?

Give an appropriate counting number answer to each question in Exercises 37–40. (Find the least counting number that will work.)

37. *Pages to Store Trading Cards*   A plastic page designed to hold trading cards will hold up to 9 cards. How many pages will be needed to store 563 cards?

38. *Drawers for DVDs*   A sliding drawer designed to hold DVD cases has 20 compartments. If Chris wants to house his collection of 408 Disney DVDs, how many such drawers will he need?

39. *Containers for African Violets*   A gardener wants to fertilize 800 African violets. Each container of fertilizer will supply up to 60 plants. How many containers will she need to do the job?

40. *Fifth-Grade Teachers Needed*   False River Academy has 155 fifth-grade students. The principal has decided that each fifth-grade teacher should have a maximum of 24 students. How many fifth-grade teachers does he need?

In Exercises 41–46, use estimation to determine the choice closest to the correct answer.

41. *Price per Acre of Land*   To build a "millennium clock" on Mount Washington in Nevada that would tick once each year, chime once each century, and last at least 10,000 years, the nonprofit Long Now Foundation purchased 80 acres of land for $140,000. Which one of the following is the closest estimate to the price per acre?

   A. $1000   **B.** $2000   **C.** $4000   **D.** $11,200

42. *Time of a Round Trip*   The distance from Seattle, Washington, to Springfield, Missouri, is 2009 miles. About how many hours would a round trip from Seattle to Springfield and back take a bus that averages 50 miles per hour for the entire trip?

   A. 60   **B.** 70   **C.** 80   **D.** 90

43. *People per Square Mile*   Baton Rouge, LA has a population of 230,058 and covers 76.9 square miles. About how many people per square mile live in Baton Rouge?

   A. 3000   **B.** 300   **C.** 30   **D.** 30,000

44. *Revolutions of Mercury*   The planet Mercury takes 88.0 Earth days to revolve around the sun once. Pluto takes 90,824.2 days to do the same. When Pluto has revolved around the sun once, about how many times will Mercury have revolved around the sun?

   A. 100,000   **B.** 10,000   **C.** 1000   **D.** 100

45. *Reception Average*   In 2013, A. J. Green of the Cincinnati Bengals caught 98 passes for 1426 yards. His approximate number of yards gained per catch was _____.

   A. $\frac{1}{14}$ **B.** 0.07   **C.** 139,748   **D.** 14

46. *Area of the Sistine Chapel*   The Sistine Chapel in Vatican City measures 40.5 meters by 13.5 meters.

Which is the closest approximation to its area?

   A. 110 meters   **B.** 55 meters
   C. 110 square meters   **D.** 600 square meters

Foreign-Born Americans   *Approximately 37.5 million people living in the United States in a recent year were born in other countries. The circle graph gives the share from each region of birth for these people. Use the graph to answer each question in Exercises 47–50.*

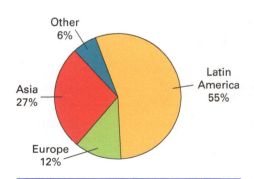

U.S. Foreign-Born Population by Region of Birth

Other 6%
Latin America 55%
Asia 27%
Europe 12%

Source: U.S. Census Bureau.

47. What share was from other regions?

48. What share was from Latin America or Asia?

49. How many people (in millions) were born in Europe?

50. How many more people (in millions) were born in Latin America than in Asia?

Milk Production The bar graph shows total U.S. milk production (in billions of pounds) for the years 2004 through 2010. Use the bar graph to work Exercises 51–54.

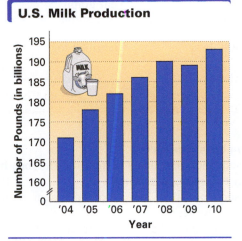

U.S. Milk Production

Source: U.S. Department of Agriculture.

51. In what years was U.S. milk production greater than 185 billion pounds?

52. In what years was U.S. milk production about the same?

53. Estimate U.S. milk production in 2004 and 2010.

54. Describe the change from 2004 to 2010.

U.S. Car Imports The line graph shows the number of new and used passenger cars (in millions) imported into the United States over the years 2005 through 2010. Use the line graph to work Exercises 55–58.

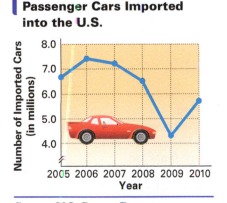

Passenger Cars Imported into the U.S.

Source: U.S. Census Bureau.

55. Over which two consecutive years did the number of imported cars increase the most? About how much was this increase?

56. Estimate the number of cars imported during 2007, 2008, and 2009.

57. Describe the trend in car imports from 2006 to 2009.

58. During which year(s) were fewer than 6 millions cars imported into the United States?

Use effective writing skills to address the following.

59. *Mathematics Web Sites* The following Web sites provide a fascinating list of mathematics-related topics. Investigate, choose a topic that interests you, and report on it according to the guidelines provided by your instructor.

www.mathworld.wolfram.com

world.std.com/~reinhold/mathmovies.html

www.maths.surrey.ac.uk/hosted-sites/R.Knott/

http://dir.yahoo.com/Science/Mathematics/

www.cut-the-knot.org

www.ics.uci.edu/~eppstein/recmath.html

www.coolmathguy.com

ptri1.tripod.com

mathforum.org

rosettacode.org

plus.maths.org

60. *The Simpsons* The longest-running animated television series is *The Simpsons,* having begun in 1989. The Web site www.simpsonsmath.com explores the occurrence of mathematics in the episodes on a season-by-season basis. Watch several episodes and elaborate on the mathematics found in them.

61. *Donald in Mathmagic Land* One of the most popular mathematical films of all time is *Donald in Mathmagic Land,* a 1959 Disney short that is available on DVD. Spend an entertaining half-hour watching this film, and write a report on it according to the guidelines provided by your instructor.

62. *Mathematics in Hollywood* A theme of mathematics-related scenes in movies and television is found throughout this text. Prepare a report on one or more such scenes, and determine whether the mathematics involved is correct or incorrect. If correct, show why. If incorrect, find the correct answer.

CHAPTER 1	SUMMARY

KEY TERMS

1.1

conjecture
inductive reasoning
counterexample
deductive reasoning
natural (counting) numbers
base
exponent
premise
conclusion
logical argument

1.2

number sequence
terms of a sequence
arithmetic sequence
common difference
geometric sequence
common ratio
method of successive
   differences
mathematical
   induction

triangular, square, and
   pentagonal numbers
figurate number
subscript
Kaprekar constant

1.3

perfect square
square root
radical symbol

1.4

numeracy
four-function calculator
scientific calculator
graphing calculator
circle graph (pie chart)
bar graph
line graph
journal
learning log

TEST YOUR WORD POWER

See how well you have learned the vocabulary in this chapter.

1. A **conjecture** is
  A. a statement that has been proved to be true.
  B. an educated guess based on repeated observations.
  C. an example that shows that a general statement is false.
  D. an example of deductive reasoning.

2. An example of a **natural number** is
  A. 0.   **B.** $\frac{1}{2}$.   **C.** -1.   **D.** 1.

3. An **arithmetic sequence** is
  A. a sequence that has a common difference between any two successive terms.
  B. a sequence that has a common sum of any two successive terms.
  C. a sequence that has a common ratio between any two successive terms.
  D. a sequence that can begin 1, 1, 2, 3, 5. . . .

4. A **geometric sequence** is
  A. a sequence that has a common difference between any two successive terms.
  B. a sequence that has a common sum of any two successive terms.
  C. a sequence that has a common ratio between any two successive terms.
  D. A sequence that can begin 1, 1, 2, 3, 5, ...

5. The symbol T_n, which uses the **subscript** n, is read
  A. "T to the nth power."   **B.** "T times n."
  C. "T of n."   **D.** "T sub n."

ANSWERS
1. B **2.** D **3.** A **4.** C **5.** D

QUICK REVIEW

Concepts	Examples
1.1 **Solving Problems by Inductive Reasoning**	
Inductive Reasoning Inductive reasoning is characterized by drawing a general conclusion (making a conjecture) from repeated observations of specific examples. The conjecture may or may not be true.	Consider the following: *When I square the first twenty numbers ending in 5, the result always ends in 25. Therefore, I make the conjecture that this happens in the twenty-first case.*
A general conclusion from inductive reasoning can be shown to be false by providing a single counterexample.	This is an example of inductive reasoning because a general conclusion follows from repeated observations.

Concepts	Examples

Deductive Reasoning
Deductive reasoning is characterized by applying general principles to specific examples.

Consider the following:
The formula for finding the perimeter P of a rectangle with length L and width W is P = 2L + 2W. Therefore, the perimeter P is

$$2(5) + 2(3) = 16.$$

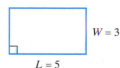

This is an example of deductive reasoning because a specific conclusion follows from a mathematical formula that is true in general.

1.2 An Application of Inductive Reasoning: Number Patterns

Sequences
A number sequence is a list of numbers having a first number, a second number, a third number, and so on, which are called the terms of the sequence.

Arithmetic Sequence
In an arithmetic sequence, each term after the first is obtained by adding the same number, called the common difference.

The arithmetic sequence that begins
$$2, 4, 6, 8$$
has common difference $4 - 2 = 2$, and the next term in the sequence is $8 + 2 = 10$.

Geometric Sequence
In a geometric sequence, each term after the first is obtained by multiplying by the same number, called the common ratio.

The geometric sequence that begins
$$4, 20, 100, 500$$
has common ratio $\frac{20}{4} = 5$, and the next term in the sequence is $500 \times 5 = 2500$.

Method of Successive Differences
The next term in a sequence can sometimes be found by computing successive differences between terms until a pattern can be established.

The sequence that begins
$$7, 15, 25, 37$$
has the following successive differences.

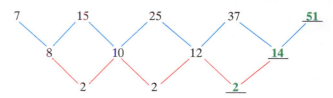

The next term in the sequence is $37 + 14 = 51$.

Figurate Numbers
Figurate numbers, such as triangular, square, and pentagonal numbers, can be represented by geometric arrangements of points.

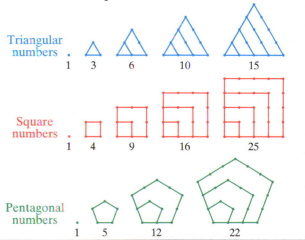

Concepts	Examples

1.3 **Strategies for Problem Solving**

Polya's Four-Step Method for Problem Solving

Step 1 Understand the problem.

Step 2 Devise a plan.

Step 3 Carry out the plan.

Step 4 Look back and check.

Problem-Solving Strategies

• Make a table or a chart.

• Look for a pattern.

• Solve a similar, simpler problem.

• Draw a sketch.

• Use inductive reasoning.

• Write an equation and solve it.

• If a formula applies, use it.

• Work backward.

• Guess and check.

• Use trial and error.

• Use common sense.

• Look for a "catch" if an answer seems too obvious or impossible.

What is the ones, or units, digit in 7^{350}?

Solution
We can observe a pattern in the table of simpler powers of 7. (Use a calculator.)

$7^1 = 7$	$7^5 = 16,807$	$7^9 = 40,353,607$
$7^2 = 49$	$7^6 = 117,649$	. . .
$7^3 = 343$	$7^7 = 823,543$	. . .
$7^4 = 2401$	$7^8 = 5,764,801$	. . .

The ones digit appears in a pattern of four digits over and over: 7, 9, 3, 1, 7, 9, 3, 1, ... If the exponent is divided by 4, the remainder helps predict the ones digit. If we divide the exponent 350 by 4, the quotient is 87 and the remainder is 2, just as it is in the second row above for 7^2 and 7^6, where the units digit is **9**. So the units digit in 7^{350} is 9.

How many ways are there to make change equivalent to one dollar using only nickels, dimes, and quarters? You do not need at least one coin of each denomination. (December 26, 2013)

Solution (Verify each of the following by trial and error.)

If we start with 4 quarters, there is 1 way to make change for a dollar.

If we start with 3 quarters, there are 3 ways.

If we start with 2 quarters, there are 6 ways.

If we start with 1 quarter, there are 8 ways.

If we start with 0 quarters, there are 11 ways.

Thus, there are
$1 + 3 + 6 + 8 + 11$
$= 29$ ways in all.

1.4 **Numeracy in Today's World**

There are a variety of types of calculators available; four-function, scientific, and graphing calculators are some of them.

In practical applications, it is often convenient to simply approximate to get an idea of an answer to a problem.

Circle graphs (pie charts), bar graphs, and line graphs are used in today's media to illustrate data in a compact way.

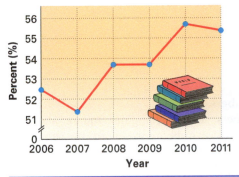

Percents of Students Who Return for Second Year (2-Year Public Institutions)

Source: ACT.

Over which two consecutive years did the percent stay the same? Estimate this percent.

Solution
The graph is horizontal between 2008 and 2009, at about 53.7%.

CHAPTER 1   TEST

In Exercises 1 and 2, decide whether the reasoning involved is an example of inductive or deductive reasoning.

1. Michelle is a sales representative for a publishing company. For the past 16 years, she has exceeded her annual sales goal, primarily by selling mathematics textbooks. Therefore, she will also exceed her annual sales goal this year.

2. For all natural numbers n, n^2 is also a natural number. 176 is a natural number. Therefore, 176^2 is a natural number.

3. ***Counting Puzzle (Rectangles)*** How many rectangles of any size are in the figure shown? (September 10, 2001)

4. Use the list of equations and inductive reasoning to predict the next equation, and then verify your conjecture.

    $65,359,477,124,183 \times 17 = 1,111,111,111,111,111$
    $65,359,477,124,183 \times 34 = 2,222,222,222,222,222$
    $65,359,477,124,183 \times 51 = 3,333,333,333,333,333$

5. Use the method of successive differences to find the next term in the sequence

    $$3, 11, 31, 69, 131, 223, \ldots.$$

6. Find the sum $1 + 2 + 3 + \cdots + 250$.

7. Consider the following equations, where the left side of each is an octagonal number.

    $$1 = 1$$
    $$8 = 1 + 7$$
    $$21 = 1 + 7 + 13$$
    $$40 = 1 + 7 + 13 + 19$$

    Use the pattern established on the right sides to predict the next octagonal number. What is the next equation in the list?

8. Use the result of **Exercise 7** and the method of successive differences to find the first eight octagonal numbers. Then divide each by 4 and record the remainder. What is the pattern obtained?

9. Describe the pattern used to obtain the terms of the Fibonacci sequence below. What is the next term?

    $$1, 1, 2, 3, 5, 8, 13, 21, \ldots.$$

Use problem-solving strategies to solve each problem, taken from the date indicated in the monthly calendar of Mathematics Teacher.

10. ***Building a Fraction*** Each of the four digits 2, 4, 6, and 9 is placed in one of the boxes to form a fraction. The numerator and the denominator are both two-digit whole numbers. What is the smallest value of all the common fractions that can be formed? Express your answer as a common fraction. (November 17, 2004)

11. ***Units Digit of a Power of 9*** What is the units digit (ones digit) in the decimal representation of 9^{1997}? (January 27, 1997)

12. ***Counting Puzzle (Triangles)*** How many triangles are in this figure? (January 6, 2000)

13. ***Make Them Equal*** Consider the following:

    $$1\ 2\ 3\ 4\ 5\ 6\ 7\ 8\ 9\ 0 = 100.$$

    Leaving all the numerals in the order given, insert addition and subtraction signs into the expression to make the equation true. (March 23, 2008)

14. ***Shrinkage*** Dr. Small is 36 inches tall, and Ms. Tall is 96 inches tall. If Dr. Small shrinks 2 inches per year and Ms. Tall grows $\frac{2}{3}$ of an inch per year, how tall will Ms. Tall be when Dr. Small disappears altogether? (November 2, 2007)

15. ***Units Digit of a Sum*** Find the units digit (ones digit) of the decimal numeral representing the number

    $$11^{11} + 14^{14} + 16^{16}. \text{ (February 14, 1994)}$$

16. Based on your knowledge of elementary arithmetic, describe the pattern that can be observed when the following operations are performed.

    $$9 \times 1, \quad 9 \times 2, \quad 9 \times 3, \ldots, \quad 9 \times 9$$

    (*Hint:* Add the digits in the answers. What do you notice?)

Use your calculator to evaluate each of the following. Give as many decimal places as the calculator displays.

17. $\sqrt{98.16}$

18. 3.25^3

19. *Basketball Scoring Results* During the 2012–13 NCAA women's basketball season, Brittney Griner of Baylor made 148 of her 208 free throw attempts. This means that for every 20 attempts, she made approximately _____ of them.

    A. 10 **B.** 14 **C.** 8 **D.** 11

20. *Unemployment Rate* *The line graph shows the overall unemployment rate in the U.S. civilian labor force for the years 2003 through 2010.*

  (a) Between which pairs of consecutive years did the unemployment rate decrease?

  (b) What was the general trend in the unemployment rate between 2007 and 2010?

  (c) Estimate the overall unemployment rate in 2008 and 2009. About how much did the unemployment rate increase between 2008 and 2009?

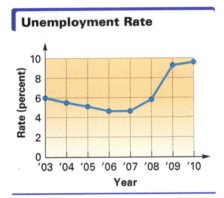

Unemployment Rate

Source: Bureau of Labor Statistics.

The Basic Concepts of Set Theory

2

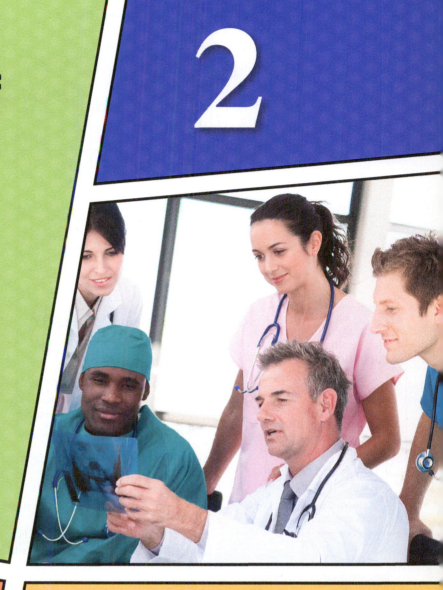

For many reasons, the job outlook in the United States is good—and improving—for all types of nursing careers, as well as other health-related categories. Students may be considering groups of different training programs; lists of certificate or degree objectives; an array of employment opportunities; the pros and cons of different opportunities relative to upward mobility, stress level, and flexibility; and many others. The properties of these arrays, groups, and collections can be better understood by treating them all as *sets* (the mathematical term) and applying the methods presented in this chapter. (See, for example, the opening discussion of **Section 2.4** on page 71.)

Refer to the table on the next page. To the nearest tenth of a percent, what percentage increase is predicted for number of jobs over the decade 2012–2022 for RNs? For LPNs? (See page 50 for the answer.)

	2012 Median Salary	Number of Jobs in 2012	Predicted Increase in Number of Jobs, 2012–2022
Registered Nurses, RNs	$65,470	2,711,500	526,800
Licensed Practical (or Vocational) Nurses, LPNs (or LVNs)	$41,540	738,400	182,900

Source of data: www.bls.gov

2.1   SYMBOLS AND TERMINOLOGY

OBJECTIVES

1 Use three methods to designate sets.

2 Understand important categories of numbers, and determine cardinal numbers of sets.

3 Distinguish between finite and infinite sets.

4 Determine whether two sets are equal.

The basic ideas of set theory were developed by the German mathematician **Georg Cantor** (1845–1918) in about 1875. Cantor created a new field of theory and at the same time continued the long debate over infinity that began in ancient times. He developed counting by one-to-one correspondence to determine how many objects are contained in a set. Infinite sets differ from finite sets by not obeying the familiar law that the whole is greater than any of its parts.

Designating Sets

A **set** is a collection of objects. The objects belonging to the set are called the **elements,** or **members,** of the set. Sets are designated using the following three methods: (1) *word description,* (2) the *listing method,* and (3) *set-builder notation.*

The set of even counting numbers less than 10   Word description

$\{2, 4, 6, 8\}$ Listing method

$\{x \mid x$ is an even counting number less than 10$\}$ Set-builder notation

The set-builder notation above is read "the set of all x such that x is an even counting number less than 10." Set-builder notation uses the algebraic idea of a *variable.* (Any symbol would do, but just as in other algebraic applications, the letter x is a common choice.)

Variable representing an element in general
↓
$\{x \mid x$ is an even counting number less than 10$\}$
↑
Criteria by which an element qualifies for membership in the set

Sets are commonly given names (usually capital letters), such as E for the set of all letters of the English alphabet.

$$E = \{a, b, c, d, e, f, g, h, i, j, k, l, m, n, o, p, q, r, s, t, u, v, w, x, y, z\}$$

The listing notation can often be shortened by establishing the pattern of elements included and using ellipsis points to indicate a continuation of the pattern.

$$E = \{a, b, c, d, \ldots, x, y, z\}, \quad \text{or} \quad E = \{a, b, c, d, e, \ldots, z\}$$

The set containing no elements is called the **empty set,** or **null set.** The symbol $\varnothing$ is used to denote the empty set, so $\varnothing$ and $\{\ \}$ have the same meaning. We do *not* denote the empty set with the symbol $\{\varnothing\}$ because this notation represents a set with one element (that element being the empty set).

EXAMPLE 1 Listing Elements of Sets

Give a complete listing of all the elements of each set.

(a) the set of counting numbers between eight and thirteen

(b) $\{5, 6, 7, \ldots, 13\}$

(c) $\{x \mid x$ is a counting number between 4 and 5$\}$

Solution

(a) This set can be denoted $\{9, 10, 11, 12\}$. (Notice that the word *between* excludes the endpoint values.)

(b) This set begins with the element 5, then 6, then 7, and so on, with each element obtained by adding 1 to the previous element in the list. This pattern stops at 13, so a complete listing is

$$\{5, 6, 7, 8, 9, 10, 11, 12, 13\}.$$

(c) There are no counting numbers between 4 and 5, so this is the empty set: $\{\ \}$, or $\varnothing$. ∎

For a set to be useful, it must be *well defined*. For example, the preceding set E of the letters of the English alphabet is well defined. Given the letter q, we know that q is an element of E. Given the Greek letter θ (theta), we know that it is not an element of set E.

However, given the set C of all good singers, and a particular singer, Adilah, it may not be possible to say whether

Adilah is an element of C or   Adilah is *not* an element of C.

The problem is the word "good"; how good is good? Because we cannot necessarily decide whether a given singer belongs to set C, set C is not well defined.

The fact that the letter q is an element of set E is denoted by using the symbol $\in$.

$$q \in E \quad \text{This is read "q is an element of set } E." $$

The letter θ is not an element of E. To show this, $\in$ with a slash mark is used.

$$\theta \notin E \quad \text{This is read "}\theta \text{ is not an element of set } E." $$

Many other mathematical symbols also have their meanings negated by use of a **slash mark.** The most common example, $\neq$, means "does not equal" or "is not equal to."

EXAMPLE 2 Applying the Symbol $\in$

Decide whether each statement is *true* or *false*.

(a) $4 \in \{1, 2, 5, 8, 13\}$ **(b)** $0 \in \{0, 1, 2, 3\}$ **(c)** $\frac{1}{5} \notin \left\{\frac{1}{3}, \frac{1}{4}, \frac{1}{6}\right\}$

Solution

(a) Because 4 is *not* an element of the set $\{1, 2, 5, 8, 13\}$, the statement is *false*.

(b) Because 0 is indeed an element of the set $\{0, 1, 2, 3\}$, the statement is *true*.

(c) This statement says that $\frac{1}{5}$ is not an element of the set $\left\{\frac{1}{3}, \frac{1}{4}, \frac{1}{6}\right\}$, which is *true*. ∎

Sets of Numbers and Cardinality

Important categories of numbers are summarized below.

Most concepts in this chapter will be illustrated using the **sets of numbers** shown here, not only to solidify understanding of these sets of numbers but also because all these sets are precisely, or "well," defined and therefore provide clear illustrations.

SETS OF NUMBERS

Natural numbers (or counting numbers) $\{1, 2, 3, 4, \dots\}$

Whole numbers $\{0, 1, 2, 3, 4, \dots\}$

Integers $\{\dots, -3, -2, -1, 0, 1, 2, 3, \dots\}$

Rational numbers $\left\{\frac{p}{q} \mid p \text{ and } q \text{ are integers, and } q \neq 0\right\}$

(*Examples:* $\frac{3}{5}$, $-\frac{7}{9}$, 5, 0. Any rational number may be written as a terminating decimal number, such as 0.25, or a repeating decimal number, such as 0.666 . . . .)

Real numbers $\{x \mid x \text{ is a number that can be expressed as a decimal}\}$

Irrational numbers $\{x \mid x \text{ is a real number and } x \text{ cannot be expressed as a quotient of integers}\}$

(*Examples:* $\sqrt{2}$, $\sqrt[3]{4}$, π. Decimal representations of irrational numbers are neither terminating nor repeating.)

The number of elements in a set is called the **cardinal number,** or **cardinality,** of the set. The symbol

$$n(A), \quad \text{which is read "n of A,"}$$

represents the cardinal number of set A. If elements are repeated in a set listing, they should not be counted more than once when determining the cardinal number of the set.

EXAMPLE 3   Finding Cardinal Numbers

Find the cardinal number of each set.

(a) $K = \{3, 9, 27, 81\}$ **(b)** $M = \{0\}$ **(c)** $B = \{1, 1, 2, 3, 2\}$

(d) $R = \{7, 8, \dots, 15, 16\}$ **(e)** $\varnothing$

Solution

(a) Set K contains four elements, so the cardinal number of set K is 4, and $n(K) = 4$.

(b) Set M contains only one element, 0, so $n(M) = 1$.

(c) Do not count repeated elements more than once. Set B has only three *distinct* elements, so $n(B) = 3$.

(d) Although only four elements are listed, the ellipsis points indicate that there are other elements in the set. Counting them all, we find that there are ten elements, so $n(R) = 10$.

(e) The empty set, $\varnothing$, contains no elements, so $n(\varnothing) = 0$.   ∎

Finite and Infinite Sets

If the cardinal number of a set is a particular whole number (0 or a counting number), as in all parts of **Example 3,** we call that set a **finite set.** Given enough time, we could finish counting all the elements of any finite set and arrive at its cardinal number.

Some sets, however, are so large that we could never finish the counting process. The counting numbers themselves are such a set. Whenever a set is so large that its cardinal number is not found among the whole numbers, we call that set an **infinite set.**

Answers to the Chapter Opener questions
RNs: 19.4%
LPNs: 24.8%

A close-up of a camera lens shows the **infinity symbol,** ∞, defined in this case as any distance greater than 1000 times the focal length of a lens.

 The sign was invented by the mathematician John Wallis in 1655. Wallis used $1/\infty$ to represent an infinitely small quantity.

EXAMPLE 4 Designating an Infinite Set

Designate all odd counting numbers by the three common methods of set notation.

Solution

The set of all odd counting numbers   Word description

$$\{1, 3, 5, 7, 9, \dots\}$$ Listing method

$$\{x \mid x \text{ is an odd counting number}\}$$ Set-builder notation   ■

Equality of Sets

SET EQUALITY

Set A is **equal** to set B provided the following two conditions are met:

 1. Every element of A is an element of B, and

 2. Every element of B is an element of A.

Two sets are equal if they contain exactly the same elements, regardless of order.

$$\{a, b, c, d\} = \{a, c, d, b\}$$ Both sets contain exactly the same elements.

Repetition of elements in a set listing does not add new elements.

$$\{1, 0, 1, 5, 3, 3\} = \{0, 1, 3, 5\}$$ Both sets contain exactly the same elements.

EXAMPLE 5 Determining Whether Two Sets Are Equal

Are $\{-4, 3, 2, 5\}$ and $\{-4, 0, 3, 2, 5\}$ equal sets?

Solution

Every element of the first set is an element of the second. However, 0 is an element of the second and not of the first. The sets do not contain exactly the same elements.

$$\{-4, 3, 2, 5\} \ne \{-4, \mathbf{0}, 3, 2, 5\}$$ The sets are not equal.   ■

Two sets are **equivalent** if they have the *same number* of elements. (See **Exercises 87–90**.) Georg Cantor extended the idea of equivalence to infinite sets, used one-to-one correspondence to establish equivalence, and showed that, surprisingly, the natural numbers, the whole numbers, the integers, and the rational numbers are all equivalent. The elements of any one of these sets will match up, one-to-one, with those of any other, with no elements left over in either set. All these sets have cardinal number $\aleph_0$ (which is read **aleph null**).

 However, the irrational numbers and the real numbers, though equivalent to one another, are of a higher infinite order than the sets mentioned above. Their cardinal number is denoted **c** (representing the **continuum** of points on a line).

EXAMPLE 6 Determining Whether Two Sets Are Equal

Decide whether each statement is *true* or *false*.

(a) $\{3\} = \{x \mid x \text{ is a counting number between 1 and 5}\}$

(b) $\{x \mid x \text{ is a negative whole number}\} = \{y \mid y \text{ is a number that is both rational and irrational}\}$

(c) $\{(0, 0), (1, 1), (2, 4)\} = \{(x, y) \mid x \text{ is a natural number less than 3, and } y = x^2\}$

Solution

(a) The set on the right contains *all* counting numbers between 1 and 5, namely 2, 3, and 4, while the set on the left contains *only* the number 3. Because the sets do not contain exactly the same elements, they are not equal. The statement is *false*.

(b) No whole numbers are negative, so the set on the left is ∅. By definition, if a number is rational, it cannot be irrational, so the set on the right is also ∅. Because each set is the empty set, the sets are equal. The statement is *true*.

(c) The first listed ordered pair in the set on the left has x-value 0, which is not a natural number. Therefore, the ordered pair $(0, 0)$ is not an element of the set on the right, even though the relationship $y = x^2$ is true for $(0, 0)$. Thus the sets are not equal. The statement is *false*.   ■

2.1 EXERCISES

Match each set in Group I with the appropriate description in Group II.

I

1. $\{1, 3, 5, 7, 9\}$

2. $\{x \mid x \text{ is an even integer greater than 4 and less than 6}\}$

3. $\{\ldots, -4, -3, -2, -1\}$

4. $\{\ldots, -5, -3, -1, 1, 3, 5, \ldots\}$

5. $\{2, 4, 8, 16, 32\}$

6. $\{\ldots, -4, -2, 0, 2, 4, \ldots\}$

7. $\{2, 4, 6, 8, 10\}$

8. $\{2, 4, 6, 8\}$

II

A. the set of all even integers

B. the set of the five least positive integer powers of 2

C. the set of even positive integers less than 10

D. the set of all odd integers

E. the set of all negative integers

F. the set of odd positive integers less than 10

G. $\varnothing$

H. the set of the five least positive integer multiples of 2

List all the elements of each set. Use set notation and the listing method to describe the set.

9. the set of all counting numbers less than or equal to 6

10. the set of all whole numbers greater than 8 and less than 18

11. the set of all whole numbers not greater than 4

12. the set of all natural numbers between 4 and 14

13. $\{6, 7, 8, \ldots, 14\}$

14. $\{3, 6, 9, 12, \ldots, 30\}$

15. $\{2, 4, 8, \ldots, 256\}$

16. $\{90, 87, 84, \ldots, 69\}$

17. $\{x \mid x \text{ is an even whole number less than 11}\}$

18. $\{x \mid x \text{ is an odd integer between } -8 \text{ and } 7\}$

Denote each set by the listing method. There may be more than one correct answer.

19. the set of all multiples of 20 that are greater than 200

20. $\{x \mid x \text{ is a negative multiple of 6}\}$

21. the set of U.S. Great Lakes

22. the set of U.S. presidents who served after Richard Nixon and before Barack Obama

23. $\{x \mid x \text{ is the reciprocal of a natural number}\}$

24. $\{x \mid x \text{ is a positive integer power of 4}\}$

25. $\{(x, y) \mid x \text{ and } y \text{ are whole numbers and } x^2 + y^2 = 25\}$

26. $\{(x, y) \mid x \text{ and } y \text{ are integers, and } x^2 = 9y^2 + 16\}$

Denote each set by set-builder notation, using x as the variable. There may be more than one correct answer.

27. the set of all rational numbers

28. the set of all even natural numbers

29. $\{1, 3, 5, \ldots, 75\}$

30. $\{35, 40, 45, \ldots, 95\}$

Give a word description for each set. There may be more than one correct answer.

31. $\{-9, -8, -7, \ldots, 7, 8, 9\}$

32. $\left\{\dfrac{1}{2}, \dfrac{2}{3}, \dfrac{3}{4}, \ldots\right\}$

33. $\{\text{Alabama, Alaska, Arizona}, \ldots, \text{Wisconsin, Wyoming}\}$

34. $\{\text{Alaska, California, Hawaii, Oregon, Washington}\}$

Identify each set as finite *or* infinite.

35. $\{2, 4, 6, \ldots, 932\}$

36. $\{6, 12, 18\}$

37. $\left\{1, \dfrac{1}{2}, \dfrac{1}{3}, \dfrac{1}{4}, \ldots\right\}$

38. $\{3, 6, 9, \ldots\}$

39. $\{x \mid x$ is a natural number greater than $50\}$

40. $\{x \mid x$ is a natural number less than $50\}$

41. $\{x \mid x$ is a rational number $\}$

42. $\{x \mid x$ is a rational number between 0 and $1\}$

Find n(A) for each set.

43. $A = \{0, 1, 2, 3, 4, 5, 6, 7\}$

44. $A = \{-3, -1, 1, 3, 5, 7, 9\}$

45. $A = \{2, 4, 6, \ldots, 1000\}$

46. $A = \{0, 1, 2, 3, \ldots, 2000\}$

47. $A = \{a, b, c, \ldots, z\}$

48. $A = \{x \mid x$ is a vowel in the English alphabet$\}$

49. $A =$ the set of integers between -20 and 20

50. $A =$ the set of sanctioned U.S. senate seats

51. $A = \left\{\dfrac{1}{3}, \dfrac{2}{4}, \dfrac{3}{5}, \dfrac{4}{6}, \ldots, \dfrac{27}{29}, \dfrac{28}{30}\right\}$

52. $A = \left\{\dfrac{1}{2}, -\dfrac{1}{2}, \dfrac{1}{3}, -\dfrac{1}{3}, \ldots, \dfrac{1}{10}, -\dfrac{1}{10}\right\}$

53. Although x is a consonant, why can we write

"x is a vowel in the English alphabet"

in **Exercise 48?**

54. Explain how **Exercise 51** can be answered without actually listing and then counting all the elements.

Identify each set as well defined *or* not well defined.

55. $\{x \mid x$ is a real number $\}$

56. $\{x \mid x$ is a good athlete $\}$

57. $\{x \mid x$ is a difficult course $\}$

58. $\{x \mid x$ is a counting number less than $2\}$

Fill each blank with either $\in$ *or* $\notin$ *to make each statement true.*

59. 3 ___ $\{2, 4, 5, 7\}$ **60.** -4 ___ $\{4, 7, 8, 12\}$

61. 8 ___ $\{3, 8, 12, 18\}$ **62.** 0 ___ $\{-2, 0, 5, 9\}$

63. 8 ___ $\{10 - 2, 10\}$ **64.** $\{6\}$ ___ $\{5 + 1, 6 + 1\}$

65. Is the statement $\{0\} = \varnothing$ true, or is it false?

66. The statement

$$3 \in \{9 - 6, 8 - 6, 7 - 6\}$$

is true even though the *symbol* 3 does not appear in the set. Explain.

Write true *or* false *for each statement.*

67. $3 \in \{2, 5, 6, 8\}$ **68.** $m \in \{l, m, n, o, p\}$

69. $c \in \{c, d, a, b\}$ **70.** $2 \in \{-2, 5, 8, 9\}$

71. $\{k, c, r, a\} = \{k, c, a, r\}$ **72.** $\{e, h, a, n\} = \{a, h, e, n\}$

73. $\{5, 8, 9\} = \{5, 8, 9, 0\}$ **74.** $\{3, 7\} = \{3, 7, 0\}$

75. $\{4\} \in \{\{3\}, \{4\}, \{5\}\}$ **76.** $4 \in \{\{3\}, \{4\}, \{5\}\}$

77. $\{x \mid x$ is a natural number less than $3\} = \{1, 2\}$

78. $\{x \mid x$ is a natural number greater than $10\}$
$= \{11, 12, 13, \ldots\}$

Write true *or* false *for each statement in Exercises 79–84.*

Let $A = \{2, 4, 6, 8, 10, 12\}$, $B = \{2, 4, 8, 10\}$,
and $C = \{4, 10, 12\}$.

79. $4 \in A$ **80.** $10 \in B$

81. $4 \notin C$ **82.** $10 \notin A$

83. Every element of C is also an element of A.

84. Every element of C is also an element of B.

85. The human mind likes to create collections. Why do you suppose this is so? In your explanation, use one or more particular "collections," mathematical or otherwise.

86. Explain the difference between a well-defined set and a set that is not well defined. Give examples, and use terms introduced in this section.

Two sets are **equal** *if they contain identical elements. Two sets are* **equivalent** *if they contain the same number of elements (but not necessarily the same elements). For each condition, give an example or explain why it is impossible.*

87. two sets that are neither equal nor equivalent

88. two sets that are equal but not equivalent

89. two sets that are equivalent but not equal

90. two sets that are both equal and equivalent

91. *Hiring Nurses* A medical organization plans to hire three nurses from the pool of applicants shown.

Name	Certification
Bernice	RN
Heather	RN
Marcy	LVN
Natalie	LVN
Susan	RN

Show all possible sets of hires that would include

(a) two RNs and one LVN.

(b) one RN and two LVNs.

(c) no LVNs.

92. *Burning Calories* Candice Cotton likes cotton candy, each serving of which contains 220 calories. To burn off unwanted calories, Candice participates in her favorite activities, shown in the next column, in increments of 1 hour and never repeats a given activity on a given day.

Activity	Symbol	Calories Burned per Hour
Volleyball	v	160
Golf	g	260
Canoeing	c	340
Swimming	s	410
Running	r	680

(a) On Monday, Candice has time for no more than two hours of activities. List all possible sets of activities that would burn off at least the number of calories obtained from three cotton candies.

(b) Assume that Candice can afford up to three hours of time for activities on Wednesday. List all sets of activities that would burn off at least the number of calories in five cotton candies.

(c) Candice can spend up to four hours in activities on Saturday. List all sets of activities that would burn off at least the number of calories in seven cotton candies.

2.2 VENN DIAGRAMS AND SUBSETS

OBJECTIVES

1 Use Venn diagrams to depict set relationships.

2 Determine the complement of a set within a universal set.

3 Determine whether one set is a subset of another.

4 Understand the distinction between a subset and a proper subset.

5 Determine the number of subsets of a given set.

Venn Diagrams

In most discussions, there is either a stated or an implied **universe of discourse.** The universe of discourse includes all things under discussion at a given time. For example, if the topic of interest is what courses to offer at a vocational school, the universe of discourse might be all students at the school, or the board of trustees of the school, or the members of a local overseer board, or the members of a state regulatory agency, or perhaps all these groups of people.

In set theory, the universe of discourse is called the **universal set,** typically designated by the letter *U.* The universal set might change from one discussion to another.

Also in set theory, we commonly use **Venn diagrams,** developed by the logician John Venn (1834–1923). In these diagrams, the universal set is represented by a rectangle, and other sets of interest within the universal set are depicted by circular regions (sometimes ovals or other shapes). See **Figure 1.**

Complement of a Set

The colored region inside *U* and outside the circle in **Figure 1** is labeled *A'* (read "*A* prime"). This set, called the *complement* of *A*, contains all elements that are contained in *U* but are not contained in *A*.

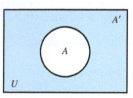

The entire region bounded by the rectangle represents the universal set *U*, and the portion bounded by the circle represents set *A*.

Figure 1

THE COMPLEMENT OF A SET

For any set *A* within a universal set *U*, the **complement** of *A*, written *A'*, is the set of elements of *U* that are not elements of *A*. That is,

$$A' = \{x \,|\, x \in U \text{ and } x \notin A\}.$$

EXAMPLE 1 **Finding Complements**

Find each set.

Let $U = \{a, b, c, d, e, f, g, h\}$, $M = \{a, b, e, f\}$, and $N = \{b, d, e, g, h\}$.

(a) M' **(b)** N'

Solution

(a) Set M' contains all the elements of set U that are *not* in set M. Because set M contains a, b, e, and f, these elements will be disqualified from belonging to set M'.

$$M' = \{c, d, g, h\}$$

(b) Set N' contains all the elements of U that are not in set N, so $N' = \{a, c, f\}$. ∎

Consider the complement of the universal set, U'. The set U' is found by selecting all the elements of U that do not belong to U. There are no such elements, so there can be no elements in set U'. This means that for any universal set U,

$$U' = \emptyset.$$

Now consider the complement of the empty set, $\emptyset'$. The set $\emptyset'$ includes all elements of U that do *not* belong to $\emptyset$. All elements of U qualify, because none of them belongs to $\emptyset$. Therefore, for any universal set U,

$$\emptyset' = U.$$

Subsets of a Set

Suppose that we are given the universal set $U = \{1, 2, 3, 4, 5\}$, while $A = \{1, 2, 3\}$. Every element of set A is also an element of set U. Because of this, set A is called a *subset* of set U, written

$$A \subseteq U.$$

("A is not a subset of set U" would be written $A \not\subseteq U$.)

A Venn diagram showing that set M is a subset of set N is shown in **Figure 2**.

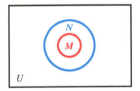

Figure 2

SUBSET OF A SET

Set A is a **subset** of set B if every element of A is also an element of B. This is written $A \subseteq B$.

EXAMPLE 2 **Determining If One Set Is a Subset of Another**

Write $\subseteq$ or $\not\subseteq$ in each blank to make a true statement.

(a) $\{3, 4, 5, 6\}$ _____ $\{3, 4, 5, 6, 8\}$ **(b)** $\{1, 2, 6\}$ _____ $\{2, 4, 6, 8\}$

(c) $\{5, 6, 7, 8\}$ _____ $\{6, 5, 8, 7\}$

Solution

(a) Because every element of $\{3, 4, 5, 6\}$ is also an element of $\{3, 4, 5, 6, 8\}$, the first set is a subset of the second, so $\subseteq$ goes in the blank.

$$\{3, 4, 5, 6\} \subseteq \{3, 4, 5, 6, 8\}$$

(b) $\{1, 2, 6\} \not\subseteq \{2, 4, 6, 8\}$ 1 does not belong to $\{2, 4, 6, 8\}$.

(c) $\{5, 6, 7, 8\} \subseteq \{6, 5, 8, 7\}$ ∎

As **Example 2(c)** suggests, every set is a subset of itself.

$$B \subseteq B, \quad \text{for any set } B.$$

SET EQUALITY (ALTERNATIVE DEFINITION)

Suppose A and B are sets. Then $A = B$ if $A \subseteq B$ and $B \subseteq A$ are both true.

Proper Subsets

Suppose that we are given the following sets.

$$B = \{5, 6, 7, 8\} \quad \text{and} \quad A = \{6, 7\}$$

A is a subset of B, but A is not all of B. There is at least one element in B that is not in A. (Actually, in this case there are two such elements, 5 and 8.) In this situation, A is called a *proper subset* of B, written $A \subset B$.

Notice the similarity of the subset symbols, $\subset$ and $\subseteq$, to the inequality symbols from algebra, $<$ and $\leq$.

PROPER SUBSET OF A SET

Set A is a **proper subset** of set B if $A \subseteq B$ and $A \neq B$. This is written $\mathbf{A \subset B}$.

EXAMPLE 3 **Determining Subsets and Proper Subsets**

Decide whether $\subset$, $\subseteq$, or both could be placed in each blank to make a true statement.

(a) $\{5, 6, 7\}$ _____ $\{5, 6, 7, 8\}$ **(b)** $\{a, b, c\}$ _____ $\{a, b, c\}$

Solution

(a) Every element of $\{5, 6, 7\}$ is contained in $\{5, 6, 7, 8\}$, so $\subseteq$ could be placed in the blank. Also, the element 8 belongs to $\{5, 6, 7, 8\}$ but not to $\{5, 6, 7\}$, making $\{5, 6, 7\}$ a proper subset of $\{5, 6, 7, 8\}$. Thus $\subset$ could also be placed in the blank.

(b) The set $\{a, b, c\}$ is a subset of $\{a, b, c\}$. Because the two sets are equal, $\{a, b, c\}$ is not a proper subset of $\{a, b, c\}$. Only $\subseteq$ may be placed in the blank. ∎

Set A is a subset of set B if every element of set A is also an element of set B. Alternatively, we say that set A is a subset of set B if there are no elements of A that are not also elements of B. Thus, the empty set is a subset of any set.

$$\varnothing \subseteq B, \quad \text{for any set } B.$$

This is true because it is not possible to find any element of $\varnothing$ that is not also in B. (There are no elements in $\varnothing$.) The empty set $\varnothing$ is a proper subset of every set except itself.

$$\varnothing \subset B \quad \text{if } B \text{ is any set other than } \varnothing.$$

Every set (except $\varnothing$) has at least two subsets, $\varnothing$ and the set itself.

EXAMPLE 4 **Listing All Subsets of a Set**

Find all possible subsets of each set.

(a) $\{7, 8\}$ **(b)** $\{a, b, c\}$

Solution

(a) By trial and error, the set $\{7, 8\}$ has four subsets: $\varnothing, \{7\}, \{8\}, \{7, 8\}$.

(b) Here, trial and error leads to eight subsets for $\{a, b, c\}$:

$$\varnothing, \{a\}, \{b\}, \{c\}, \{a, b\}, \{a, c\}, \{b, c\}, \{a, b, c\}. \quad ∎$$

One-to-one correspondence was employed by Georg Cantor to establish many controversial facts about infinite sets. For example, the correspondence

$$\{1, 2, 3, 4, \ldots, n, \ldots\}$$
$$\updownarrow \updownarrow \updownarrow \updownarrow \quad \updownarrow$$
$$\{2, 4, 6, 8, \ldots, 2n, \ldots\},$$

which can be continued indefinitely without leaving any elements in either set unpaired, shows that the counting numbers and the even counting numbers are equivalent (have the same number of elements), even though logic may seem to say that the first set has twice as many elements as the second.

Counting Subsets

In **Example 4,** the subsets of $\{7, 8\}$ and the subsets of $\{a, b, c\}$ were found by trial and error. An alternative method involves drawing a **tree diagram,** a systematic way of listing all the subsets of a given set. See **Figure 3.**

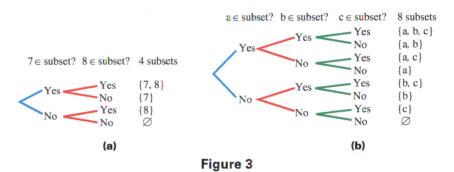

Figure 3

In **Example 4,** we determined the number of subsets of a given set by making a list of all such subsets and then counting them. The tree diagram method also produced a list of all possible subsets. To obtain a formula for finding the number of subsets, we use inductive reasoning. That is, we observe particular cases to try to discover a general pattern.

Begin with the set containing the least number of elements possible—the empty set. This set, $\varnothing$, has only one subset, $\varnothing$ itself. Next, a set with one element has only two subsets, itself and $\varnothing$. These facts, together with those obtained in **Example 4** for sets with two and three elements, are summarized here.

Number of elements	0	1	2	3
Number of subsets	1	2	4	8

This chart suggests that as the number of elements of the set increases by one, the number of subsets doubles. If so, then the number of subsets in each case might be a power of 2. Since every number in the second row of the chart is indeed a power of 2, add this information to the chart.

Number of elements	0	1	2	3
Number of subsets	$1 = 2^0$	$2 = 2^1$	$4 = 2^2$	$8 = 2^3$

This chart shows that the number of elements in each case is the same as the exponent on the base 2. Inductive reasoning gives the following generalization.

NUMBER OF SUBSETS

The number of subsets of a set with n elements is 2^n.

Because the value 2^n includes the set itself, we must subtract 1 from this value to obtain the number of proper subsets of a set containing n elements.

NUMBER OF PROPER SUBSETS

The number of proper subsets of a set with n elements is $2^n - 1.$

Powers of 2

$2^0 = 1$
$2^1 = 2$
$2^2 = 2 \cdot 2 = 4$
$2^3 = 2 \cdot 2 \cdot 2 = 8$
$2^4 = 2 \cdot 2 \cdot 2 \cdot 2 = 16$
$2^5 = 32$
$2^6 = 64$
$2^7 = 128$
$2^8 = 256$
$2^9 = 512$
$2^{10} = 1024$
$2^{11} = 2048$
$2^{12} = 4096$
$2^{15} = 32,768$
$2^{20} = 1,048,576$
$2^{25} = 33,554,432$
$2^{30} = 1,073,741,824$

As shown in **Chapter 1,** although inductive reasoning is a good way of *discovering* principles or arriving at a *conjecture,* it does not provide a proof that the conjecture is true in general. The two formulas above are true, by observation, for $n = 0, 1, 2,$ and 3. (For a general proof, see **Exercise 63** at the end of this section.)

EXAMPLE 5 Finding Numbers of Subsets and Proper Subsets

Find the number of subsets and the number of proper subsets of each set.

(a) $\{3, 4, 5, 6, 7\}$ **(b)** $\{1, 2, 3, 4, 5, 9, 12, 14\}$

Solution

(a) This set has 5 elements and $2^5 = 2 \cdot 2 \cdot 2 \cdot 2 \cdot 2 = 32$ subsets. Of these,

$$2^5 - 1 = 32 - 1 = 31 \text{ are proper subsets.}$$

(b) This set has 8 elements. There are $2^8 = 256$ subsets and 255 proper subsets. ∎

2.2 EXERCISES

Match each set or sets in Column I with the appropriate description in Column II.

I	II
1. $\{p\}, \{q\}, \{p, q\}, \varnothing$	**A.** the proper subsets of $\{p, q\}$
2. $\{p\}, \{q\}, \varnothing$	**B.** the complement of $\{c, d\}$, if $U = \{a, b, c, d\}$
3. $\{a, b\}$	**C.** the complement of U
4. $\varnothing$	**D.** the subsets of $\{p, q\}$

Insert $\subseteq$ or $\nsubseteq$ in each blank to obtain a true statement.

5. $\{-2, 0, 2\}$ ____ $\{-2, -1, 1, 2\}$

6. $\{M, W, F\}$ ____ $\{S, M, T, W, Th\}$

7. $\{2, 5\}$ ____ $\{0, 1, 5, 3, 7, 2\}$

8. $\{a, n, d\}$ ____ $\{r, a, n, d, y\}$

9. $\varnothing$ ____ $\{a, b, c, d, e\}$

10. $\varnothing$ ____ $\varnothing$

11. $\{-5, 2, 9\}$ ____ $\{x \mid x \text{ is an odd integer}\}$

12. $\left\{1, 2, \dfrac{9}{3}\right\}$ ____ the set of rational numbers

Decide whether $\subset$, $\subseteq$, both, or neither can be placed in each blank to make the statement true.

13. $\{P, Q, R\}$ ____ $\{P, Q, R, S\}$

14. $\{red, blue, yellow\}$ ____ $\{yellow, blue, red\}$

15. $\{9, 1, 7, 3, 5\}$ ____ $\{1, 3, 5, 7, 9\}$

16. $\{S, M, T, W, Th\}$ ____ $\{W, E, E, K\}$

17. $\varnothing$ ____ $\{0\}$

18. $\varnothing$ ____ $\varnothing$

19. $\{0, 1, 2, 3\}$ ____ $\{1, 2, 3, 4\}$

20. $\left\{\dfrac{5}{6}, \dfrac{9}{8}\right\}$ ____ $\left\{\dfrac{6}{5}, \dfrac{8}{9}\right\}$

For Exercises 21–36, tell whether each statement is true *or* false. *U is the universal set.*

Let $U = \{a, b, c, d, e, f, g\}$, $A = \{a, e\}$,

$B = \{a, b, e, f, g\}$, $C = \{b, f, g\}$, and $D = \{d, e\}$.

21. $A \subset U$ **22.** $C \not\subset U$

23. $D \subseteq B$ **24.** $D \nsubseteq A$

25. $A \subset B$ **26.** $B \subseteq C$

27. $\varnothing \not\subset A$ **28.** $\varnothing \subseteq D$

29. $D \nsubseteq B$ **30.** $A \nsubseteq B$

31. There are exactly 6 subsets of C.

32. There are exactly 31 subsets of B.

33. There are exactly 3 proper subsets of A.

34. There are exactly 4 subsets of D.

35. The Venn diagram below correctly represents the relationship among sets A, D, and U.

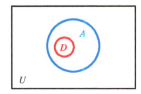

36. The Venn diagram below correctly represents the relationship among sets B, C, and U.

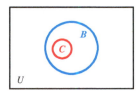

For Exercises 37–40, find **(a)** *the number of subsets and* **(b)** *the number of proper subsets of each set.*

37. $\{a, b, c, d, e, f\}$

38. the set of days of the week

39. $\{x \mid x \text{ is an odd integer between } -4 \text{ and } 6\}$

40. $\{x \mid x \text{ is an even whole number less than } 4\}$

For Exercises 41–44, let $U = \{1, 2, 3, 4, 5, 6, 7, 8, 9, 10\}$ *and find the complement of each set.*

41. U **42.** $\varnothing$

43. $\{1, 2, 3, 4, 6, 8\}$ **44.** $\{2, 5, 9, 10\}$

Vacationing in California *Terry is planning a trip with her two sons to California. In weighing her options concerning whether to fly or drive from their home in Iowa, she has listed the following considerations.*

Fly to California	Drive to California
Higher cost	Lower cost
Educational	Educational
More time to see the sights in California	Less time to see the sights in California
Cannot visit friends along the way	Can visit friends along the way

Refer to the table for Exercises 45–50.

45. Find the smallest universal set U that contains all listed considerations of both options.

Let F represent the set of considerations of the flying option and let D represent the set of considerations of the driving option. Use the universal set from ***Exercise 45.***

46. Give the set F'.   **47.** Give the set D'.

Find the set of elements common to both sets in Exercises 48–50.

48. F and D **49.** F' and D'

50. F and D'

Meeting in the Conference Room *Amie, Bruce, Corey, Dwayne, and Eric, members of an architectural firm, plan to meet in the company conference room to discuss the project coordinator's plans for their next project. Denoting these five people by A, B, C, D, and E, list all the possible sets of this group in which the given number of them can gather.*

51. five people   **52.** four people

53. three people   **54.** two people

55. one person   **56.** no people

57. Find the total number of ways that members of this group can gather. (*Hint:* Find the total number of sets in your answers to **Exercises 51–56.**)

58. How does your answer in **Exercise 57** compare with the number of subsets of a set of five elements? Interpret the answer to **Exercise 57** in terms of subsets.

59. ***Selecting a Club Delegation*** The twenty-five members of the mathematics club must send a delegation to a meeting for student groups at their school. The delegation can include as many members of the club as desired, but at least one member must attend. How many different delegations are possible? (*Mathematics Teacher* calendar problem)

60. In **Exercise 59,** suppose ten of the club members say they do not want to be part of the delegation. Now how many delegations are possible?

61. ***Selecting Bills*** Suppose you have the bills shown here.

(a) How many sums of money can you make using nonempty subsets of these bills?

(b) Repeat part (a) without the condition "nonempty."

62. *Selecting Coins* The photo shows a group of obsolete U.S. coins, consisting of one each of the penny, nickel, dime, quarter, and half dollar. Repeat **Exercise 61,** replacing "bill(s)" with "coin(s)."

63. In discovering the expression (2^n) for finding the number of subsets of a set with n elements, we observed that for the first few values of n, increasing the number of elements by one doubles the number of subsets.

Here, you can prove the formula in general by showing that the same is true for any value of n. Assume set A has n elements and s subsets. Now add one additional element, say e, to the set A. (We now have a new set, say B, with $n + 1$ elements.) Divide the subsets of B into those that do not contain e and those that do.

(a) How many subsets of B do not contain e? (*Hint:* Each of these is a subset of the original set A.)

(b) How many subsets of B do contain e? (*Hint:* Each of these would be a subset of the original set A, with the additional element e included.)

(c) What is the total number of subsets of B?

(d) What do you conclude?

64. Explain why $\varnothing$ is both a subset and an element of $\{\varnothing\}$.

OBJECTIVES

1 Determine intersections of sets.

2 Determine unions of sets.

3 Determine the difference of two sets.

4 Understand ordered pairs and their uses.

5 Determine Cartesian products of sets.

6 Analyze sets and set operations with Venn diagrams.

7 Apply De Morgan's laws for sets.

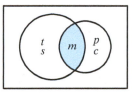

Figure 4

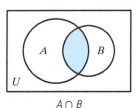

$A \cap B$

Figure 5

Intersection of Sets

Two candidates, Aimee and Darien, are running for a seat on the city council. A voter deciding for whom she should vote recalled the campaign promises, each given a code letter, made by the candidates.

Honest Aimee	**Determined Darien**
Spend less money, *m*	**Spend less money, *m***
Emphasize traffic law enforcement, *t*	Crack down on crooked politicians, *p*
Increase service to suburban areas, *s*	Increase service to the city, *c*

The only promise common to both candidates is promise m, to spend less money. Suppose we take each candidate's promises to be a set. The promises of Aimee give the set $\{m, t, s\}$, while the promises of Darien give $\{m, p, c\}$. The common element m belongs to the *intersection* of the two sets, as shown in color in the Venn diagram in **Figure 4**.

$$\{m, t, s\} \cap \{m, p, c\} = \{m\} \quad \cap \text{ represents set intersection.}$$

The intersection of two sets is itself a set.

> **INTERSECTION OF SETS**
>
> The **intersection** of sets A and B, written $A \cap B$, is the set of elements common to both A and B.
>
> $$A \cap B = \{x \mid x \in A \text{ and } x \in B\}$$

Form the intersection of sets A and B by taking all the elements included in both sets, as shown in color in **Figure 5**.

White light can be viewed as the intersection of the three primary colors.

EXAMPLE 1 **Finding Intersections**

Find each intersection.

(a) $\{3, 4, 5, 6, 7\} \cap \{4, 6, 8, 10\}$ (b) $\{9, 14, 25, 30\} \cap \{10, 17, 19, 38, 52\}$

(c) $\{5, 9, 11\} \cap \varnothing$

Solution

(a) The elements common to both sets are 4 and 6.

$$\{3, 4, 5, 6, 7\} \cap \{4, 6, 8, 10\} = \{4, 6\}$$

(b) These two sets have no elements in common.

$$\{9, 14, 25, 30\} \cap \{10, 17, 19, 38, 52\} = \varnothing$$

(c) There are no elements in $\varnothing$, so there can be no elements belonging to both $\{5, 9, 11\}$ and $\varnothing$.

$$\{5, 9, 11\} \cap \varnothing = \varnothing$$

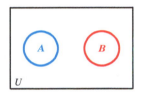

Disjoint sets

Figure 6

Examples 1(b) and 1(c) show two sets that have no elements in common. Sets with no elements in common are called **disjoint sets**. (See **Figure 6**.) A set of dogs and a set of cats would be disjoint sets.

Sets A and B are disjoint if $A \cap B = \varnothing$.

Union of Sets

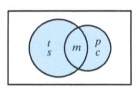

Figure 7

Referring again to the lists of campaign promises, suppose a pollster wants to summarize the types of promises made by the candidates. The pollster would need to study *all* the promises made by *either* candidate, or the set

$$\{m, t, s, p, c\}.$$

This set is the *union* of the sets of promises, as shown in color in the Venn diagram in **Figure 7**.

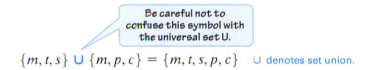

Be careful not to confuse this symbol with the universal set U.

$$\{m, t, s\} \cup \{m, p, c\} = \{m, t, s, p, c\}$$ ∪ denotes set union.

Again, the union of two sets is a set.

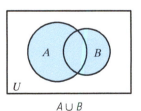

$A \cup B$

Figure 8

UNION OF SETS

The **union** of sets A and B, written $A \cup B$, is the set of all elements belonging to either A or B.

$$A \cup B = \{x \mid x \in A \text{ or } x \in B\}$$

Form the union of sets A and B by first taking every element of set A and then also including every element of set B that is not already listed. See Figure 8.

EXAMPLE 2 **Finding Unions**

Find each union.

(a) $\{2, 4, 6\} \cup \{4, 6, 8, 10, 12\}$ (b) $\{a, b, d, f, g, h\} \cup \{c, f, g, h, k\}$

(c) $\{3, 4, 5\} \cup \varnothing$

Solution

(a) Start by listing all the elements from the first set, 2, 4, and 6. Then list all the elements from the second set that are not in the first set, 8, 10, and 12. The union is made up of *all* these elements.

$$\{2, 4, 6\} \cup \{4, 6, 8, 10, 12\} = \{2, 4, 6, 8, 10, 12\}$$

(b) $\{a, b, d, f, g, h\} \cup \{c, f, g, h, k\} = \{a, b, c, d, f, g, h, k\}$

(c) Because there are no elements in $\varnothing$, the union of $\{3, 4, 5\}$ and $\varnothing$ contains only the elements 3, 4, and 5.

$$\{3, 4, 5\} \cup \varnothing = \{3, 4, 5\} \qquad \blacksquare$$

Recall from the previous section that A' represents the *complement* of set A. **Set A' is formed by taking every element of the universal set U that is not in set A.**

EXAMPLE 3 **Finding Intersections and Unions of Complements**

Find each set. Let

$$U = \{1, 2, 3, 4, 5, 6, 9\}, \quad A = \{1, 2, 3, 4\}, \quad B = \{2, 4, 6\}, \quad \text{and} \quad C = \{1, 3, 6, 9\}.$$

(a) $A' \cap B$ (b) $B' \cup C'$ (c) $A \cap (B \cup C')$ (d) $(B \cup C)'$

Solution

(a) First identify the elements of set A', the elements of U that are not in set A.

$$A' = \{5, 6, 9\}$$

Now, find $A' \cap B$, the set of elements belonging both to A' and to B.

$$A' \cap B = \{5, 6, 9\} \cap \{2, 4, 6\} = \{6\}$$

(b) $B' \cup C' = \{1, 3, 5, 9\} \cup \{2, 4, 5\} = \{1, 2, 3, 4, 5, 9\}$

(c) First find the set inside the parentheses.

$$B \cup C' = \{2, 4, 6\} \cup \{2, 4, 5\} = \{2, 4, 5, 6\}$$

Now, find the intersection of this set with A.

$$A \cap (B \cup C') = A \cap \{2, 4, 5, 6\}$$
$$= \{1, 2, 3, 4\} \cap \{2, 4, 5, 6\}$$
$$= \{2, 4\}$$

(d) $B \cup C = \{2, 4, 6\} \cup \{1, 3, 6, 9\} = \{1, 2, 3, 4, 6, 9\}$,   so

$$(B \cup C)' = \{5\}. \qquad \blacksquare$$

Comparing **Examples 3(b) and 3(d),** we see that, interestingly, $(B \cup C)'$ is not the same as $B' \cup C'$. This fact will be investigated further later in this section.

FOR FURTHER THOUGHT

Comparing Properties

The arithmetic operations of addition and multiplication, when applied to numbers, have some familiar properties. If a, b, and c are *real numbers*, then the **commutative property of addition** says that the order of the numbers being added makes no difference:

$$a + b = b + a.$$

(Is there a **commutative property of multiplication?**) The **associative property of addition** says that when three numbers are added, the grouping used makes no difference:

$$(a + b) + c = a + (b + c).$$

(Is there an **associative property of multiplication?**) The number 0 is called the **identity element for addition** since adding it to any number does not change that number:

$$a + 0 = a.$$

(What is the **identity element for multiplication?**) Finally, the **distributive property of multiplication over addition** says that

$$a(b + c) = ab + ac.$$

(Is there a distributive property of addition over multiplication?)

For Group or Individual Investigation

Now consider the operations of union and intersection, applied to sets. By recalling definitions, trying examples, or using Venn diagrams, answer the following questions.

1. Is set union commutative?   Set intersection?
2. Is set union associative?   Set intersection?
3. Is there an identity element for set union? If so, what is it? How about set intersection?
4. Is set intersection distributive over set union? Is set union distributive over set intersection?

EXAMPLE 4 Describing Sets in Words

Describe each set in words.

(a) $A \cap (B \cup C')$ **(b)** $(A' \cup C') \cap B'$

Solution

(a) This set might be described as "the set of all elements that are in A, and also are in B or not in C."

(b) One possibility is "the set of all elements that are not in A or not in C, and also are not in B." ∎

Difference of Sets

Suppose that $A = \{1, 2, 3, \ldots, 10\}$ and $B = \{2, 4, 6, 8, 10\}$. If the elements of B are excluded (or taken away) from A, the set $C = \{1, 3, 5, 7, 9\}$ is obtained. C is called the *difference* of sets A and B.

DIFFERENCE OF SETS

The **difference** of sets A and B, written $A - B$, is the set of all elements belonging to set A and not to set B.

$$A - B = \{x \,|\, x \in A \text{ and } x \notin B\}$$

Assume a universal set U containing both A and B. Then because $x \notin B$ has the same meaning as $x \in B'$, the set difference $A - B$ can also be described as

$$\{x \,|\, x \in A \text{ and } x \in B'\}, \quad \text{or} \quad A \cap B'.$$

Figure 9 illustrates the idea of set difference. The region in color represents $A - B$.

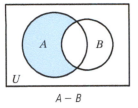

$A - B$

Figure 9

EXAMPLE 5 **Finding Set Differences**

Find each set.

$$\text{Let} \quad U = \{1, 2, 3, 4, 5, 6, 7\}, \quad A = \{1, 2, 3, 4, 5, 6\},$$
$$B = \{2, 3, 6\}, \quad \text{and} \quad C = \{3, 5, 7\}.$$

(a) $A - B$ **(b)** $B - A$ **(c)** $(A - B) \cup C'$

Solution

(a) Begin with set A and exclude any elements found also in set B.

$$A - B = \{1, 2, 3, 4, 5, 6\} - \{2, 3, 6\} = \{1, 4, 5\}$$

(b) To be in $B - A$, an element must be in set B and not in set A. But all elements of B are also in A. Thus, $B - A = \varnothing$.

(c) From part (a), $A - B = \{1, 4, 5\}$. Also, $C' = \{1, 2, 4, 6\}$.

$$(A - B) \cup C' = \{1, 2, 4, 5, 6\}$$ ∎

The results in **Examples 5(a) and 5(b)** illustrate that, in general,

$$A - B \neq B - A.$$

Ordered Pairs

When writing a set that contains several elements, the order in which the elements appear is not relevant. For example,

$$\{1, 5\} = \{5, 1\}.$$

However, there are many instances in mathematics where, when two objects are paired, the order in which the objects are written is important. This leads to the idea of *ordered pair.* When writing ordered pairs, use parentheses rather than braces, which are reserved for writing sets.

ORDERED PAIRS

In the **ordered pair** (a, b), a is called the **first component** and b is called the **second component.** In general, $(a, b) \neq (b, a)$.

Two ordered pairs (a, b) and (c, d) are **equal** provided that their first components are equal and their second components are equal.

$$(a, b) = (c, d) \quad \textit{if and only if} \quad a = c \quad \textit{and} \quad b = d.$$

EXAMPLE 6 **Determining Equality of Sets and of Ordered Pairs**

Decide whether each statement is *true* or *false.*

(a) $(3, 4) = (5 - 2, 1 + 3)$ **(b)** $\{3, 4\} \neq \{4, 3\}$ **(c)** $(7, 4) = (4, 7)$

Solution

(a) Because $3 = 5 - 2$ and $4 = 1 + 3$, the first components are equal and the second components are equal. The statement is *true.*

(b) Because these are sets and not ordered pairs, the order in which the elements are listed is not important. Because these sets are equal, the statement is *false.*

(c) The ordered pairs $(7, 4)$ and $(4, 7)$ are not equal because their corresponding components are not equal. The statement is *false.* ∎

Cartesian Product of Sets

A set may contain ordered pairs as elements. If A and B are sets, then each element of A can be paired with each element of B, and the results can be written as ordered pairs. The set of all such ordered pairs is called the *Cartesian product* of A and B, which is written $A \times B$ and read **"A cross B."** The name comes from that of the French mathematician René Descartes, profiled in **Chapter 8.**

CARTESIAN PRODUCT OF SETS

The **Cartesian product** of sets A and B is defined as follows.

$$A \times B = \{(a, b) \mid a \in A \text{ and } b \in B\}$$

EXAMPLE 7 **Finding Cartesian Products**

Let $A = \{1, 5, 9\}$ and $B = \{6, 7\}$. Find each set.

(a) $A \times B$ **(b)** $B \times A$

Solution

(a) Pair each element of A with each element of B. Write the results as ordered pairs, with the element of A written first and the element of B written second. Write as a set.

$$A \times B = \{(1, 6), (1, 7), (5, 6), (5, 7), (9, 6), (9, 7)\}$$

(b) Because B is listed first, this set will consist of ordered pairs that have their components interchanged when compared to those in part (a).

$$B \times A = \{(6, 1), (7, 1), (6, 5), (7, 5), (6, 9), (7, 9)\} \qquad \blacksquare$$

The order in which the ordered pairs themselves are listed is not important. For example, another way to write $B \times A$ in **Example 7(b)** would be

$$\{(6, 1), (6, 5), (6, 9), (7, 1), (7, 5), (7, 9)\}.$$

From **Example 7** it can be seen that, in general,

$$A \times B \neq B \times A,$$

because they do not contain exactly the same ordered pairs. However, each set contains the same *number* of elements, six. Furthermore, $n(A) = 3$, $n(B) = 2$, and $n(A \times B) = n(B \times A) = 6$. Because $3 \cdot 2 = 6$, one might conclude that the cardinal number of the Cartesian product of two sets is equal to the product of the cardinal numbers of the sets. In general, this conclusion is correct.

CARDINAL NUMBER OF A CARTESIAN PRODUCT

If $n(A) = a$ and $n(B) = b$, then the following is true.

$$n(A \times B) = n(B \times A) = n(A) \cdot n(B) = n(B) \cdot n(A) = ab = ba$$

EXAMPLE 8   **Finding Cardinal Numbers of Cartesian Products**

Find $n(A \times B)$ and $n(B \times A)$ from the given information.

(a) $A = \{a, b, c, d, e, f, g\}$ and $B = \{2, 4, 6\}$ **(b)** $n(A) = 24$ and $n(B) = 5$

Solution

(a) Because $n(A) = 7$ and $n(B) = 3$, $n(A \times B)$ and $n(B \times A)$ both equal $7 \cdot 3$, or 21.

(b) $n(A \times B) = n(B \times A) = 24 \cdot 5 = 5 \cdot 24 = 120$ ∎

An **operation** is a rule or procedure by which one or more objects are used to obtain another object. The most common operations on sets are summarized in the following box.

SET OPERATIONS

Let A and B be any sets within a universal set U.

The **complement** of A, written A', is

$$A' = \{x \mid x \in U \text{ and } x \notin A\}.$$

The **intersection** of A and B is

$$A \cap B = \{x \mid x \in A \text{ and } x \in B\}.$$

The **union** of A and B is

$$A \cup B = \{x \mid x \in A \text{ or } x \in B\}.$$

The **difference** of A and B is

$$A - B = \{x \mid x \in A \text{ and } x \notin B\}.$$

The **Cartesian product** of A and B is

$$A \times B = \{(x, y) \mid x \in A \text{ and } y \in B\}.$$

More on Venn Diagrams

It is often helpful to use numbers in Venn diagrams, as in **Figures 10, 11, and 12,** depending on whether the discussion involves one, two, or three (distinct) sets, respectively. In each case, the numbers are neither elements nor cardinal numbers, but simply arbitrary labels for the various regions within the diagram.

In **Figure 11,** region 3 includes the elements belonging to both A and B, while region 4 includes those elements (if any) belonging to B but not to A. How would you describe region 7 in **Figure 12?**

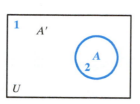

Figure 10

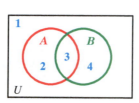

Figure 11

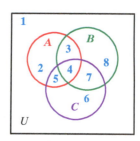

Figure 12

EXAMPLE 9 Shading Venn Diagrams to Represent Sets

Draw a Venn diagram similar to **Figure 11,** and shade the region or regions representing each set.

(a) $A' \cap B$ **(b)** $A' \cup B'$

Solution

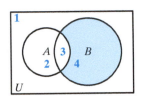

Figure 13

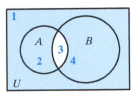

Figure 14

(a) See **Figure 11.** Set A' contains all the elements outside of set A—in other words, the elements in regions 1 and 4. Set B contains the elements in regions 3 and 4. The intersection of sets A' and B is made up of the elements in the region common to (1 and 4) and (3 and 4), which is region 4. Thus, $A' \cap B$ is represented by region 4, shown in color in **Figure 13.** This region can also be described as $B - A$.

(b) Again, set A' is represented by regions 1 and 4, and B' is made up of regions 1 and 2. The union of A' and B', the set $A' \cup B'$, is made up of the elements belonging to the union of (1 and 4) with (1 and 2)—that is, regions 1, 2, and 4, shown in color in **Figure 14.**   ■

EXAMPLE 10 Locating Elements in a Venn Diagram

Place the elements of the sets in their proper locations in a Venn diagram.

Let $U = \{q, r, s, t, u, v, w, x, y, z\}$, $A = \{r, s, t, u, v\}$,   and $B = \{t, v, x\}$.

Solution

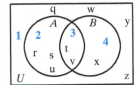

Figure 15

Because $A \cap B = \{t, v\}$, elements t and v are placed in region 3 in **Figure 15.** The remaining elements of A, that is, r, s, and u, go in region 2. The figure shows the proper placement of all other elements.   ■

EXAMPLE 11 Shading a Set in a Venn Diagram

Shade the set $(A' \cap B') \cap C$ in a Venn diagram similar to the one in **Figure 12.**

Solution

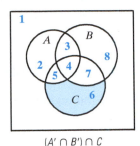

$(A' \cap B') \cap C$

Figure 16

Work first inside the parentheses. Set A' is made up of the regions outside set A, or regions 1, 6, 7, and 8. Set B' is made up of regions 1, 2, 5, and 6. The intersection of these sets is given by the overlap of regions 1, 6, 7, 8 and 1, 2, 5, 6, or regions 1 and 6.

For the final Venn diagram, find the intersection of regions 1 and 6 with set C. Set C is made up of regions 4, 5, 6, and 7. The overlap of regions 1, 6 and 4, 5, 6, 7 is region 6, the region shown in color in **Figure 16.**   ■

EXAMPLE 12 Verifying a Statement Using a Venn Diagram

Suppose $A, B \subseteq U$. Is the statement $(A \cap B)' = A' \cup B'$ true for every choice of sets A and B?

Solution

Use the regions labeled in **Figure 11.** Set $A \cap B$ is made up of region 3, so $(A \cap B)'$ is made up of regions 1, 2, and 4. These regions are shown in color in **Figure 17(a)** on the next page.

To identify set $A' \cup B'$, proceed as in **Example 9(b).** The result, shown in **Figure 14,** is repeated in **Figure 17(b)** on the next page.

(A ∩ B)′ is shaded.   A′ ∪ B′ is shaded.
(a)   (b)

Figure 17

The fact that the same regions are in color in both Venn diagrams suggests that

$$(A \cap B)' = A' \cup B'. \qquad \blacksquare$$

De Morgan's Laws

The result of **Example 12** can be stated in words.

> *The complement of the intersection of two sets is equal to the union of the complements of the two sets.*

Interchanging the words "intersection" and "union" produces another true statement.

> *The complement of the union of two sets is equal to the intersection of the complements of the two sets.*

Both of these "laws" were established by the British logician Augustus De Morgan (1806–1871), profiled on **page 22.** They are stated in set symbols as follows.

DE MORGAN'S LAWS FOR SETS

For any sets A and B, where $A, B \subseteq U$,

$$(A \cap B)' = A' \cup B' \quad \text{and} \quad (A \cup B)' = A' \cap B'.$$

The Venn diagrams in **Figure 17** strongly suggest the truth of the first of De Morgan's laws. They provide a *conjecture*. Actual proofs of De Morgan's laws would require methods used in more advanced courses in set theory.

EXAMPLE 13   **Describing Venn Diagram Regions Using Symbols**

For the Venn diagrams, write several symbolic descriptions of the region in color, using $A, B, C, \cap, \cup, -,$ and $'$ as necessary.

(a)
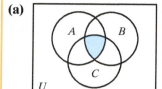

(b)
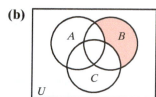

Solution

(a) The region in color can be described as belonging to all three sets, A and B and C. Therefore, the region corresponds to

$$(A \cap B) \cap C, \quad \text{or} \quad A \cap (B \cap C), \quad \text{or} \quad A \cap B \cap C.$$

(b) The region in color is in set B and is not in A and is not in C. Because it is not in A, it is in A', and similarly it is in C'. The region can be described as

$$B \cap A' \cap C', \quad \text{or} \quad B - (A \cup C), \quad \text{or} \quad B \cap (A \cup C)'. \qquad \blacksquare$$

2.3 EXERCISES

Match each term in Group I with the appropriate description from A–F in Group II. Assume that A and B are sets.

I

1. the intersection of A and B

2. the union of A and B

3. the difference of A and B

4. the complement of A

5. the Cartesian product of A and B

6. the difference of B and A

II

A. the set of elements in A that are not in B

B. the set of elements common to both A and B

C. the set of elements in the universal set that are not in A

D. the set of elements in B that are not in A

E. the set of ordered pairs such that each first element is from A and each second element is from B, with every element of A paired with every element of B

F. the set of elements that are in A or in B or in both A and B

Perform the indicated operations, and designate each answer using the listing method.

Let $U = \{a, b, c, d, e, f, g\}$, $X = \{a, c, e, g\}$, $Y = \{a, b, c\}$, and $Z = \{b, c, d, e, f\}$.

7. $X \cap Y$

8. $X \cup Y$

9. $Y \cup Z$

10. $Y \cap Z$

11. X'

12. Y'

13. $X' \cap Y'$

14. $X' \cap Z$

15. $X \cup (Y \cap Z)$

16. $Y \cap (X \cup Z)$

17. $X - Y$

18. $Y - X$

19. $(Z \cup X')' \cap Y$

20. $(Y \cap X')' \cup Z'$

21. $X \cap (X - Y)$

22. $Y \cup (Y - X)$

23. $X' - Y$

24. $Y' - (X \cap Z)$

Describe each set in words.

25. $A \cup (B' \cap C')$

26. $(A \cap B') \cup (B \cap A')$

27. $(C - B) \cup A$

28. $(A' \cap B') \cup C'$

Adverse Effects of Tobacco and Alcohol *The table lists some common adverse effects of prolonged tobacco and alcohol use.*

Tobacco	Alcohol
Emphysema, e	Liver damage, l
Heart damage, h	Brain damage, b
Cancer, c	Heart damage, h

In Exercises 29–32, let T be the set of listed effects of tobacco and A be the set of listed effects of alcohol. Find each set.

29. the smallest possible universal set U that includes all the effects listed

30. $T \cap A$

31. $T \cup A$

32. $T \cap A'$

An accountant is sorting tax returns in her files that require attention in the next week.

Describe in words each set in Exercises 33–36.

Let $U =$ the set of all tax returns in the file,

$A =$ the set of all tax returns with itemized deductions,

$B =$ the set of all tax returns showing business income,

$C =$ the set of all tax returns filed in 2014,

$D =$ the set of all tax returns selected for audit.

33. $C - A$

34. $D \cup A'$

35. $(A \cup B) - D$

36. $(C \cap A) \cap B'$

For Exercises 37–40, assume that A and B represent any two sets. Identify each statement as either always true *or* not always true.

37. $(A \cap B) \subseteq A$

38. $A \subseteq (A \cap B)$

39. $n(A \cup B) = n(A) + n(B)$

40. $n(A \cup B) = n(A) + n(B) - n(A \cap B)$

For Exercises 41–44, use your results in parts (a) and (b) to answer part (c).

Let $U = \{1, 2, 3, 4, 5\}$, $X = \{1, 3, 5\}$, $Y = \{1, 2, 3\}$, and $Z = \{3, 4, 5\}$.

41. (a) Find $X \cup Y$.　　**(b)** Find $Y \cup X$.

　　(c) State a conjecture.

42. (a) Find $X \cap Y$.　　**(b)** Find $Y \cap X$.

　　(c) State a conjecture.

43. (a) Find $X \cup (Y \cup Z)$.　　**(b)** Find $(X \cup Y) \cup Z$.

　　(c) State a conjecture.

44. (a) Find $X \cap (Y \cap Z)$.　　**(b)** Find $(X \cap Y) \cap Z$.

　　(c) State a conjecture.

Decide whether each statement is true *or* false.

45. $(3, 2) = (5 - 2, 1 + 1)$

46. $(2, 13) = (13, 2)$

47. $\{6, 3\} = \{3, 6\}$

48. $\{(5, 9), (4, 8), (4, 2)\} = \{(4, 8), (5, 9), (2, 4)\}$

Find $A \times B$ and $B \times A$, for A and B defined as follows.

49. $A = \{d, o, g\}$, $B = \{p, i, g\}$

50. $A = \{3, 6, 9, 12\}$, $B = \{6, 8\}$

Use the given information to find $n(A \times B)$ and $n(B \times A)$.

51. $n(A) = 35$ and $n(B) = 6$

52. $n(A) = 13$ and $n(B) = 5$

Find the cardinal number specified.

53. If $n(A \times B) = 72$ and $n(A) = 12$, find $n(B)$.

54. If $n(A \times B) = 300$ and $n(B) = 30$, find $n(A)$.

Place the elements of these sets in the proper locations in the given Venn diagram.

55. Let $U = \{a, b, c, d, e, f, g\}$,

　　　$A = \{b, d, f, g\}$,

　　　$B = \{a, b, d, e, g\}$.

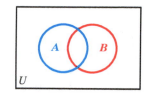

56. Let $U = \{5, 6, 7, 8, 9, 10, 11, 12, 13\}$,

　　　$M = \{5, 8, 10, 11\}$,

　　　$N = \{5, 6, 7, 9, 10\}$.

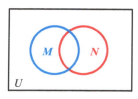

Use a Venn diagram similar to the one shown below to shade each set.

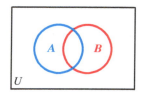

57. $A' \cup B$　　　　　　**58.** $A' \cap B'$

59. $B \cap A'$　　　　　　**60.** $A \cup B$

61. $B' \cap B$　　　　　　**62.** $A' \cup A$

63. $B' \cup (A' \cap B')$　　　**64.** $(A - B) \cup (B - A)$

In Exercises 65 and 66, place the elements of the sets in the proper locations in a Venn diagram similar to the one shown below.

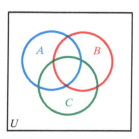

65. Let $U = \{m, n, o, p, q, r, s, t, u, v, w\}$,

　　　$A = \{m, n, p, q, r, t\}$,

　　　$B = \{m, o, p, q, s, u\}$,

　　　$C = \{m, o, p, r, s, t, u, v\}$.

66. Let $U = \{1, 2, 3, 4, 5, 6, 7, 8, 9\}$,

　　　$A = \{1, 3, 5, 7\}$,

　　　$B = \{1, 3, 4, 6, 8\}$,

　　　$C = \{1, 4, 5, 6, 7, 9\}$.

Use a Venn diagram to shade each set.

67. $(A \cap B) \cap C$　　　　**68.** $(A' \cap B) \cap C$

69. $(A' \cap B') \cap C$　　　**70.** $(A \cap C') \cap B$

71. $(A \cap B') \cap C'$　　　**72.** $(A \cap B)' \cup C$

Write a description of each shaded area. Use the symbols A, B, C, ∩, ∪, −, and '. Different answers are possible.

73.
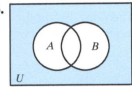

74.
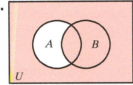

75.
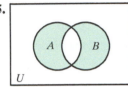

76.
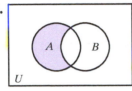

77.
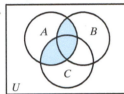

78.
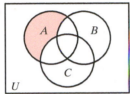

Suppose A and B are sets. Describe the conditions under which each statement would be true.

79. $A = A - B$

80. $A = B - A$

81. $A = A - \varnothing$

82. $A \cap \varnothing = \varnothing$

83. $A \cup B = A$

84. $A \cap B = B$

For Exercises 85–88, draw two Venn diagrams to decide whether the statement is always true *or* not always true.

85. $(A \cap B) \subseteq A$

86. $(A \cup B) \subseteq A$

87. If $A \subseteq B$, then $A \cup B = A$.

88. If $A \subseteq B$, then $A \cap B = A$.

89. Explain why, if A and B are sets, it is not necessarily true that $n(A - B) = n(A) - n(B)$. Give a counterexample.

90. The five set operations listed on **page 66** are applied to subsets of U (that is, to A and/or B). Is the result always a subset of U also? Explain why or why not.

2.4 SURVEYS AND CARDINAL NUMBERS

OBJECTIVES

1 Analyze survey results.

2 Apply the cardinal number formula.

3 Interpret information from tables.

Surveys

As suggested in the chapter opener, the techniques of set theory can be applied to many different groups of people (or objects). The problems addressed sometimes require analyzing known information about certain subsets to obtain cardinal numbers of other subsets. In this section, we apply three problem-solving strategies to such problems:

Venn diagrams,   cardinal number formulas,   and   tables.

The "known information" is quite often (although not always) obtained by conducting a survey.

Suppose a group of attendees at an educational seminar, all of whom desire to become registered nurses (RNs), are asked their preferences among the three traditional ways to become an RN, and the following information is produced.

23 would consider pursuing the Bachelor of Science in Nursing (BSN).

16 would consider pursuing the Associate Degree in Nursing (ADN).

7 would consider a Diploma program (Diploma).

10 would consider both the BSN and the ADN.

5 would consider both the BSN and the Diploma.

3 would consider both the ADN and the Diploma.

2 would consider all three options.

4 are looking for an option other than these three.

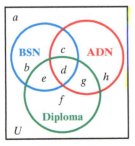

Figure 18

To determine the total number of attendees surveyed, we cannot just add the eight numbers above because there is some overlap. For example, in **Figure 18,** the 23 who like the BSN option should not be positioned in region b but, rather, should be distributed among regions b, c, d, and e in a way that is consistent with all of the data. (Region b actually contains those who like the BSN but neither of the other two options.)

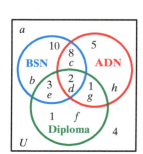

Figure 19

Because, at the start, we do not know how to distribute the 23 who like the BSN, we look first for some more manageable data. The smallest total listed, the 2 who like all three options, can be placed in region d (the intersection of the three sets). The 4 who like none of the three must go into region a. Then, the 10 who like the BSN and the ADN must go into regions c and d. Because region d already contains 2 attendees, we must place

$$10 - 2 = 8 \quad \text{in region } c.$$

Because 5 like BSN and Diploma (regions d and e), we place

$$5 - 2 = 3 \quad \text{in region } e.$$

Now that regions c, d, and e contain 8, 2, and 3, respectively, we must place

$$23 - 8 - 2 - 3 = 10 \quad \text{in region } b.$$

By similar reasoning, all regions are assigned their correct numbers. See **Figure 19.**

EXAMPLE 1   Analyzing a Survey

Using the survey data on personal preferences for nursing preparation, as summarized in **Figure 19,** answer each question.

(a) How many persons like the BSN option only?

(b) How many persons like exactly two of these three options?

(c) How many persons were surveyed?

Solution

(a) A person who likes BSN only does not like ADN and does not like Diploma. These persons are inside the regions for BSN and outside the regions for ADN and Diploma. Region b is the appropriate region in **Figure 19,** and we see that ten persons like BSN only.

(b) The persons in regions c, e, and g like exactly two of the three options. The total number of such persons is

$$8 + 3 + 1 = 12.$$

(c) Each person surveyed has been placed in exactly one region of **Figure 19,** so the total number surveyed is the sum of the numbers in all eight regions:

$$4 + 10 + 8 + 2 + 3 + 1 + 1 + 5 = 34. \qquad \blacksquare$$

Cardinal Number Formula

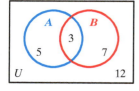

Figure 20

If the numbers shown in **Figure 20** are the cardinal numbers of the individual regions, then

$$n(A) = 5 + 3 = 8, \quad n(B) = 3 + 7 = 10, \quad n(A \cap B) = 3,$$

and

$$n(A \cup B) = 5 + 3 + 7 = 15.$$

Notice that $n(A \cup B) = n(A) + n(B) - n(A \cap B)$ because $15 = 8 + 10 - 3$. This relationship is true for any two sets A and B.

CARDINAL NUMBER FORMULA

For any two sets A and B, the following is true.

$$n(A \cup B) = n(A) + n(B) - n(A \cap B)$$

The cardinal number formula can be rearranged to find any one of its four terms when the others are known.

EXAMPLE 2   Applying the Cardinal Number Formula

Find $n(A)$ if $n(A \cup B) = 22$, $n(A \cap B) = 8$, and $n(B) = 12$.

Solution

We solve the cardinal number formula for $n(A)$.

$$n(A) = n(A \cup B) - n(B) + n(A \cap B)$$
$$= 22 - 12 + 8$$
$$= 18$$

Sometimes, even when information is presented as in **Example 2,** it is more convenient to fit that information into a Venn diagram as in **Example 1.**

WHEN   Will I   Ever USE   This   ?

Suppose you run a small construction company, building a few "spec" homes at a time. This week's work will require the following jobs:

Hanging drywall (D),   Installing roofing (R),   Doing electrical work (E).

You will assign 9 workers, with job skills as described here.

6 of the 9 can do D	5 can do both D and E
4 can do R	4 can do both R and E
7 can do E	4 can do all three

1. Construct a Venn diagram to decide how many of your workers have

   (a) exactly two of the three skills

   (b) none of the three skills

   (c) no more than one of the three skills.

2. Which skill is common to the greatest number of workers, and how many possess that skill?

Answers: 1. (a) 1 (b) 1 (c) 4 2. Electrical work; 7

EXAMPLE 3 **Analyzing Data in a Report**

Scott, who leads a group of software engineers who investigate illegal activities on social networking sites, reported the following information.

T = the set of group members following patterns on Twitter

F = the set of group members following patterns on Facebook

L = the set of group members following patterns on LinkedIn

$$n(T) = 13 \qquad n(T \cap F) = 9 \qquad n(T \cap F \cap L) = 5$$
$$n(F) = 16 \qquad n(F \cap L) = 10 \qquad n(T' \cap F' \cap L') = 3$$
$$n(L) = 13 \qquad n(T \cap L) = 6$$

How many engineers are in Scott's group?

Solution

The data supplied by Scott are reflected in **Figure 21.** The sum of the numbers in the diagram gives the total number of engineers in the group.

$$3 + 3 + 1 + 2 + 5 + 5 + 4 + 2 = 25 \qquad \blacksquare$$

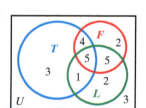

Figure 21

Tables

Sometimes information appears in a table rather than a Venn diagram, but the basic ideas of union and intersection still apply.

EXAMPLE 4 **Analyzing Data in a Table**

Melanie, the officer in charge of the cafeteria on a military base, wanted to know if the beverage that enlisted men and women preferred with lunch depended on their ages. On a given day, Melanie categorized her lunch patrons according to age and preferred beverage, recording the results in a table.

		Beverage			
		Cola (*C*)	Iced Tea (*I*)	Sweet Tea (*S*)	Totals
Age	18–25 (*Y*)	45	10	35	90
	26–33 (*M*)	20	25	30	75
	Over 33 (*O*)	5	30	20	55
	Totals	70	65	85	220

Using the letters in the table, find the number of people in each set.

(a) $Y \cap C$ **(b)** $O' \cup I$

Solution

(a) The set Y includes all personnel represented across the top row of the table (90 in all), while C includes the 70 down the left column. The intersection of these two sets is just the upper left entry, 45 people.

(b) The set O' excludes the bottom row, so it includes the first and second rows. The set I includes the middle column only. The union of the two sets represents

$$45 + 10 + 35 + 20 + 25 + 30 + 30 = 195 \text{ people.} \qquad \blacksquare$$

2.4 EXERCISES

Use the numerals representing cardinalities in the Venn diagrams to give the cardinality of each set specified.

1.
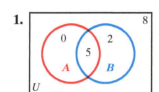

(a) $A \cap B$ (b) $A \cup B$
(c) $A \cap B'$ (d) $A' \cap B$
(e) $A' \cap B'$

2.
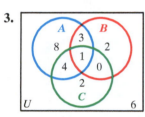

(a) $A \cap B$ (b) $A \cup B$
(c) $A \cap B'$ (d) $A' \cap B$
(e) $A' \cap B'$

3.
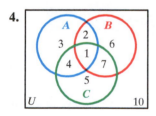

(a) $A \cap B \cap C$ (b) $A \cap B \cap C'$
(c) $A \cap B' \cap C$ (d) $A' \cap B \cap C$
(e) $A' \cap B' \cap C$ (f) $A \cap B' \cap C'$
(g) $A' \cap B \cap C'$ (h) $A' \cap B' \cap C'$

4.

(a) $A \cap B \cap C$ (b) $A \cap B \cap C'$
(c) $A \cap B' \cap C$ (d) $A' \cap B \cap C$
(e) $A' \cap B' \cap C$ (f) $A \cap B' \cap C'$
(g) $A' \cap B \cap C'$ (h) $A' \cap B' \cap C'$

In Exercises 5–10, make use of an appropriate formula.

5. Find the value of $n(A \cup B)$ if $n(A) = 12$, $n(B) = 14$, and $n(A \cap B) = 5$.

6. Find the value of $n(A \cup B)$ if $n(A) = 16$, $n(B) = 28$, and $n(A \cap B) = 5$.

7. Find the value of $n(A \cap B)$ if $n(A) = 20$, $n(B) = 12$, and $n(A \cup B) = 25$.

8. Find the value of $n(A \cap B)$ if $n(A) = 20$, $n(B) = 24$, and $n(A \cup B) = 30$.

9. Find the value of $n(A)$ if $n(B) = 35$, $n(A \cap B) = 15$, and $n(A \cup B) = 55$.

10. Find the value of $n(B)$ if $n(A) = 20$, $n(A \cap B) = 6$, and $n(A \cup B) = 30$.

Draw a Venn diagram and use the given information to fill in the number of elements in each region.

11. $n(A) = 19$, $n(B) = 13$, $n(A \cup B) = 25$, $n(A') = 11$

12. $n(U) = 43$, $n(A) = 25$, $n(A \cap B) = 5$, $n(B') = 30$

13. $n(A') = 25$, $n(B) = 28$, $n(A' \cup B') = 40$, $n(A \cap B) = 10$

14. $n(A \cup B) = 15$, $n(A \cap B) = 8$, $n(A) = 13$, $n(A' \cup B') = 11$

15. $n(A) = 57$, $n(A \cap B) = 35$, $n(A \cup B) = 81$, $n(A \cap B \cap C) = 15$, $n(A \cap C) = 21$, $n(B \cap C) = 25$, $n(C) = 49$, $n(B') = 52$

16. $n(A) = 24$, $n(B) = 24$, $n(C) = 26$, $n(A \cap B) = 10$, $n(B \cap C) = 8$, $n(A \cap C) = 15$, $n(A \cap B \cap C) = 6$, $n(U) = 50$

17. $n(A) = 15$, $n(A \cap B \cap C) = 5$, $n(A \cap C) = 13$, $n(A \cap B') = 9$, $n(B \cap C) = 8$, $n(A' \cap B' \cap C') = 21$, $n(B \cap C') = 3$, $n(B \cup C) = 32$

18. $n(A \cap B) = 21$, $n(A \cap B \cap C) = 6$, $n(A \cap C) = 26$, $n(B \cap C) = 7$, $n(A \cap C') = 20$, $n(B \cap C') = 25$, $n(C) = 40$, $n(A' \cap B' \cap C') = 2$

Use Venn diagrams to work each problem.

19. *Writing and Producing Music*   Joe Long worked on 9 music projects last year.

Joe Long, Bob Gaudio, Tommy DeVito, and Frankie Valli
The Four Seasons

He wrote and produced 3 projects.
He wrote a total of 5 projects.
He produced a total of 7 projects.

(a) How many projects did he write but not produce?

(b) How many projects did he produce but not write?

20. *Compact Disc Collection* Gitti is a fan of the music of Paul Simon and Art Garfunkel. In her collection of 25 compact discs, she has the following:

> 5 on which both Simon and Garfunkel sing
>
> 7 on which Simon sings
>
> 8 on which Garfunkel sings
>
> 15 on which neither Simon nor Garfunkel sings.

(a) How many of her compact discs feature only Paul Simon?

(b) How many of her compact discs feature only Art Garfunkel?

(c) How many feature at least one of these two artists?

(d) How many feature at most one of these two artists?

21. *Student Response to Classical Composers* The 65 students in a classical music lecture class were polled, with the following results:

> 37 like Wolfgang Amadeus Mozart
>
> 36 like Ludwig van Beethoven
>
> 31 like Franz Joseph Haydn
>
> 14 like Mozart and Beethoven
>
> 21 like Mozart and Haydn
>
> 14 like Beethoven and Haydn
>
> 8 like all three composers.

How many of these students like:

(a) exactly two of these composers?

(b) exactly one of these composers?

(c) none of these composers?

(d) Mozart, but neither Beethoven nor Haydn?

(e) Haydn and exactly one of the other two?

(f) no more than two of these composers?

22. *Financial Aid for Students* At a southern university, half of the 48 mathematics majors were receiving federal financial aid.

> 5 had Pell Grants
>
> 14 participated in the College Work Study Program
>
> 4 had TOPS scholarships
>
> 2 had TOPS scholarships and participated in Work Study.
>
> Those with Pell Grants had no other federal aid.

How many of the 48 math majors had:

(a) no federal aid?

(b) more than one of these three forms of aid?

(c) federal aid other than these three forms?

(d) a TOPS scholarship or Work Study?

(e) exactly one of these three forms of aid?

(f) no more than one of these three forms of aid?

23. *Animated Movies* A middle school counselor, attempting to correlate school performance with leisure interests, found the following information for a group of students:

> 34 had seen *Despicable Me*
>
> 29 had seen *Epic*
>
> 26 had seen *Turbo*
>
> 16 had seen *Despicable Me* and *Epic*
>
> 12 had seen *Despicable Me* and *Turbo*
>
> 10 had seen *Epic* and *Turbo*
>
> 4 had seen all three of these films
>
> 5 had seen none of the three films.

(a) How many students had seen *Turbo* only?

(b) How many had seen exactly two of the films?

(c) How many students were surveyed?

24. *Non-Mainline Religious Beliefs* 140 U.S. adults were surveyed.

Let A = the set of respondents who believe in astrology,

R = the set of respondents who believe in reincarnation,

Y = the set of respondents who believe in the spirituality of yoga.

The survey revealed the following information:

$$n(A) = 35 \qquad n(R \cap Y) = 8$$
$$n(R) = 36 \qquad n(A \cap Y) = 10$$
$$n(Y) = 32 \qquad n(A \cap R \cap Y) = 6$$
$$n(A \cap R) = 19$$

How many of the respondents believe in:

(a) astrology but not reincarnation?

(b) at least one of these three things?

(c) reincarnation but neither of the others?

(d) exactly two of these three things?

(e) none of the three?

25. Survey on Attitudes toward Religion Researchers interviewed a number of people and recorded the following data. Of all the respondents:

> 240 think Hollywood is unfriendly toward religion
>
> 160 think the media are unfriendly toward religion
>
> 181 think scientists are unfriendly toward religion
>
> 145 think both Hollywood and the media are unfriendly toward religion
>
> 122 think both scientists and the media are unfriendly toward religion
>
> 80 think exactly two of these groups are unfriendly toward religion
>
> 110 think all three groups are unfriendly toward religion
>
> 219 think none of these three groups is unfriendly toward religion.

How many respondents:

(a) were surveyed?

(b) think exactly one of these three groups is unfriendly toward religion?

26. Student Goals Sofia, who sells college textbooks, interviewed first-year students on a community college campus to find out the main goals of today's students.

Let W = the set of those who want to be wealthy,

F = the set of those who want to raise a family,

E = the set of those who want to become experts in their fields.

Sofia's findings are summarized here.

$$n(W) = 160 \qquad n(E \cap F) = 90$$
$$n(F) = 140 \qquad n(W \cap F \cap E) = 80$$
$$n(E) = 130 \qquad n(E') = 95$$
$$n(W \cap F) = 95 \qquad n[(W \cup F \cup E)'] = 10$$

Find the total number of students interviewed.

27. Hospital Patient Symptoms A survey was conducted among 75 patients admitted to a hospital cardiac unit during a two-week period.

Let B = the set of patients with high blood pressure,

C = the set of patients with high cholesterol levels,

S = the set of patients who smoke cigarettes.

The survey produced the following data.

$$n(B) = 47 \qquad n(B \cap S) = 33$$
$$n(C) = 46 \qquad n(B \cap C) = 31$$
$$n(S) = 52 \qquad n(B \cap C \cap S) = 21$$
$$n[(B \cap C) \cup (B \cap S) \cup (C \cap S)] = 51$$

Find the number of these patients who:

(a) had either high blood pressure or high cholesterol levels, but not both

(b) had fewer than two of the indications listed

(c) were smokers but had neither high blood pressure nor high cholesterol levels

(d) did not have exactly two of the indications listed.

28. Song Themes It was once said that country-western songs emphasize three basic themes: love, prison, and trucks. A survey of the local country-western radio station produced the following data.

> 12 songs about a truck driver who is in love while in prison
>
> 13 about a prisoner in love
>
> 28 about a person in love
>
> 18 about a truck driver in love
>
> 3 about a truck driver in prison who is not in love
>
> 2 about people in prison who are not in love and do not drive trucks
>
> 8 about people who are out of prison, are not in love, and do not drive trucks
>
> 16 about truck drivers who are not in prison

(a) How many songs were surveyed?

Find the number of songs about:

(b) truck drivers

(c) prisoners

(d) truck drivers in prison

(e) people not in prison

(f) people not in love.

29. Use the figure to find the numbers of the regions belonging to each set.

(a) $A \cap B \cap C \cap D$

(b) $A \cup B \cup C \cup D$

(c) $(A \cap B) \cup (C \cap D)$

(d) $(A' \cap B') \cap (C \cup D)$

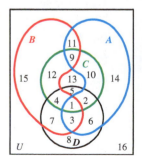

30. Sports Viewing A survey of 130 TV viewers was taken.

> 52 watch football
>
> 56 watch basketball
>
> 62 watch tennis
>
> 60 watch golf
>
> 21 watch football and basketball
>
> 19 watch football and tennis
>
> 22 watch basketball and tennis
>
> 27 watch football and golf
>
> 30 watch basketball and golf
>
> 21 watch tennis and golf
>
> 3 watch football, basketball, and tennis
>
> 15 watch football, basketball, and golf
>
> 10 watch football, tennis, and golf
>
> 10 watch basketball, tennis, and golf
>
> 3 watch all four of these sports
>
> 5 don't watch any of these four sports

Use a Venn diagram to answer these questions.

(a) How many of these viewers watch football, basketball, and tennis, but not golf?

(b) How many watch exactly one of these four sports?

(c) How many watch exactly two of these four sports?

Solve each problem.

31. Basketball Positions Sakda runs a basketball program in California. On the first day of the season, 60 young women showed up and were categorized by age level and by preferred basketball position, as shown in the following table.

	Position			
	Guard (G)	Forward (F)	Center (N)	Totals
Junior High (J)	9	6	4	19
Age **Senior High (S)**	12	5	9	26
College (C)	5	8	2	15
Totals	26	19	15	60

Using the set labels (letters) in the table, find the number of players in each of the following sets.

(a) $J \cap G$

(b) $S \cap N$

(c) $N \cup (S \cap F)$

(d) $S' \cap (G \cup N)$

(e) $(S \cap N') \cup (C \cap G')$

(f) $N' \cap (S' \cap C')$

32. Army Housing A study of U.S. Army housing trends categorized personnel as commissioned officers (C), warrant officers (W), or enlisted (E), and categorized their living facilities as on-base (B), rented off-base (R), or owned off-base (O). One survey yielded the following data.

		Facilities			
		B	**R**	**O**	**Totals**
	C	12	29	54	95
Personnel	**W**	4	5	6	15
	E	374	71	285	730
	Totals	390	105	345	840

Find the number of personnel in each of the following sets.

(a) $W \cap O$

(b) $C \cup B$

(c) $R' \cup W'$

(d) $(C \cup W) \cap (B \cup R)$

(e) $(C \cap B) \cup (E \cap O)$

(f) $B \cap (W \cup R)'$

33. Could the information of **Example 4** have been presented in a Venn diagram similar to those in **Examples 1 and 3**? If so, construct such a diagram. Otherwise, explain the essential difference of **Example 4.**

34. Explain how a cardinal number formula can be derived for the case where *three* sets occur. Specifically, give a formula relating $n(A \cup B \cup C)$ to

$$n(A), \ n(B), \ n(C), \ n(A \cap B), \ n(A \cap C),$$
$$n(B \cap C), \ \text{and} \ n(A \cap B \cap C).$$

Illustrate with a Venn diagram.

CHAPTER 2 **SUMMARY**

KEY TERMS

2.1

integers
rational numbers
set | real numbers
elements | irrational numbers
members | cardinal number
empty (null) set | (cardinality)
natural (counting) numbers | finite set
whole numbers | infinite set

2.2

universal set
Venn diagram
complement (of a set)
subset (of a set)
proper subset (of a set)
tree diagram

2.3

intersection (of sets)
disjoint sets
union (of sets)
difference (of sets)
ordered pairs
Cartesian product (of sets)
operation (on sets)

NEW SYMBOLS

$\varnothing$	empty set (or null set)	$\not\subseteq$	is not a subset of
$\in$	is an element of	$\subset$	is a proper subset of
$\notin$	is not an element of	$\cap$	intersection (of sets)
$n(A)$	cardinal number of set A	$\cup$	union (of sets)
$\aleph_0$	aleph null	$-$	difference (of sets)
U	universal set	(a, b)	ordered pair
A'	complement of set A	$\times$	Cartesian product
$\subseteq$	is a subset of		

TEST YOUR WORD POWER

See how well you have learned the vocabulary in this chapter.

1. In an **ordered pair,**
   A. the components are always numbers of the same type.
   B. the first component must be less than the second.
   C. the components can be any kinds of objects.
   D. the first component is a subset of the second component.

2. The **complement** of a set
   A. contains only some, but not all, of that set's elements.
   B. contains the same number of elements as the given set.
   C. always contains fewer elements than the given set.
   D. cannot be determined until a universal set is given.

3. Any **subset** of set A
   A. must have fewer elements than A.
   B. has fewer elements than A only if it is a proper subset.
   C. is an element of A.
   D. contains all the elements of A, plus at least one additional element.

4. Examples of **operations on sets** are
   A. union, intersection, and subset.
   B. Cartesian product, difference, and proper subset.
   C. complement, supplement, and intersection.
   D. complement, union, and Cartesian product.

5. The **set difference** $A - B$ must have cardinality
   A. $n(A) - n(B)$.
   B. less than or equal to $n(A)$.
   C. greater than or equal to $n(B)$.
   D. less than $n(A)$.

6. The **cardinal number formula** says that
   A. $n(A \cup B) = n(A) + n(B) - n(A \cap B)$.
   B. $n(A \cup B) = n(A) + n(B) + n(A \cap B)$.
   C. $n(A \cup B) = n(A) + n(B)$.
   D. $n(A \cup B) = n(A) \cdot n(B)$.

ANSWERS
1. C **2.** D **3.** B **4.** D **5.** B **6.** A

QUICK REVIEW

Concepts	Examples

2.1 Symbols and Terminology

Designating Sets

Sets are designated using the following methods:

(1) word descriptions,

(2) the listing method,

(3) set-builder notation.

The **cardinal number** of a set is the number of elements it contains.

Equal sets have exactly the same elements.

The set of odd counting numbers less than 7

$\{1, 3, 5\}$

$\{x \mid x$ is an odd counting number and $x < 7\}$

All are equal.

If $A = \{10, 20, 30, ..., 80\}$, then $n(A) = 8$.

$\{a, e, i, o, u\} = \{i, o, u, a, e\}, \quad \{q, r, s, t\} \neq \{q, p, s, t\}$

2.2 Venn Diagrams and Subsets

Sets are normally discussed within the context of a designated **universal set, U.**

The **complement** of a set A contains all elements in U that are not in A.

Set B is a **subset** of set A if every element of B is also an element of A.

Set B is a **proper subset** of A if $B \subseteq A$ and $B \neq A$

If a set has cardinal number n, it has 2^n subsets and $2^n - 1$ proper subsets.

Let $U = \{x \mid x$ is a whole number$\}$

and

$A = \{x \mid x$ is an even whole number$\}$.

Then $A' = \{x \mid x$ is an odd whole number$\}$.

Let $U = \{2, 3, 5, 7\}$, $A = \{3, 5, 7\}$, and $B = \{5\}$. Then $A' = \{2\}$ and $B \subseteq A$.

For the sets A and B given above, $B \subset A$.

If $D = \{x, y, z\}$, then $n(D) = 3$.
D has $2^3 = 8$ subsets and $2^3 - 1 = 8 - 1 = 7$ proper subsets.

2.3 Set Operations

Intersection of sets

$A \cap B = \{x \mid x \in A \text{ and } x \in B\}$

Union of sets

$A \cup B = \{x \mid x \in A \text{ or } x \in B\}$

Difference of sets

$A - B = \{x \mid x \in A \text{ and } x \notin B\}$

Cartesian product of sets

$A \times B = \{(a, b) \mid a \in A \text{ and } b \in B\}$

Cardinal number of a Cartesian product

If $n(A) = a$ and $n(B) = b$, then

$n(A \times B) = n(A) \cdot n(B) = ab.$

$\{1, 2, 7\} \cap \{5, 7, 9, 11\} = \{7\}$

$\{20, 40, 60\} \cup \{40, 60, 80\} = \{20, 40, 60, 80\}$

$\{5, 6, 7, 8\} - \{1, 3, 5, 7\} = \{6, 8\}$

$\{1, 2\} \times \{30, 40, 50\}$
$= \{(1, 30), (1, 40), (1, 50), (2, 30), (2, 40), (2, 50)\}$

If $A = \{2, 4, 6, 8\}$ and $B = \{3, 7\}$, then

$n(A \times B) = n(A) \cdot n(B) = 4 \cdot 2 = 8.$

Verify by listing the elements of $A \times B$.

$\{(2, 3), (2, 7), (4, 3), (4, 7), (6, 3), (6, 7), (8, 3), (8, 7)\}$

It, indeed, has 8 elements.

Concepts

Examples

Numbering the regions in a Venn diagram facilitates identification of various sets and relationships among them.

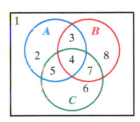

A consists of regions 2, 3, 4, and 5.

$A \cap B$ consists of regions 3 and 4.

$B \cup C$ consists of regions 3, 4, 5, 6, 7, and 8.

$(A \cup B) \cap C'$ consists of regions 2, 3, and 8.

$A \cap B \cap C$ consists of region 4.

$(A \cup B \cup C)'$ consists of region 1.

$A - C$ consists of regions 2 and 3.

De Morgan's Laws

For any sets A and B, where $A, B \subseteq U$, the following are true.

$$(A \cap B)' = A' \cup B'$$

$$(A \cup B)' = A' \cap B'$$

Let $U = \{1, 2, 3, 4, \ldots, 9\}$, $A = \{2, 4, 6, 8\}$, and $B = \{4, 5, 6, 7\}$.

Then $A \cap B = \{4, 6\}$,

so $(A \cap B)' = \{1, 2, 3, 5, 7, 8, 9\}$.

Also $A' = \{1, 3, 5, 7, 9\}$ and $B' = \{1, 2, 3, 8, 9\}$, ─ Same

so $A' \cup B' = \{1, 2, 3, 5, 7, 8, 9\}$.

Using the same sets A and B as above,

$$A \cup B = \{2, 4, 5, 6, 7, 8\},$$

so $(A \cup B)' = \{1, 3, 9\}$.

Also, $A' = \{1, 3, 5, 7, 9\}$ and $B' = \{1, 2, 3, 8, 9\}$, ─ Same

so $A' \cap B' = \{1, 3, 9\}$.

2.4 Surveys and Cardinal Numbers

Cardinal Number Formula

For any two sets A and B, the following is true.

$$n(A \cup B) = n(A) + n(B) - n(A \cap B)$$

Enter known facts in a Venn diagram to find desired facts.

Suppose $n(A) = 5, n(B) = 12$, and $n(A \cup B) = 10$. Then, to find $n(A \cap B)$, first solve the formula for that term.

$$n(A \cap B) = n(A) + n(B) - n(A \cup B)$$
$$= 5 + 12 - 10 = 7$$

Given $n(A) = 12, n(B) = 27, n(A \cup B) = 32$, and $n(U) = 50$, find $n(A - B)$ and $n[(A \cap B)']$.

$$n(A \cap B) = n(A) + n(B) - n(A \cup B)$$
$$= 12 + 27 - 32$$
$$= 7$$

Then the cardinalities of all regions can be entered in the diagram, starting with 7 in the center.

Now observe that $n(A - B) = 5$ and $n[(A \cap B)'] = 18 + 5 + 20 = 43$.

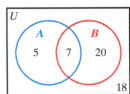

CHAPTER 2	**TEST**

In Exercises 1–14, let

$U = \{a, b, c, d, e, f, g, h\}, \quad A = \{a, b, c, d\},$
$B = \{b, e, a, d\}, \quad and \quad C = \{e, a\}.$

Find each set.

1. $A \cup C$ **2.** $B \cap A$

3. B' **4.** $A - (B \cap C')$

Identify each statement as true *or* false.

5. $e \in A$ **6.** $C \subseteq B$

7. $B \subset (A \cup C)$ **8.** $c \notin C$

9. $n[(A \cup B) - C] = 4$ **10.** $\varnothing \not\subset C$

11. $A \cap B'$ is equivalent to $B \cap A'$

12. $(A \cup B)' = A' \cap B'$

Find each of the following.

13. $n(B \times A)$

14. the number of proper subsets of B

Give a word description for each set.

15. $\{-3, -1, 1, 3, 5, 7, 9\}$

16. $\{\text{Sun, Mon, Tue}, \ldots, \text{Sat}\}$

Express each set in set-builder notation.

17. $\{-1, -2, -3, -4, \ldots\}$

18. $\{24, 32, 40, 48, \ldots, 88\}$

Place $\subset$, $\subseteq$, both, *or* neither *in each blank to make a true statement.*

19. $\varnothing$ _____ $\{x \mid x$ is a counting number between 20 and 21$\}$

20. $\{3, 5, 7\}$ _____ $\{4, 5, 6, 7, 8, 9, 10\}$

Shade each set in an appropriate Venn diagram.

21. $X \cup Y'$ **22.** $X' \cap Y'$

23. $(X \cup Y) - Z$

24. $[(X \cap Y) \cup (X \cap Z)] - (Y \cap Z)$

25. State De Morgan's laws for sets in words rather than symbols.

Facts about Inventions *The table lists ten inventions, together with other pertinent data.*

Invention	Date	Inventor	Nation
Adding machine	1642	Pascal	France
Baking powder	1843	Bird	England
Electric razor	1917	Schick	U.S.
Fiber optics	1955	Kapany	England
Geiger counter	1913	Geiger	Germany
Pendulum clock	1657	Huygens	Holland
Radar	1940	Watson-Watt	Scotland
Telegraph	1837	Morse	U.S.
Thermometer	1593	Galileo	Italy
Zipper	1891	Judson	U.S.

Let $U =$ the set of all ten inventions,
$A =$ the set of items invented in the United States,
and $T =$ the set of items invented in the twentieth century.

List the elements of each set.

26. $A \cap T$ **27.** $(A \cup T)'$ **28.** $T' - A'$

29. The numerals in the Venn diagram indicate the number of elements in each particular subset. Determine the number of elements in each set.

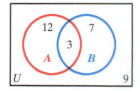

(a) $A \cup B$ (b) $A \cap B'$

(c) $(A \cap B)'$

30. ***Financial Aid to College Students*** Three sources of financial aid are government grants, private scholarships, and the colleges themselves. Susan, a financial aid director of a private college, surveyed the records of 100 sophomores and found the following:

49 receive government grants

55 receive private scholarships

43 receive aid from the college

23 receive government grants and private scholarships

18 receive government grants and aid from the college

28 receive private scholarships and aid from the college

8 receive help from all three sources.

How many of the students in the survey:

(a) have government grants only?

(b) have scholarships but not government grants?

(c) receive financial aid from only one of these sources?

(d) receive aid from exactly two of these sources?

(e) receive no financial aid from any of these sources?

(f) receive no aid from the college or the government?

Introduction to Logic

3

To capture the imagination of their readers, authors of fantasy literature (such as J. R. R. Tolkien, J. K. Rowling, and Lewis Carroll) have long used magical imagery, including riddles. Such a riddle confronts a hero named Humphrey, who finds himself trapped in a magical maze with only two exits.

> *One door to freedom opens wide,*
> *The other naught but dungeon hides.*
> *Twin watchers, one by truth is bound,*
> *The other never truth will sound.*
>
> *The watcher here shall speak when he*
> *Spies a man who would be free.*
> *Name him Lying Troll or Truthful,*
> *Rightly named he shall be useful.*
>
> *If he moves, then be not late,*
> *Choose a door and seal thy fate.*

As Humphrey approaches the troll, it speaks:

> *"If truly Truthful Troll I be,*
> *then go thou east, and be thou free."*

Logic, the topic of this chapter, is useful to authors in the creation of such puzzles—and also to readers in solving them.

3.1 STATEMENTS AND QUANTIFIERS

OBJECTIVES

1 Distinguish between statements and non-statements.

2 Compose negations of statements.

3 Translate between words and symbols.

4 Interpret statements with quantifiers and form their negations.

5 Find truth values of statements involving quantifiers and number sets.

Statements

This section introduces the study of **symbolic logic,** which uses letters to represent statements, and symbols for words such as *and, or, not.* Logic is used to determine the **truth value** (that is, the truth or falsity) of statements with multiple parts. The truth value of such statements depends on their components.

Many kinds of sentences occur in ordinary language, including factual statements, opinions, commands, and questions. Symbolic logic discusses only statements that involve facts. A **statement** is a declarative sentence that is either true or false, but not both simultaneously.

Electronic mail provides a means of communication. } Statements
$$12 + 6 = 13$$ Each is either true or false.

Access the file.   Not statements
Did the Seahawks win the Super Bowl?   Each cannot be
Dustin Pedroia is a better baseball player than Miguel Cabrera. identified as being
This sentence is false.   either true or false.

Of the sentences that are not statements, the first is a command, and the second is a question. The third is an opinion. "This sentence is false" is a paradox: If we assume it is true, then it is false, and if we assume it is false, then it is true.

A **compound statement** may be formed by combining two or more statements. The statements making up a compound statement are called **component statements.** Various **logical connectives,** or simply **connectives,** such as *and, or, not,* and *if . . . then,* can be used in forming compound statements. (Although a statement such as "Today is not Tuesday" does not consist of two component statements, for convenience it is considered compound, because its truth value is determined by noting the truth value of a different statement, "Today is Tuesday.")

EXAMPLE 1 Deciding Whether a Statement Is Compound

Decide whether each statement is compound. If so, identify the connective.

(a) Lord Byron wrote sonnets, and the poem exhibits iambic pentameter.

(b) You can pay me now, or you can pay me later.

(c) If it's on the Internet, then it must be true.

(d) My pistol was made by Smith and Wesson.

Solution

(a) This statement is compound, because it is made up of the component statements "Lord Byron wrote sonnets" and "the poem exhibits iambic pentameter." The connective is *and.*

(b) The connective here is *or.* The statement is compound.

(c) The connective here is *if . . . then*, discussed in more detail in **Section 3.3.** The statement is compound.

(d) Although the word "and" is used in this statement, it is not used as a *logical* connective. It is part of the name of the manufacturer. The statement is not compound. ∎

Negations

Gottfried Leibniz (1646–1716) was a wide-ranging philosopher and a universalist who tried to patch up Catholic–Protestant conflicts. He promoted cultural exchange between Europe and the East. Chinese ideograms led him to search for a universal symbolism. He was an early inventor of **symbolic logic.**

The sentence "Anthony Mansella has a red truck" is a statement. The **negation** of this statement is "Anthony Mansella does not have a red truck." *The negation of a true statement is false, and the negation of a false statement is true.*

EXAMPLE 2 Forming Negations

Form the negation of each statement.

(a) That city has a mayor. **(b)** The moon is not a planet.

Solution

(a) To negate this statement, we introduce *not* into the sentence: "That city does not have a mayor."

(b) The negation is "The moon is a planet." ∎

One way to detect incorrect negations is to check truth values. *A negation must have the opposite truth value from the original statement.*

The next example uses some of the inequality symbols in **Table 1.** In the case of an inequality involving a variable, the negation must have the opposite truth value for *any* replacement of the variable.

```
TEST  LOGIC
1:=
2:≠
3:>
4:≥
5:<
6:≤
```

The TEST menu of the TI-83/84 Plus calculator allows the user to test the truth value of statements involving $=$, $\neq$, $>$, $\geq$, $<$, and $\leq$. If a statement is true, it returns a 1. If a statement is false, it returns a 0.

Table 1

Symbolism	Meaning	Examples	
$a < b$	a is less than b	$4 < 9$	$\frac{1}{2} < \frac{3}{4}$
$a > b$	a is greater than b	$6 > 2$	$-5 > -11$
$a \leq b$	a is less than or equal to b	$8 \leq 10$	$3 \leq 3$
$a \geq b$	a is greater than or equal to b	$-2 \geq -3$	$-5 \geq -5$

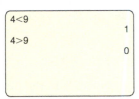

$4 < 9$ is true, as indicated by the 1.
$4 > 9$ is false, as indicated by the 0.

EXAMPLE 3 Negating Inequalities

Give a negation of each inequality. Do *not* use a slash symbol.

(a) $x < 9$ **(b)** $7x + 11y \geq 77$

Solution

(a) The negation of "x is less than 9" is "x is *not* less than 9." Because we cannot use "not," which would require writing $x \not< 9$, phrase the negation as "x is greater than or equal to 9," or

$$x \geq 9.$$

(b) The negation, with no slash, is

$$7x + 11y < 77.$$ ∎

Symbols

The study of logic uses symbols. Statements are represented with letters, such as p, q, or r. Several symbols for connectives are shown in **Table 2**.

Table 2

Connective	Symbol	Type of Statement
and	$\wedge$	Conjunction
or	$\vee$	Disjunction
not	$\sim$	Negation

The symbol $\sim$ represents the connective *not*. If p represents the statement "Barack Obama was president in 2014," then $\sim p$ represents "Barack Obama was *not* president in 2014."

EXAMPLE 4 **Translating from Symbols to Words**

Let p represent the statement "Nursing informatics is a growing field," and let q represent "Critical care will always be in demand." Translate each statement from symbols to words.

(a) $p \vee q$ **(b)** $\sim p \wedge q$ **(c)** $\sim(p \vee q)$ **(d)** $\sim(p \wedge q)$

Solution

(a) From the table, $\vee$ symbolizes *or*. Thus, $p \vee q$ represents

Nursing informatics is a growing field, or critical care will always be in demand.

(b) Nursing informatics is not a growing field and critical care will always be in demand.

(c) It is not the case that nursing informatics is a growing field or critical care will always be in demand. (This is usually translated as **"Neither p nor q."**)

(d) It is not the case that nursing informatics is a growing field and critical care will always be in demand. ∎

Quantifiers

Quantifiers are used to indicate *how many* cases of a particular situation exist. The words *all, each, every,* and *no(ne)* are **universal quantifiers,** while words and phrases such as *some, there exists,* and *(for) at least one* are **existential quantifiers.**

Be careful when forming the negation of a statement involving quantifiers. A statement and its negation must have opposite truth values in all possible cases. Consider this statement.

All girls in the group are named Mary.

Many people would write the negation of this statement as "No girls in the group are named Mary" or "All girls in the group are not named Mary." But neither of these is correct. To see why, look at the three groups below.

Group I: Mary Jane Payne, Mary Meyer, Mary O'Hara

Group II: Mary Johnson, Lisa Pollak, Margaret Watson

Group III: Donna Garbarino, Paula Story, Rhonda Alessi, Kim Falgout

These groups contain all possibilities that need to be considered. In Group I, *all* girls are named Mary. In Group II, *some* girls are named Mary (and some are not). In Group III, *no* girls are named Mary.

Consider **Table 3**. Keep in mind that "some" means "at least one (and possibly all)."

Table 3 Truth Value as Applied to:

	Group I	Group II	Group III
(1) All girls in the group are named Mary. **(Given)**	T	F	F
(2) No girls in the group are named Mary. **(Possible negation)**	F	F	T
(3) All girls in the group are not named Mary. **(Possible negation)**	F	F	T
(4) Some girls in the group are not named Mary. **(Possible negation)**	F	T	T

The negation of the given statement (1) must have opposite truth values in *all* cases. It can be seen that statements (2) and (3) do not satisfy this condition (for Group II), but statement (4) does. It may be concluded that the correct negation for "All girls in the group are named Mary" is "Some girls in the group are not named Mary." Other ways of stating the negation include the following.

Not all girls in the group are named Mary.

It is not the case that all girls in the group are named Mary.

At least one girl in the group is not named Mary.

Table 4 shows how to find the negation of a statement involving quantifiers.

Table 4 Negations of Quantified Statements

Statement	Negation
All do.	Some do not. (Equivalently: Not all do.)
Some do.	None do. (Equivalently: All do not.)

The negation of the negation of a statement is simply the statement itself. For instance, the negations of the statements in the Negation column are simply the corresponding original statements in the Statement column.

EXAMPLE 5 Forming Negations of Quantified Statements

Form the negation of each statement.

(a) Some cats have fleas.   **(b)** Some cats do not have fleas.

(c) No cats have fleas.

Solution

(a) Because *some* means "at least one," the statement "Some cats have fleas" is really the same as "At least one cat has fleas." The negation of this is

"No cat has fleas."

(b) The statement "Some cats do not have fleas" claims that at least one cat, somewhere, does not have fleas. The negation of this is

"All cats have fleas."

(c) The negation is "Some cats have fleas." ——— Avoid the incorrect answer "All cats have fleas."

Quantifiers and Number Sets

Earlier we introduced sets of numbers.

SETS OF NUMBERS

Natural numbers (or counting numbers) $\{1, 2, 3, 4, \dots\}$

Whole numbers $\{0, 1, 2, 3, 4, \dots\}$

Integers $\{\dots, -3, -2, -1, 0, 1, 2, 3, \dots\}$

Rational numbers $\left\{\frac{p}{q} \mid p \text{ and } q \text{ are integers, and } q \neq 0\right\}$

Real numbers $\{x \mid x \text{ is a number that can be written as a decimal}\}$

Irrational numbers $\{x \mid x \text{ is a real number and } x \text{ cannot be written as a quotient of integers}\}$

EXAMPLE 6 **Deciding Whether Quantified Statements Are True or False**

Decide whether each statement involving a quantifier is *true* or *false*.

(a) There exists a whole number that is not a natural number.

(b) Every integer is a natural number.

(c) Every natural number is a rational number.

(d) There exists an irrational number that is not real.

Solution

(a) Because there is such a whole number (it is 0), this statement is true.

(b) This statement is false, because we can find at least one integer that is not a natural number. For example, -1 is an integer but is not a natural number.

(c) Because every natural number can be written as a fraction with denominator 1, this statement is true.

(d) In order to be an irrational number, a number must first be real. Because we cannot give an irrational number that is not real, this statement is false. (Had we been able to find at least one, the statement would have been true.) ■

3.1 EXERCISES

Decide whether each is a statement or is not a statement.

1. February 2, 2009, was a Monday.

2. The ZIP code for Oscar, Louisiana, is 70762.

3. Listen, my children, and you shall hear of the midnight ride of Paul Revere.

4. Did you yield to oncoming traffic?

5. $5 + 9 \neq 14$ and $4 - 1 = 12$

6. $5 + 9 \neq 12$ or $4 - 2 = 5$

7. Some numbers are positive.

8. Grover Cleveland was president of the United States in 1885 and 1897.

9. Accidents are the main cause of deaths of children under the age of 7.

10. It is projected that in the United States between 2010 and 2020, there will be over 500,000 job openings per year for elementary school teachers, with median annual salaries of about $51,000.

11. Where are you going tomorrow?

12. Behave yourself and sit down.

13. Kevin "Catfish" McCarthy once took a prolonged continuous shower for 340 hours, 40 minutes.

14. One gallon of milk weighs more than 3 pounds.

Decide whether each statement is compound.

15. I read the *Detroit Free Press,* and I read the *Sacramento Bee.*

16. My brother got married in Copenhagen.

17. Tomorrow is Saturday.

18. Jing is younger than 18 years of age, and so is her friend Shu-fen.

19. Jay's wife loves Ben and Jerry's ice cream.

20. The sign on the back of the car read "Canada or bust!"

21. If Lorri sells her quota, then Michelle will be happy.

22. If Bobby is a politician, then Mitch is a crook.

Write a negation for each statement.

23. Her aunt's name is Hermione.

24. No rain fell in southern California today.

25. Some books are longer than this book.

26. All students present will get another chance.

27. No computer repairman can play blackjack.

28. Some people have all the luck.

29. Everybody loves somebody sometime.

30. Everyone needs a friend.

31. The trash needs to be collected.

32. Every architect who wants a job can find one.

Give a negation of each inequality. Do not use a slash symbol.

33. $x > 12$

34. $x < -6$

35. $x \geq 5$

36. $x \leq 19$

37. Try to negate the sentence "The exact number of words in this sentence is ten" and see what happens. Explain the problem that arises.

38. Explain why the negation of "$x > 5$" is not "$x < 5$."

Let p represent the statement "She has green eyes" *and let q represent the statement* "He is 60 years old." *Translate each symbolic compound statement into words.*

39. $\sim p$

40. $\sim q$

41. $p \land q$

42. $p \lor q$

43. $\sim p \lor q$

44. $p \land \sim q$

45. $\sim p \lor \sim q$

46. $\sim p \land \sim q$

47. $\sim(\sim p \land q)$

48. $\sim(p \lor \sim q)$

Let p represent the statement "Tyler collects DVDs" *and let q represent the statement* "Josh is an art major." *Convert each compound statement into symbols.*

49. Tyler collects DVDs and Josh is not an art major.

50. Tyler does not collect DVDs or Josh is not an art major.

51. Tyler does not collect DVDs or Josh is an art major.

52. Josh is an art major and Tyler does not collect DVDs.

53. Neither Tyler collects DVDs nor Josh is an art major.

54. Either Josh is an art major or Tyler collects DVDs, and it is not the case that both Josh is an art major and Tyler collects DVDs.

55. Incorrect use of quantifiers often is heard in everyday language. Suppose you hear that a local electronics chain is having a 40% off sale, and the radio advertisement states "All items are not available in all stores." Do you think that, literally translated, the ad really means what it says? What do you think is really meant? Explain your answer.

56. Repeat **Exercise 55** for the following: "All people don't have the time to devote to maintaining their vehicles properly."

Refer to the groups of art labeled A, B, and C, and identify by letter the group or groups that satisfy the given statements involving quantifiers in Exercises 57–64.

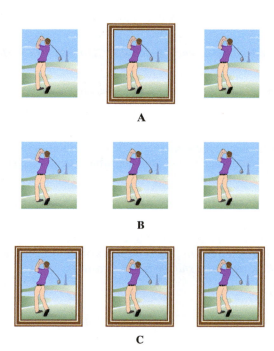

A

B

C

57. All pictures have frames.

58. No picture has a frame.

59. At least one picture does not have a frame.

60. Not every picture has a frame.

61. At least one picture has a frame.

62. No picture does not have a frame.

63. All pictures do not have frames.

64. Not every picture does not have a frame.

Decide whether each statement in Exercises 65–74 involving a quantifier is true *or* false.

65. Every whole number is an integer.

66. Every integer is a whole number.

67. There exists a natural number that is not an integer.

68. There exists an integer that is not a natural number.

69. All rational numbers are real numbers.

70. All irrational numbers are real numbers.

71. Some rational numbers are not integers.

72. Some whole numbers are not rational numbers.

73. Each whole number is a positive number.

74. Each rational number is a positive number.

75. Explain the difference between the statements "All students did not pass the test" and "Not all students passed the test."

76. The statement "For some real number x, $x^2 \geq 0$" is true. However, your friend does not understand why, because he claims that $x^2 \geq 0$ is true for *all* real numbers x (and not *some*). How would you explain his misconception to him?

77. Write the following statement using "every": There is no one here who has not made mistakes before.

78. Only one of these statements is true. Which one is it?

   A. For some real number x, $x \not< 0$.

   B. For all real numbers x, $x^3 > 0$.

   C. For all real numbers x less than 0, x^2 is also less than 0.

   D. For some real number x, $x^2 < 0$.

Symbolic logic also uses symbols for quantifiers. The symbol for the existential quantifier is

$$\exists \quad (a \text{ rotated } \text{E}),$$

and the symbol for the universal quantifier is

$$\forall \quad (an \text{ inverted } \text{A}).$$

The statement "For some x, p is true" can be symbolized

$$(\exists x)(p).$$

The statement "For all x, p is true" can be symbolized

$$(\forall x)(p).$$

The negation of $(\exists x)(p)$ is

$$(\forall x)(\sim p),$$

and the negation of $(\forall x)(p)$ is

$$(\exists x)(\sim p).$$

79. Refer to **Example 5.** If we let c represent "cat" and f represent "The cat has fleas," then the statement "Some cats have fleas" can be represented by $(\exists c)(f)$. Use symbols to express the negation of this statement.

80. Use symbols to express the statements for parts (b) and (c) of **Example 5** and their negations. Verify that the symbolic expressions translate to the negations found in the text.

3.2 TRUTH TABLES AND EQUIVALENT STATEMENTS

OBJECTIVES

1 Find the truth value of a conjunction.

2 Find the truth value of a disjunction.

3 Find the truth values for compound mathematical statements.

4 Construct truth tables for compound statements.

5 Understand and determine equivalence of statements.

6 Use De Morgan's laws to find negations of compound statements.

Conjunctions

Truth values of component statements are used to find truth values of compound statements. To begin, we must decide on truth values of the **conjunction p and q,** symbolized $p \wedge q$. Here, the connective *and* implies the idea of "both." The following statement is true, because each component statement is true.

Monday immediately follows Sunday, and March immediately follows February.

True

On the other hand, the following statement is false, even though part of the statement (Monday immediately follows Sunday) is true.

Monday immediately follows Sunday, and March immediately follows January.

False

For the conjunction $p \wedge q$ to be true, both p and q must be true. This result is summarized by a table, called a **truth table,** which shows truth values of $p \wedge q$ for all four possible combinations of truth values for the component statements p and q.

TRUTH TABLE FOR THE CONJUNCTION p and q

p and q

p	q	$p \wedge q$
T	T	T
T	F	F
F	T	F
F	F	F

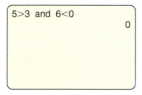

5>3 and 6<0

0

The calculator returns a 0 for

5 > 3 *and* 6 < 0,

indicating that the statement is false.

EXAMPLE 1   Finding the Truth Value of a Conjunction

Let p represent "$5 > 3$" and let q represent "$6 < 0$." Find the truth value of $p \wedge q$.

Solution

Here p is true and q is false. The second row of the conjunction truth table shows that $p \wedge q$ is false in this case. ∎

In some cases, the logical connective *but* is used in compound statements.

He wants to go to the mountains but she wants to go to the beach.

Here, *but* is used in place of *and* to give a different emphasis to the statement. We consider this statement as we would consider the conjunction using the word *and*. The truth table for the conjunction, given above, would apply.

Disjunctions

In ordinary language, the word *or* can be ambiguous. The expression "this or that" can mean either "this or that or both," or "this or that but not both." For example, consider the following statement.

I will paint the wall or I will paint the ceiling.

This statement probably means: "I will paint the wall or I will paint the ceiling or I will paint both."

On the other hand, consider the following statement.

I will wear my glasses or my contact lenses.

It probably means "I will wear my glasses, or I will wear my contacts, but I will not wear both."

The symbol ∨ represents the first *or* described. That is,

> **$p ∨ q$ means "p or q or both."** Disjunction

With this meaning of *or*, $p ∨ q$ is called the **inclusive disjunction,** or just the **disjunction** of p and q. In everyday language, the disjunction implies the idea of "either." For example, consider the following disjunction.

I have a quarter or I have a dime.

It is true whenever I have either a quarter, a dime, or both. The only way this disjunction could be false would be if I had neither coin. *The disjunction $p ∨ q$ is false only if both component statements are false.*

TRUTH TABLE FOR THE DISJUNCTION *p* or *q*

		p or *q*
p	*q*	*p* ∨ *q*
T	T	T
T	F	T
F	T	T
F	F	F

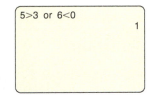

The calculator returns a 1 for

$5 > 3$ *or* $6 < 0$,

indicating that the statement is true.

EXAMPLE 2 Finding the Truth Value of a Disjunction

Let p represent "$5 > 3$" and let q represent "$6 < 0$." Find the truth value of $p ∨ q$.

Solution

Here, as in **Example 1,** p is true and q is false. The second row of the disjunction truth table shows that $p ∨ q$ is true. ∎

The symbol ≥ is read **"is greater than or equal to,"** while ≤ is read **"is less than or equal to."** If a and b are real numbers, then $a ≤ b$ is true if $a < b$ or $a = b$. See **Table 5**.

Table 5

Statement	Reason It Is True
$8 ≥ 8$	$8 = 8$
$3 ≥ 1$	$3 > 1$
$-5 ≤ -3$	$-5 < -3$
$-4 ≤ -4$	$-4 = -4$

Negations

The **negation** of a statement p, symbolized **$∼p$,** must have the opposite truth value from the statement p itself. This leads to the truth table for the negation.

TRUTH TABLE FOR THE NEGATION not *p*

	not *p*
p	*∼p*
T	F
F	T

EXAMPLE 3 Finding the Truth Value of a Compound Statement

Suppose p is false, q is true, and r is false. What is the truth value of the compound statement $\sim p \wedge (q \vee \sim r)$?

Solution

Here parentheses are used to group q and $\sim r$ together. Work first inside the parentheses. Because r is false, $\sim r$ will be true. Because $\sim r$ is true and q is true, the truth value of $q \vee \sim r$ is T, as shown in the first row of the *or* truth table.

Because p is false, $\sim p$ is true, and the truth value of $\sim p \wedge (q \vee \sim r)$ is found in the top row of the *and* truth table. The statement

$$\sim p \wedge (q \vee \sim r) \quad \text{is true.}$$

We can use a shortcut symbolic method that involves replacing the statements with their truth values, letting T represent a true statement and F represent a false statement.

$$\sim p \wedge (q \vee \sim r)$$
$$\sim F \wedge (T \vee \sim F) \quad \text{← Work within parentheses first.}$$
$$T \wedge (T \vee T) \quad \sim\!F \text{ gives T.}$$
$$T \wedge T \quad T \vee T \text{ gives T.}$$

The compound statement is true. → **T** $T \wedge T$ gives T. ∎

Mathematical Statements

We can use truth tables to determine the truth values of compound mathematical statements.

EXAMPLE 4 Deciding Whether Compound Mathematical Statements Are True or False

Let p represent the statement $3 > 2$, q represent $5 < 4$, and r represent $3 < 8$. Decide whether each statement is *true* or *false*.

(a) $\sim p \wedge \sim q$ **(b)** $\sim(p \wedge q)$ **(c)** $(\sim p \wedge r) \vee (\sim q \wedge \sim p)$

Solution

(a) Because p is true, $\sim p$ is false. By the *and* truth table, if one part of an "and" statement is false, the entire statement is false.

$$\sim p \wedge \sim q \quad \text{is false.}$$

(b) For $\sim(p \wedge q)$, first work within the parentheses. Because p is true and q is false, $p \wedge q$ is false by the *and* truth table. Next, apply the negation. The negation of a false statement is true.

$$\sim(p \wedge q) \quad \text{is true.}$$

(c) Here p is true, q is false, and r is true. This makes $\sim p$ false and $\sim q$ true. By the *and* truth table, $\sim p \wedge r$ is false, and $\sim q \wedge \sim p$ is also false. By the *or* truth table,

$$(\sim p \wedge r) \vee (\sim q \wedge \sim p) \quad \text{is false.}$$
$$\downarrow \qquad\qquad \downarrow \qquad\qquad \text{(Alternatively, see \textbf{Example 8(b)}.)}$$
$$\text{F} \quad \vee \quad \text{F}$$

∎

When a quantifier is used with a conjunction or a disjunction, we must be careful in determining the truth value, as shown in the following example.

---

not(3>2) and not (5<4)

not((3>2) and (5<4))

0

1

Example 4(a) explains why

$$\sim(3 > 2) \wedge \sim(5 < 4)$$

is false. The calculator returns a 0. For a true statement such as

$$\sim[(3 > 2) \wedge (5 < 4)]$$

from **Example 4(b)**, it returns a 1.

George Boole (1815–1864) grew up in poverty. His father, a London tradesman, gave him his first mathematics lessons and taught him to make optical instruments. Boole was largely self-educated. At 16 he worked in an elementary school and by age 20 had opened his own school. He studied mathematics in his spare time. He died of lung disease at age 49.

Boole's ideas have been used in the design of computers, telephone systems, and search engines.

EXAMPLE 5 **Deciding Whether Quantified Mathematical Statements Are True or False**

Decide whether each statement is *true* or *false*.

(a) For some real number x, $x < 5$ and $x > 2$.

(b) For every real number x, $x > 0$ or $x < 1$.

(c) For all real numbers x, $x^2 > 0$.

Solution

(a) Because "some" is an *existential* quantifier, we need only find one real number x that makes both component statements true, and $x = 3$ is such a number. The statement is true by the *and* truth table.

(b) No matter which real number might be tried as a replacement for x, at least one of the two statements

$$x > 0, \quad x < 1$$

will be true. Because an "or" statement is true if one or both component statements are true, the entire statement as given is true.

(c) Because "for all" is a *universal* quantifier, we need only find one case in which the inequality is false to make the entire statement false. Can we find a real number whose square is not positive (that is, not greater than 0)? Yes, we can—0 is such a number. In fact 0 is the *only* real number whose square is not positive. This statement is false.   ∎

FOR FURTHER THOUGHT

Whose Picture Am I Looking At?

Raymond Smullyan is one of today's foremost writers of logic puzzles. This professor of mathematics and philosophy is now retired from Indiana University and has written several books on recreational logic, including *What Is the Name of This Book?*, *The Lady or the Tiger?*, and *The Gödelian Puzzle Book*. The first of these includes the following puzzle, which has been around for many years.

For Group or Individual Investigation

A man was looking at a portrait. Someone asked him, "Whose picture are you looking at?" He replied: "Brothers and sisters, I have none, but this man's father is my father's son." ("This man's father" means, of course, the father of the man in the picture.)

Whose picture was the man looking at? (The answer is on **page 96**.)

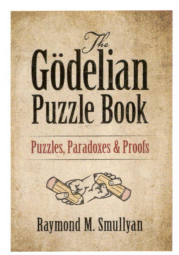

Truth Tables

In the preceding examples, the truth value for a given statement was found by going back to the basic truth tables. It is generally easier to first create a complete truth table for the given statement itself. Then final truth values can be read directly from this table.

In this book we use the standard format shown in the margin for listing the possible truth values in compound statements involving two component statements.

p	q	Compound Statement
T	T	
T	F	
F	T	
F	F	

EXAMPLE 6   Constructing a Truth Table

Consider the statement $(\sim p \wedge q) \vee \sim q$.

(a) Construct a truth table.

(b) Suppose both p and q are true. Find the truth value of the compound statement.

Solution

(a) As shown below, begin by listing all possible combinations of truth values for p and q. Then list the truth values of $\sim p$, which are the opposite of those of p.

Use the "$\sim p$" and "q" columns, along with the *and* truth table, to find the truth values of $\sim p \wedge q$. List them in a separate column.

Next include a column for $\sim q$.

p	q	$\sim p$	$\sim p \wedge q$	$\sim q$
T	T	F	F	F
T	F	F	F	T
F	T	T	T	F
F	F	T	F	T

Finally, make a column for the entire compound statement. To find the truth values, use *or* to combine $\sim p \wedge q$ with $\sim q$ and refer to the *or* truth table.

p	q	$\sim p$	$\sim p \wedge q$	$\sim q$	$(\sim p \wedge q) \vee \sim q$
T	T	F	F	F	F
T	F	F	F	T	T
F	T	T	T	F	T
F	F	T	F	T	T

(b) Look in the first row of the final truth table above, where both p and q have truth value T. Read across the row to find that the compound statement is false.   ∎

EXAMPLE 7   Constructing a Truth Table

Construct the truth table for $p \wedge (\sim p \vee \sim q)$.

Solution

p	q	$\sim p$	$\sim q$	$\sim p \vee \sim q$	$p \wedge (\sim p \vee \sim q)$
T	T	F	F	F	F
T	F	F	T	T	T
F	T	T	F	T	F
F	F	T	T	T	F

∎

If a compound statement involves three component statements p, q, and r, we will use the following standard format in setting up the truth table.

p	q	r	Compound Statement
T	T	T	
T	T	F	
T	F	T	
T	F	F	
F	T	T	
F	T	F	
F	F	T	
F	F	F	

Emilie, Marquise du Châtelet
(1706–1749) participated in the scientific activity of the generation after Newton and Leibniz. Educated in science, music, and literature, she was studying mathematics at the time (1733) she began a long intellectual relationship with the philosopher **François Voltaire** (1694–1778). She and Voltaire competed independently in 1738 for a prize offered by the French Academy on the subject of fire. Although du Châtelet did not win, her dissertation was published by the academy in 1744.

EXAMPLE 8 **Constructing a Truth Table**

Consider the statement $(\sim p \wedge r) \vee (\sim q \wedge \sim p)$.

(a) Construct a truth table.

(b) Suppose p is true, q is false, and r is true. Find the truth value of this statement.

Solution

(a) There are three component statements: p, q, and r. The truth table thus requires eight rows to list all possible combinations of truth values of p, q, and r. The final truth table can be found in much the same way as the ones earlier.

p	q	r	$\sim p$	$\sim p \wedge r$	$\sim q$	$\sim q \wedge \sim p$	$(\sim p \wedge r) \vee (\sim q \wedge \sim p)$
T	T	T	F	F	F	F	F
T	T	F	F	F	F	F	F
T	F	T	F	F	T	F	F
T	F	F	F	F	T	F	F
F	T	T	T	T	F	F	T
F	T	F	T	F	F	F	F
F	F	T	T	T	T	T	T
F	F	F	T	F	T	T	T

(b) By the third row of the truth table in part (a), the compound statement is false. (This is an alternative method for working part (c) of **Example 4**.) ∎

Problem-Solving Strategy

One strategy for problem solving is to notice a pattern and then use inductive reasoning. This strategy is applied in the next example.

EXAMPLE 9 **Using Inductive Reasoning**

Suppose that n is a counting number, and a logical statement is composed of n component statements. How many rows will appear in the truth table for the compound statement?

Solution

We examine some of the earlier truth tables in this section. The truth table for the negation has one statement and two rows. The truth tables for the conjunction and the disjunction have two component statements, and each has four rows. The truth table in **Example 8(a)** has three component statements and eight rows.

Summarizing these in **Table 6** (seen in the margin) reveals a pattern encountered earlier. Inductive reasoning leads us to the conjecture that if a logical statement is composed of n component statements, it will have 2^n rows. This can be proved using more advanced concepts. ∎

Table 6

Number of Statements	Number of Rows
1	$2 = 2^1$
2	$4 = 2^2$
3	$8 = 2^3$

The result of **Example 9** is reminiscent of the formula for the number of subsets of a set having n elements.

NUMBER OF ROWS IN A TRUTH TABLE

A logical statement having n component statements will have 2^n rows in its truth table.

Alternative Method for Constructing Truth Tables

After making a reasonable number of truth tables, some people prefer the shortcut method shown in **Example 10,** which repeats **Examples 6 and 8.**

Ada Lovelace (1815–1852) was born Augusta Ada Byron. Her talents as a mathematician and logician led to her work with **Charles Babbage** (1791–1871) on his Analytical Engine, the first programmable mechanical computer. Lovelace's notes on this machine include what is regarded by many as the first computer program and reveal her visionary belief that computing machines would have applications beyond numerical calculations.

EXAMPLE 10 Constructing Truth Tables

Construct the truth table for each compound statement.

(a) $(\sim p \wedge q) \vee \sim q$ **(b)** $(\sim p \wedge r) \vee (\sim q \wedge \sim p)$

Solution

(a) Start by inserting truth values for $\sim p$ and for q. Then use the *and* truth table to obtain the truth values for $\sim p \wedge q$.

p	q	(~p	∧	q)	∨	~q
T	T	F		T		
T	F	F		F		
F	T	T		T		
F	F	T		F		

p	q	(~p	∧	q)	∨	~q
T	T	F	F	T		
T	F	F	F	F		
F	T	T	T	T		
F	F	T	F	F		

Now disregard the two preliminary columns of truth values for $\sim p$ and for q, and insert truth values for $\sim q$. Finally, use the *or* truth table.

p	q	(~p ∧ q)	∨	~q
T	T	F		F
T	F	F		T
F	T	T		F
F	F	F		T

p	q	(~p ∧ q)	∨	~q
T	T	F	F	F
T	F	F	T	T
F	T	T	T	F
F	F	F	T	T

These steps can be summarized as follows.

p	q	(~p	∧	q)	∨	~q
T	T	F	F	T	F	F
T	F	F	F	F	T	T
F	T	T	T	T	T	F
F	F	T	F	F	T	T
		①	②	①	④	③

> The circled numbers indicate the order in which the various columns of the truth table were found.

(b) Work as follows.

p	q	r	(~p	∧	r)	∨	(~q	∧	~p)
T	T	T	F	F	T	F	F	F	F
T	T	F	F	F	F	F	F	F	F
T	F	T	F	F	T	F	T	F	F
T	F	F	F	F	F	F	T	F	F
F	T	T	T	T	T	T	F	F	T
F	T	F	T	F	F	F	F	F	T
F	F	T	T	T	T	T	T	T	T
F	F	F	T	F	F	T	T	T	T
			①	②	①	⑤	③	④	③

> The circled numbers indicate the order.

Equivalent Statements and De Morgan's Laws

Two statements are **equivalent** if they have the same truth value in *every* possible situation. The columns of the two truth tables that were the last to be completed will be the same for equivalent statements.

EXAMPLE 11 Deciding Whether Two Statements Are Equivalent

Are the following two statements equivalent?

$$\sim p \wedge \sim q \quad \text{and} \quad \sim (p \vee q)$$

Solution

Construct a truth table for each statement.

p	q	$\sim p \wedge \sim q$
T	T	F
T	F	F
F	T	F
F	F	T

p	q	$\sim (p \vee q)$
T	T	F
T	F	F
F	T	F
F	F	T

Because the truth values are the same in all cases, as shown in the columns in color, the statements $\sim p \wedge \sim q$ and $\sim (p \vee q)$ are equivalent.

$$\sim p \wedge \sim q \equiv \sim (p \vee q) \qquad \text{The symbol} \equiv \text{denotes equivalence.} \qquad \blacksquare$$

In the same way, the statements $\sim p \vee \sim q$ and $\sim (p \wedge q)$ are equivalent. We call these equivalences *De Morgan's laws*.

Compare **De Morgan's Laws** for logical statements with the set theoretical version on **page 68**.

DE MORGAN'S LAWS FOR LOGICAL STATEMENTS

For any statements p and q, the following equivalences are valid.

$$\sim (p \vee q) \equiv \sim p \wedge \sim q \quad \text{and} \quad \sim (p \wedge q) \equiv \sim p \vee \sim q$$

EXAMPLE 12 Applying De Morgan's Laws

Find a negation of each statement by applying De Morgan's laws.

(a) I got an A or I got a B. **(b)** She won't try and he will succeed.

(c) $\sim p \vee (q \wedge \sim p)$

Solution

(a) If p represents "I got an A" and q represents "I got a B," then the compound statement is symbolized $p \vee q$. The negation of $p \vee q$ is $\sim (p \vee q)$. By one of De Morgan's laws, this is equivalent to $\sim p \wedge \sim q$, or, in words,

I didn't get an A and I didn't get a B.

This negation is reasonable—the original statement says that I got either an A or a B. The negation says that I didn't get *either* grade.

(b) From De Morgan's laws, $\sim (p \wedge q) \equiv \sim p \vee \sim q$, so the negation becomes

She will try or he won't succeed.

(c) Negate both component statements and change $\vee$ to $\wedge$.

$$\sim [\sim p \vee (q \wedge \sim p)] \equiv p \wedge \sim (q \wedge \sim p)$$

$$\equiv p \wedge (\sim q \vee \sim (\sim p)) \qquad \text{Apply De Morgan's law again.}$$

$$\equiv p \wedge (\sim q \vee p)$$

A truth table will show that the statements

$$\sim p \vee (q \wedge \sim p) \quad \text{and} \quad p \wedge (\sim q \vee p) \quad \text{are negations of each other.} \qquad \blacksquare$$

FOR FURTHER THOUGHT

The Logic Behind Computers

Computer designers use *logic gates* at the foundation of digital circuits. Logic gates treat the binary digits 0 and 1 as logical values, and give a single output of 1 or 0 based on the input value(s), which also have a value of 1 or 0. Since 1 and 0 are the only digits in the binary number system, that system is the perfect bridge between logic and numerical computation (as Leibniz realized).

There are three basic logic gates: the AND gate, the OR gate, and the NOT gate, whose symbols and truth tables follow.

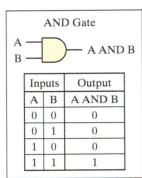

AND Gate		OR Gate		

A — [ ] — A AND B

Inputs		Output
A	B	A AND B
0	0	0
0	1	0
1	0	0
1	1	1

A — [ ] — A OR B

Inputs		Output
A	B	A OR B
0	0	0
0	1	1
1	0	1
1	1	1

NOT Gate

A — [ ]o— NOT A

Input	Output
A	NOT A
0	1
1	0

Other logic gates include NAND, NOR (the negations of AND and OR, respectively), XOR (exclusive OR—see **Exercise 77** in this section), and XNOR (the negation of XOR). Their symbols are shown below.

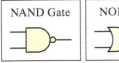

NAND Gate	NOR Gate	XOR Gate	XNOR Gate

Early logic gates were constructed with electrical relays and vacuum tubes, but the development of semiconductors and the transistor have made possible the placement of millions of logic gates in integrated circuits, microchips that perform billions of logical calculations per second. More recently, logic gates have even been fashioned from biological organisms and DNA.

There are some excellent Android Apps (such as Logic Simulator, free in the Android Play Store) for simulating logic circuits. The simulator shown allows the user to add logic gates, switches (connected to inputs), and light bulbs (connected to outputs). Red highlighting indicates where current is flowing. These four screenshots show the truth values for the XOR logic gate.

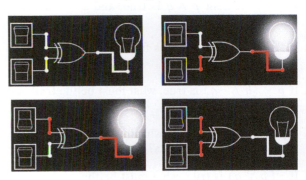

For Group or Individual Investigation

1. Create the truth table for A XOR B using the accompanying screenshots. (Remember that the red nodes are "hot," which is indicated by a 1 or a T in the truth table.)

2. Charles Peirce (1839–1914) showed that all of the logic gates we have mentioned could be constructed using only NAND gates. Construct a truth table for the logic circuit below to show that it is equivalent to the XOR gate. *Hint:* The output of the circuit shown is

   [A NAND (A NAND B)] NAND [(A NAND B) NAND B],

   which is equivalent to

   $\sim(\sim(A \wedge \sim(A \wedge B)) \wedge \sim(\sim(A \wedge B) \wedge B))$.

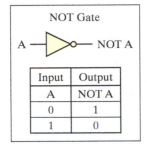

3. Download and install an app to simulate logic gates, and use it to verify your truth table from **Exercise 2**. One image of the simulated circuit is shown below.

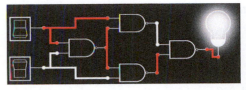

4. Use the simulator to build circuits to verify De Morgan's laws for logic.

Use the concepts introduced in this section to answer Exercises 1–8.

1. If q is false, what must be the truth value of the statement $(p \wedge \sim q) \wedge q$?

2. If q is true, what must be the truth value of the statement $q \vee (q \wedge \sim p)$?

3. If the statement $p \wedge q$ is true, and p is true, then q must be _____.

4. If the statement $p \vee q$ is false, and p is false, then q must be _____.

5. If $p \vee (q \wedge \sim q)$ is true, what must be the truth value of p?

6. If $p \wedge \sim (q \vee r)$ is true, what must be the truth value of the component statements?

7. If $\sim (p \vee q)$ is true, what must be the truth values of the component statements?

8. If $\sim (p \wedge q)$ is false, what must be the truth values of the component statements?

Let p represent a false statement and let q represent a true statement. Find the truth value of the given compound statement.

9. $\sim p$

10. $\sim q$

11. $p \vee q$

12. $p \wedge q$

13. $p \vee \sim q$

14. $\sim p \wedge q$

15. $\sim p \vee \sim q$

16. $p \wedge \sim q$

17. $\sim (p \wedge \sim q)$

18. $\sim (\sim p \vee \sim q)$

19. $\sim [\sim p \wedge (\sim q \vee p)]$

20. $\sim [(\sim p \wedge \sim q) \vee \sim q]$

21. Is the statement $6 \geq 2$ a conjunction or a disjunction? Why?

22. Why is the statement $8 \geq 3$ true? Why is $5 \geq 5$ true?

Let p represent a true statement, and let q and r represent false statements. Find the truth value of the given compound statement.

23. $(p \wedge r) \vee \sim q$

24. $(q \vee \sim r) \wedge p$

25. $p \wedge (q \vee r)$

26. $(\sim p \wedge q) \vee \sim r$

27. $\sim (p \wedge q) \wedge (r \vee \sim q)$

28. $(\sim r \wedge \sim q) \vee (\sim r \wedge q)$

29. $\sim [(\sim p \wedge q) \vee r]$

30. $\sim [r \vee (\sim q \wedge \sim p)]$

31. $\sim [\sim q \vee (r \wedge \sim p)]$

32. $\sim (p \vee q) \wedge \sim (p \wedge q)$

Let p represent the statement $16 < 8$, let q represent the statement $5 \not> 4$, and let r represent the statement $17 \leq 17$. Find the truth value of the given compound statement.

33. $p \wedge r$

34. $p \vee \sim q$

35. $\sim q \vee \sim r$

36. $\sim p \wedge \sim r$

37. $(p \wedge q) \vee r$

38. $\sim p \vee (\sim r \vee \sim q)$

39. $(\sim r \wedge q) \vee \sim p$

40. $\sim (p \vee \sim q) \vee \sim r$

Give the number of rows in the truth table for each compound statement.

41. $p \vee \sim r$

42. $p \wedge (r \wedge \sim s)$

43. $(\sim p \wedge q) \vee (\sim r \vee \sim s) \wedge r$

44. $[(p \vee q) \wedge (r \wedge s)] \wedge (t \vee \sim p)$

45. $[(\sim p \wedge \sim q) \wedge (\sim r \wedge s \wedge \sim t)] \wedge (\sim u \vee \sim v)$

46. $[(\sim p \wedge \sim q) \vee (\sim r \vee \sim s)]$
    $\vee [(\sim m \wedge \sim n) \wedge (u \wedge \sim v)]$

47. If the truth table for a certain compound statement has 64 rows, how many distinct component statements does it have?

48. Is it possible for the truth table of a compound statement to have exactly 54 rows? Why or why not?

Construct a truth table for each compound statement.

49. $\sim p \wedge q$

50. $\sim p \vee \sim q$

51. $\sim (p \wedge q)$

52. $p \vee \sim q$

53. $(q \vee \sim p) \vee \sim q$

54. $(p \wedge \sim q) \wedge p$

55. $(p \vee \sim q) \wedge (p \wedge q)$

56. $(\sim p \wedge \sim q) \vee (\sim p \vee q)$

57. $(\sim p \wedge q) \wedge r$

58. $r \vee (p \wedge \sim q)$

59. $(\sim p \wedge \sim q) \vee (\sim r \vee \sim p)$

60. $(\sim r \vee \sim p) \wedge (\sim p \vee \sim q)$

61. $\sim (\sim p \wedge \sim q) \vee (\sim r \vee \sim s)$

62. $(\sim r \vee s) \wedge (\sim p \wedge q)$

Use one of De Morgan's laws to write the negation of each statement.

63. You can pay me now or you can pay me later.

64. I am not going or she is going.

65. It is summer and there is no snow.

66. $\frac{1}{2}$ is a positive number and −9 is less than zero.

67. I said yes but she said no.

68. Dan tried to sell the software, but he was unable to do so.

69. $6 - 1 = 5$ and $9 + 13 \neq 7$

70. $8 < 10$ or $5 \neq 2$

71. Prancer or Vixen will lead Santa's reindeer sleigh next Christmas.

72. The lawyer and the client appeared in court.

Identify each statement as true *or* false.

73. For every real number x, $x < 14$ or $x > 6$.

74. For every real number x, $x > 9$ or $x < 9$.

75. There exists an integer n such that $n > 0$ and $n < 0$.

76. For some integer n, $n \geq 3$ and $n \leq 3$.

77. Complete the truth table for *exclusive disjunction*. The symbol $\underline{\vee}$ represents "one or the other is true, but not both."

p	q	$p \underline{\vee} q$
T	T	
T	F	
F	T	
F	F	

Exclusive disjunction

78. Attorneys sometimes use the phrase "and/or." This phrase corresponds to which usage of the word *or:* inclusive or exclusive disjunction?

Decide whether each compound statement is true *or* false. *Remember that* $\underline{\vee}$ *is the exclusive disjunction of* **Exercise 77.**

79. $3 + 1 = 4 \underline{\vee} 2 + 5 = 7$

80. $3 + 1 = 4 \underline{\vee} 2 + 5 = 10$

81. $3 + 1 = 6 \underline{\vee} 2 + 5 = 7$

82. $3 + 1 = 12 \underline{\vee} 2 + 5 = 10$

83. In his book *The Lady or the Tiger and Other Logic Puzzles*, Raymond Smullyan proposes the following problem. It is taken from the classic Frank Stockton short story, in which a prisoner must make a choice between two doors: behind one is a beautiful lady, and behind the other is a hungry tiger.

What if each door has a sign, and the man knows that only one sign is true?

The sign on Door 1 reads:

IN THIS ROOM THERE IS A LADY AND IN THE OTHER ROOM THERE IS A TIGER.

The sign on Door 2 reads:

IN ONE OF THESE ROOMS THERE IS A LADY AND IN ONE OF THESE ROOMS THERE IS A TIGER.

With this information, the man is able to choose the correct door. Can you?

84. (a) Build truth tables for

$$p \vee (q \wedge r) \quad \text{and} \quad (p \vee q) \wedge (p \vee r).$$

Decide whether it can be said that "OR distributes over AND." Explain.

(b) Build truth tables for

$$p \wedge (q \vee r) \quad \text{and} \quad (p \wedge q) \vee (p \wedge r).$$

Decide whether it can be said that "AND distributes over OR." Explain.

(c) Describe how the logical equivalences developed in parts (a) and (b) are related to the set-theoretical equations

$$X \cup (Y \cap Z) = (X \cup Y) \cap (X \cup Z)$$

and

$$X \cap (Y \cup Z) = (X \cap Y) \cup (X \cap Z).$$

85. De Morgan's law

$$\sim(p \vee q) \equiv \sim p \wedge \sim q$$

can be stated verbally, "The negation of a disjunction is equivalent to the conjunction of the negations." Give a similar verbal statement of

$$\sim(p \wedge q) \equiv \sim p \vee \sim q.$$

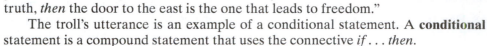

3.3   THE CONDITIONAL AND CIRCUITS

OBJECTIVES

1 Understand the structure of the conditional statement.

2 Determine the truth values of conditional statements.

3 Express a conditional statement as a disjunction.

4 Express the negation of a conditional statement.

5 Use circuits to model conditional statements.

Conditionals

"If truly Truthful Troll I be,
then go thou east, and be thou free."

This of course is the statement uttered by the troll on the second page of this chapter. A more modern paraphrase would be, "*If* I am the troll who always tells the truth, *then* the door to the east is the one that leads to freedom."

The troll's utterance is an example of a conditional statement. A **conditional** statement is a compound statement that uses the connective *if . . . then.*

> *If* I read for too long, *then* I get tired.
>
> *If* looks could kill, *then* I would be dead.
>
> *If* he doesn't get back soon, *then* you should go look for him.

Conditional statements

In each of these conditional statements, the component coming after the word *if* gives a condition (but not necessarily the only condition) under which the statement coming after *then* will be true. For example, "If it is over 90°, then I'll go to the mountains" tells one possible condition under which I will go to the mountains—if the temperature is over 90°.

The conditional is written with an arrow and symbolized as follows.

$$p \rightarrow q \qquad \text{If } p, \text{ then } q.$$

We read $p \rightarrow q$ as "**p implies q**" or "**If p, then q.**" In the conditional $p \rightarrow q$, the statement p is the **antecedent,** while q is the **consequent.**

The conditional connective may not always be explicitly stated. That is, it may be "hidden" in an everyday expression. For example, consider the following statement.

> Quitters never win.

It can be written in *if . . . then* form as

> *If* you're a quitter, *then* you will never win.

As another example, consider this statement.

> It is difficult to study when you are distracted.

It can be written

> *If* you are distracted, *then* it is difficult to study.

In the quotation "If you aim at nothing, you will hit it every time," the word "then" is not stated but understood from the context of the statement. "You aim at nothing" is the antecedent, and "you will hit it every time" is the consequent.

The conditional truth table is more difficult to define than the tables in the previous section. To see how to define the conditional truth table, imagine you have bought a used car (with financing from the car dealer), and the used-car salesman says,

> *If* you fail to make your payment on time, *then* your car will be taken.

Let p represent "You fail to make your payment on time," and let q represent "Your car will be taken." There are four combinations of truth values for the two component statements.

In his April 21, 1989, five-star review of **Field of Dreams,** the *Chicago Sun-Times* movie critic Roger Ebert gave an explanation of why the movie has become an American classic.

There is a speech in this movie about baseball that is so simple and true that it is heartbreaking. And the whole attitude toward the players reflects that attitude. Why do they come back from the great beyond and play in this cornfield? Not to make any kind of vast, earthshattering statement, but simply to hit a few and field a few, and remind us of a good and innocent time.

The photo above was taken in 2007 in Dyersville, Iowa, at the actual scene of the filming. The carving "Ray Loves Annie" in the bleacher seats can be seen in a quick shot during the movie. It has weathered over time.

What famous **conditional statement** inspired Ray to build a baseball field in his cornfield?

$\sqrt[3]{250}$ 90° $(0, -3)$

θ $45.5 \div 2^{-1}$ ∞

$x = (4+8)-3$ $|a|$

$y = -x + 2$ $\frac{1}{4}$

10^2 $\geq$ $f(x) =$

The importance of **symbols** was emphasized by the American philosopher-logician **Charles Sanders Peirce** (1839–1914), who asserted the nature of humans as symbol-using or sign-using organisms. Symbolic notation is half of mathematics, Bertrand Russell once said.

You Lie! (or Do You?) Granted, the T for Case 3 is less obvious than the F for Case 2. However, the laws of symbolic logic permit only one of two truth values. Since no lie can be established in Case 3, we give the salesman the benefit of the doubt. Likewise, *any* conditional statement is declared to be true whenever its antecedent is false.

As we consider these four possibilities, it is helpful to ask,

"Did the salesman lie?"

If so, then the conditional statement is considered false. Otherwise, the conditional statement is considered true.

Possibility	Failed to Pay on Time?	Car Taken?	
1	Yes	Yes	p is T, q is T.
2	Yes	No	p is T, q is F.
3	No	Yes	p is F, q is T.
4	No	No	p is F, q is F.

The four possibilities are as follows.

1. In the first case, assume you failed to make your payment on time, and your car *was* taken (p is T, q is T). The salesman told the truth, so place T in the first row of the truth table. (We do not claim that your car was taken *because* you failed to pay on time. It may be gone for a completely different reason.)

2. In the second case, assume that you failed to make your payment on time, and your car was *not* taken (p is T, q is F). The salesman lied (gave a false statement), so place an F in the second row of the truth table.

3. In the third case, assume that you paid on time, but your car was taken anyway (p is F, q is T). The salesman did *not* lie. He only said what would happen if you were late on payments, not what would happen if you paid on time. (The crime here is stealing, not lying.) Since we cannot say that the salesman lied, place a T in the third row of the truth table.

4. Finally, assume that you made timely payment, and your car was not taken (p is F, q is F). This certainly does not contradict the salesman's statement, so place a T in the last row of the truth table.

The completed truth table for the conditional is defined as follows.

TRUTH TABLE FOR THE CONDITIONAL   If p, then q

If p, then q

p	q	$p \rightarrow q$
T	T	T
T	F	F
F	T	T
F	F	T

The use of the conditional connective in no way implies a cause-and-effect relationship. Any two statements may have an arrow placed between them to create a compound statement. Consider this example.

If I pass mathematics, then the sun will rise the next day.

It is true, because the consequent is true. (See the special characteristics following **Example 1** on the next page.) There is, however, no cause-and-effect connection between my passing mathematics and the rising of the sun. The sun will rise no matter what grade I get.

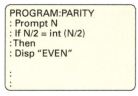

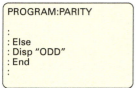

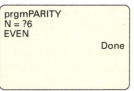

Conditional statements are useful in writing programs. The short program in the first two screens determines whether an integer is even. Notice the lines that begin with *If* and *Then*.

EXAMPLE 1 Finding the Truth Value of a Conditional

Given that p, q, and r are all false, find the truth value of the following statement.

$$(p \rightarrow \sim q) \rightarrow (\sim r \rightarrow q)$$

Solution

Using the shortcut method explained in **Example 3** of the previous section, we can replace p, q, and r with F (since each is false) and proceed as before, using the negation and conditional truth tables as necessary.

$$
\begin{aligned}
(p \rightarrow \sim q) &\rightarrow (\sim r \rightarrow q) \\
(F \rightarrow \sim F) &\rightarrow (\sim F \rightarrow F) \\
(F \rightarrow T) &\rightarrow (T \rightarrow F) \quad \text{Use the negation truth table.} \\
T &\rightarrow F \quad \text{Use the conditional truth table.} \\
&\mathbf{F}
\end{aligned}
$$

The statement $(p \rightarrow \sim q) \rightarrow (\sim r \rightarrow q)$ is false when p, q, and r are all false. ∎

SPECIAL CHARACTERISTICS OF CONDITIONAL STATEMENTS

1. $p \rightarrow q$ is false only when the antecedent is *true* and the consequent is *false*.

2. If the antecedent is *false*, then $p \rightarrow q$ is automatically *true*.

3. If the consequent is *true*, then $p \rightarrow q$ is automatically *true*.

EXAMPLE 2 Determining Whether Conditionals Are True or False

Write *true* or *false* for each statement. Here T represents a true statement, and F represents a false statement.

(a) $T \rightarrow (7 = 3)$ **(b)** $(8 < 2) \rightarrow F$ **(c)** $(4 \neq 3 + 1) \rightarrow T$

Solution

(a) Because the antecedent is true, while the consequent, $7 = 3$, is false, the given statement is false by the first point mentioned above.

(b) The antecedent is false, so the given statement is true by the second observation.

(c) The consequent is true, making the statement true by the third characteristic of conditional statements. ∎

EXAMPLE 3 Constructing Truth Tables

Construct a truth table for each statement.

(a) $(\sim p \rightarrow \sim q) \rightarrow (\sim p \wedge q)$ **(b)** $(p \rightarrow q) \rightarrow (\sim p \vee q)$

Solution

(a) Insert the truth values of $\sim p$ and $\sim q$. Find the truth values of $\sim p \rightarrow \sim q$.

p	q	$\sim p$	$\sim q$	$\sim p \rightarrow \sim q$
T	T	F	F	T
T	F	F	T	T
F	T	T	F	F
F	F	T	T	T

Next use $\sim p$ and q to find the truth values of $\sim p \wedge q$.

p	q	$\sim p$	$\sim q$	$\sim p \rightarrow \sim q$	$\sim p \wedge q$
T	T	F	F	T	F
T	F	F	T	T	F
F	T	T	F	F	T
F	F	T	T	T	F

Now find the truth values of $(\sim p \rightarrow \sim q) \rightarrow (\sim p \wedge q)$.

p	q	$\sim p$	$\sim q$	$\sim p \rightarrow \sim q$	$\sim p \wedge q$	$(\sim p \rightarrow \sim q) \rightarrow (\sim p \wedge q)$
T	T	F	F	T	F	F
T	F	F	T	T	F	F
F	T	T	F	F	T	T
F	F	T	T	T	F	F

(b) For $(p \rightarrow q) \rightarrow (\sim p \vee q)$, go through steps similar to the ones above.

p	q	$p \rightarrow q$	$\sim p$	$\sim p \vee q$	$(p \rightarrow q) \rightarrow (\sim p \vee q)$
T	T	T	F	T	T
T	F	F	F	F	T
F	T	T	T	T	T
F	F	T	T	T	T

As the truth table in **Example 3(b)** shows, the statement

$$(p \rightarrow q) \rightarrow (\sim p \vee q)$$

is always true, no matter what the truth values of the components. Such a statement is called a **tautology.** Several other examples of tautologies (as can be checked by forming truth tables) are

$$p \vee \sim p, \quad p \rightarrow p, \quad \text{and} \quad (\sim p \vee \sim q) \rightarrow \sim (p \wedge q). \quad \text{Tautologies}$$

The truth tables in **Example 3** also could have been found by the alternative method shown in **Section 3.2.**

Writing a Conditional as a Disjunction

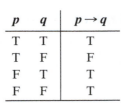

p	q	$p \rightarrow q$
T	T	T
T	F	F
F	T	T
F	F	T

Recall that the truth table for the conditional (repeated in the margin) shows that $p \rightarrow q$ is false only when p is true and q is false. But we also know that the disjunction is false for only one combination of component truth values, and it is easy to see that $\sim p \vee q$ will be false only when p is true and q is false, as the following truth table indicates.

p	q	$\sim p$	$\vee$	q
T	T	F	**T**	T
T	F	F	**F**	F
F	T	T	**T**	T
F	F	T	**T**	F

Thus we see that the disjunction $\sim p \vee q$ is equivalent to the conditional $p \rightarrow q$.

> **WRITING A CONDITIONAL AS A DISJUNCTION**
>
> $p \rightarrow q$ is equivalent to $\sim p \vee q$.

We now know that

$$p \rightarrow q \equiv \sim p \vee q,$$

so the negation of the conditional is

$$\sim(p \rightarrow q) \equiv \sim(\sim p \vee q).$$

Applying De Morgan's law to the right side of the above equivalence gives the negation as a conjunction.

NEGATION OF $p \rightarrow q$

The negation of $p \rightarrow q$ is $p \wedge \sim q$.

EXAMPLE 4 Determining Negations

Determine the negation of each statement.

(a) If I'm hungry, I will eat.   **(b)** All dogs have fleas.

Solution

(a) If p represents "I'm hungry" and q represents "I will eat," then the given statement can be symbolized by $p \rightarrow q$. The negation of $p \rightarrow q$, as shown earlier, is $p \wedge \sim q$, so the negation of the statement is

> I'm hungry and I will not eat.

(b) First, we must restate the given statement in *if . . . then* form.

> If it is a dog, then it has fleas.

Based on our earlier discussion, the negation is

> It is a dog and it does not have fleas.   ∎

> Do not try to negate a conditional with another conditional.

As seen in **Example 4,** the negation of a conditional statement is written as a conjunction.

EXAMPLE 5 Determining Statements Equivalent to Conditionals

Write each conditional as an equivalent statement without using *if . . . then*.

(a) If the Indians win the pennant, then Johnny will go to the World Series.

(b) If it's Borden's, it's got to be good.

Solution

(a) Because the conditional $p \rightarrow q$ is equivalent to $\sim p \vee q$, let p represent "The Indians win the pennant" and q represent "Johnny will go to the World Series." Restate the conditional as

> The Indians do not win the pennant or Johnny will go to the World Series.

(b) If p represents "it's Borden's" and if q represents "it's got to be good," the conditional may be restated as

> It's not Borden's or it's got to be good.   ∎

Figure 1

Series
circuit

Figure 2

Parallel circuit

Figure 3

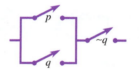

Figure 4

Switch Position	Current	Truth Value
Closed	flows	T
Open	stops	F

Circuits

One of the first nonmathematical applications of symbolic logic was seen in the master's thesis of Claude Shannon in 1937. Shannon showed how logic could be used to design electrical circuits. His work was immediately used by computer designers. Then in the developmental stage, computers could be simplified and built for less money using the ideas of Shannon.

To see how Shannon's ideas work, look at the electrical switch shown in **Figure 1.** We assume that current will flow through this switch when it is closed and not when it is open.

Figure 2 shows two switches connected in *series.* In such a circuit, current will flow only when both switches are closed. Note how closely a series circuit corresponds to the conjunction $p \wedge q$. We know that $p \wedge q$ is true only when both p and q are true.

A circuit corresponding to the disjunction $p \vee q$ can be found by drawing a *parallel* circuit, as in **Figure 3.** Here, current flows if either p *or* q is closed or if both p *and* q are closed.

The circuit in **Figure 4** corresponds to the statement $(p \vee q) \wedge {\sim}q$, which is a compound statement involving both a conjunction and a disjunction.

Simplifying an electrical circuit depends on the idea of equivalent statements from **Section 3.2.** Recall that two statements are equivalent if they have the same truth table final column. The symbol $\equiv$ is used to indicate that the two statements are equivalent. Some equivalent statements are shown in the following box.

EQUIVALENT STATEMENTS USED TO SIMPLIFY CIRCUITS

$$p \vee (q \wedge r) \equiv (p \vee q) \wedge (p \vee r) \qquad p \vee p \equiv p$$

$$p \wedge (q \vee r) \equiv (p \wedge q) \vee (p \wedge r) \qquad p \wedge p \equiv p$$

$$p \rightarrow q \equiv {\sim}q \rightarrow {\sim}p \qquad {\sim}(p \wedge q) \equiv {\sim}p \vee {\sim}q$$

$$p \rightarrow q \equiv {\sim}p \vee q \qquad {\sim}(p \vee q) \equiv {\sim}p \wedge {\sim}q$$

If T represents any true statement and F represents any false statement, then

$$p \vee \text{T} \equiv \text{T} \qquad p \vee {\sim}p \equiv \text{T}$$

$$p \wedge \text{F} \equiv \text{F} \qquad p \wedge {\sim}p \equiv \text{F.}$$

Circuits can be used as models of compound statements, with a closed switch corresponding to T (current flowing) and an open switch corresponding to F (current not flowing).

Figure 5

EXAMPLE 6 Simplifying a Circuit

Simplify the circuit of **Figure 5.**

Solution

At the top of **Figure 5,** p and q are connected in series, and at the bottom, p and r are connected in series. These are interpreted as the compound statements $p \wedge q$ and $p \wedge r$, respectively. These two conjunctions are connected in parallel, as indicated by the figure treated as a whole.

Write the disjunction of the two conjunctions.

$$(p \wedge q) \vee (p \wedge r)$$

Figure 6

(Think of the two switches labeled "p" as being controlled by the same lever.) By one of the pairs of equivalent statements in the preceding box,

$$(p \wedge q) \vee (p \wedge r) \equiv p \wedge (q \vee r),$$

which has the circuit of **Figure 6.** This circuit is logically equivalent to the one in **Figure 5,** and yet it contains only three switches instead of four—which might well lead to a large savings in manufacturing costs. ∎

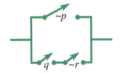

Figure 7

EXAMPLE 7 **Drawing a Circuit for a Conditional Statement**

Draw a circuit for $p \rightarrow (q \wedge \sim r)$.

Solution

From the list of equivalent statements in the box, $p \rightarrow q$ is equivalent to $\sim p \vee q$. This equivalence gives $p \rightarrow (q \wedge \sim r) \equiv \sim p \vee (q \wedge \sim r)$, which has the circuit diagram in **Figure 7.** ∎

WHEN Will I Ever USE This ?

Suppose you are a home-monitoring and control system designer. A home-owner wants the capability of turning his air-conditioning system on or off at home via his smartphone (in case he forgets before leaving for vacation). You will need to install a control module for the AC unit that receives signals from your customer's smartphone.

Draw a circuit that will allow the AC unit to turn on when both a master switch and the control module are activated, and when either (or both) of two thermostats is (or are) triggered. Then write a logical statement for the circuit.

Since the master switch (m) and the control module (c) both need to be activated, they must be connected in series, followed by two thermostats T_1 and T_2 in parallel (because only one needs to be triggered). The circuit is shown below, and the corresponding logical statement is $m \wedge c \wedge (T_1 \vee T_2)$.

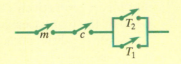

3.3 EXERCISES

Rewrite each statement using the if . . . then *connective. Rearrange the wording or add words as necessary.*

1. You can do it if you just believe.

2. It must be bad for you if it's sweet.

3. Every even integer divisible by 5 is divisible by 10.

4. No perfect square integers have units digit 2, 3, 7, or 8.

5. No grizzly bears live in California.

6. No guinea pigs get lonely.

7. Surfers can't stay away from the beach.

8. Running Bear loves Little White Dove.

Decide whether each statement is true *or* false.

9. If the antecedent of a conditional statement is false, the conditional statement is true.

10. If the consequent of a conditional statement is true, the conditional statement is true.

11. If q is true, then $(p \land (q \rightarrow r)) \rightarrow q$ is true.

12. If p is true, then $\sim p \rightarrow (q \lor r)$ is true.

13. The negation of "If pigs fly, I'll believe it" is "If pigs don't fly, I won't believe it."

14. The statements "If it flies, then it's a bird" and "It does not fly or it's a bird" are logically equivalent.

15. Given that $\sim p$ is true and q is false, the conditional $p \rightarrow q$ is true.

16. Given that $\sim p$ is false and q is false, the conditional $p \rightarrow q$ is true.

17. Explain why the statement "If $3 > 5$, then $4 < 6$" is true.

18. In a few sentences, explain how to determine the truth value of a conditional statement.

Tell whether each conditional is true (T) *or* false (F).

19. $T \rightarrow (7 < 3)$ **20.** $F \rightarrow (4 \neq 8)$

21. $F \rightarrow (5 \neq 5)$ **22.** $(8 \geq 8) \rightarrow F$

23. $(5^2 \neq 25) \rightarrow (8 - 8 = 16)$

24. $(5 = 12 - 7) \rightarrow (9 > 0)$

Let s represent "She sings for a living," let p represent "he fixes cars," and let m represent "they collect classics." Express each compound statement in words.

25. $\sim m \rightarrow p$ **26.** $p \rightarrow \sim m$

27. $s \rightarrow (m \land p)$ **28.** $(s \land p) \rightarrow m$

29. $\sim p \rightarrow (\sim m \lor s)$ **30.** $(\sim s \lor \sim m) \rightarrow \sim p$

Let b represent "I take my ball," let s represent "it is sunny," and let p represent "the park is open." Write each compound statement in symbols.

31. If I take my ball, then the park is open.

32. If I do not take my ball, then it is not sunny.

33. The park is open, and if it is sunny then I do not take my ball.

34. I take my ball, or if the park is open then it is sunny.

35. It is sunny if the park is open.

36. I'll take my ball if it is not sunny.

Find the truth value of each statement. Assume that p and r are false, and q is true.

37. $\sim r \rightarrow q$ **38.** $q \rightarrow p$

39. $p \rightarrow q$ **40.** $\sim r \rightarrow p$

41. $\sim p \rightarrow (q \wedge r)$ **42.** $(\sim r \vee p) \rightarrow p$

43. $\sim q \rightarrow (p \wedge r)$

44. $(\sim p \wedge \sim q) \rightarrow (p \wedge \sim r)$

45. $(p \rightarrow \sim q) \rightarrow (\sim p \wedge \sim r)$

46. $[(p \rightarrow \sim q) \wedge (p \rightarrow r)] \rightarrow r$

47. Explain why we know that

$$[r \vee (p \vee s)] \rightarrow [(p \vee q) \vee \sim p]$$

is true, even if we are not given the truth values of p, q, r, and s.

48. Construct a true statement involving a conditional, a conjunction, a disjunction, and a negation (not necessarily in that order) that consists of component statements p, q, and r, with all of these component statements false.

Construct a truth table for each statement. Identify any tautologies.

49. $\sim q \rightarrow p$

50. $(\sim q \rightarrow \sim p) \rightarrow \sim q$

51. $(\sim p \rightarrow q) \rightarrow p$

52. $(p \wedge q) \rightarrow (p \vee q)$

53. $(p \vee q) \rightarrow (q \vee p)$

54. $(\sim p \rightarrow \sim q) \rightarrow (p \wedge q)$

55. $[(r \vee p) \wedge \sim q] \rightarrow p$

56. $[(r \wedge p) \wedge (p \wedge q)] \rightarrow p$

57. $(\sim r \rightarrow s) \vee (p \rightarrow \sim q)$

58. $(\sim p \wedge \sim q) \rightarrow (s \rightarrow r)$

59. What is the minimum number of Fs that must appear in the final column of a truth table for us to be assured that the statement is not a tautology?

60. If all truth values in the final column of a truth table are F, how can we easily transform the statement into a tautology?

Write the negation of each statement. Remember that the negation of $p \rightarrow q$ is $p \wedge \sim q$.

61. If that is an authentic Coach bag, I'll be surprised.

62. If Muley Jones hits that note, he will shatter glass.

63. If the bullfighter doesn't get going, he's going to get gored.

64. If you don't say "I do," then you'll regret it for the rest of your life.

65. "If you want to be happy for the rest of your life, never make a pretty woman your wife." *Jimmy Soul*

66. "If I had a hammer, I'd hammer in the morning." *Lee Hayes and Pete Seeger*

Write each statement as an equivalent statement that does not use the if . . . then *connective. Remember that*

$$p \rightarrow q \quad \text{is equivalent to} \quad \sim p \vee q.$$

67. If you give your plants tender, loving care, they flourish.

68. If you scratch my back, I'll scratch yours.

69. If she doesn't, he will.

70. If I say "black," she says "white."

71. All residents of Pensacola are residents of Florida.

72. All women were once girls.

Use truth tables to decide which of the pairs of statements are equivalent.

73. $p \rightarrow q; \quad \sim p \vee q$ **74.** $\sim (p \rightarrow q); \quad p \wedge \sim q$

75. $p \rightarrow q; \quad \sim q \rightarrow \sim p$ **76.** $p \rightarrow q; \quad q \rightarrow p$

77. $p \rightarrow \sim q; \quad \sim p \vee \sim q$ **78.** $\sim p \wedge q; \quad \sim p \rightarrow q$

79. $q \to {\sim}p; \quad p \to {\sim}q$

80. ${\sim}(p \lor q) \to r; \quad (p \lor q) \lor r$

Write a logical statement representing each of the following circuits. Simplify each circuit when possible.

81.

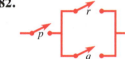

82.

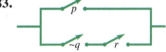

83.

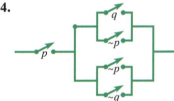

84.

85.

86.

Draw circuits representing the following statements as they are given. Simplify if possible.

87. $p \land (q \lor {\sim}p)$ **88.** $({\sim}p \land {\sim}q) \land {\sim}r$

89. $(p \lor q) \land ({\sim}p \land {\sim}q)$

90. $({\sim}q \land {\sim}p) \lor ({\sim}p \lor q)$

91. $[(p \lor q) \land r] \land {\sim}p$

92. $[({\sim}p \land {\sim}r) \lor {\sim}q] \land ({\sim}p \land r)$

93. ${\sim}q \to ({\sim}p \to q)$ **94.** ${\sim}p \to ({\sim}p \lor {\sim}q)$

95. Refer to **Figures 5 and 6** in **Example 6.** Suppose the cost of the use of one switch for an hour is $0.06. By using the circuit in **Figure 6** rather than the circuit in **Figure 5,** what is the savings for a year of 365 days, assuming that the circuit is in continuous use?

96. Explain why the circuit shown will always have exactly one open switch. What does this circuit simplify to?

97. Refer to the "For Further Thought" at the end of **Section 3.2.** Verify that the logic circuit shown below is equivalent to the conditional statement $A \to B$.

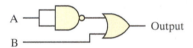

| 3.4 | **THE CONDITIONAL AND RELATED STATEMENTS** |

OBJECTIVES

1 Determine the converse, inverse, and contrapositive of a conditional statement.

2 Translate conditional statements into alternative forms.

3 Understand the structure of the biconditional.

4 Summarize the truth tables of compound statements.

Converse, Inverse, and Contrapositive

Many mathematical properties and theorems are stated in *if . . . then* form. Any conditional statement $p \to q$ is made up of an antecedent p and a consequent q. If they are interchanged, negated, or both, a new conditional statement is formed. Suppose that we begin with a conditional statement.

> If you stay, then I go.   Conditional statement

By interchanging the antecedent ("you stay") and the consequent ("I go"), we obtain a new conditional statement.

> If I go, then you stay.   Converse

This new conditional is called the **converse** of the given conditional statement.

By negating both the antecedent and the consequent, we obtain the **inverse** of the given conditional statement.

> If you do not stay, then I do not go.   Inverse

Alfred North Whitehead (1861–1947) and Bertrand Russell worked together on *Principia Mathematica.* During that time, Whitehead was teaching mathematics at Cambridge University and had written *Universal Algebra.* In 1910 he went to the University of London, exploring not only the philosophical basis of science but also the "aims of education" (as he called one of his books). It was as a philosopher that he was invited to Harvard University in 1924. Whitehead died at the age of 86 in Cambridge, Massachusetts.

If the antecedent and the consequent are both interchanged *and* negated, the **contrapositive** of the given conditional statement is formed.

If I do not go, then you do not stay.   Contrapositive

These three related statements for the conditional $p \rightarrow q$ are summarized below.

RELATED CONDITIONAL STATEMENTS

Conditional Statement	$p \rightarrow q$	(If p, then q.)
Converse	$q \rightarrow p$	(If q, then p.)
Inverse	$\sim p \rightarrow \sim q$	(If not p, then not q.)
Contrapositive	$\sim q \rightarrow \sim p$	(If not q, then not p.)

Notice that the inverse is the contrapositive of the converse.

EXAMPLE 1 Determining Related Conditional Statements

Determine each of the following, given the conditional statement

If I am running, then I am moving.

(a) the converse   **(b)** the inverse   **(c)** the contrapositive

Solution

(a) Let p represent "I am running" and q represent "I am moving." Then the given statement may be written $p \rightarrow q$. The converse, $q \rightarrow p$, is

If I am moving, then I am running.

The converse is not necessarily true, even though the given statement is true.

(b) The inverse of $p \rightarrow q$ is $\sim p \rightarrow \sim q$. Thus the inverse is

If I am not running, then I am not moving.

Again, this is not necessarily true.

(c) The contrapositive, $\sim q \rightarrow \sim p$, is

If I am not moving, then I am not running.

The contrapositive, like the given conditional statement, is true.   ∎

Example 1 shows that the converse and inverse of a true statement need not be true. They *can* be true, but they need not be. The relationships between the related conditionals are shown in the truth table that follows.

		Conditional	Converse	Inverse	Contrapositive
p	q	$p \rightarrow q$	$q \rightarrow p$	$\sim p \rightarrow \sim q$	$\sim q \rightarrow \sim p$
T	T	T	T	T	T
T	F	F	T	T	F
F	T	T	F	F	T
F	F	T	T	T	T

Equivalent (Conditional and Contrapositive)

Equivalent (Converse and Inverse)

As this truth table shows,

1. *A conditional statement and its contrapositive always have the same truth value,* making it possible to replace any statement with its contrapositive without affecting the logical meaning.

2. *The converse and inverse always have the same truth value.*

Bertrand Russell (1872–1970) was a student of Whitehead's before they wrote the *Principia*. Like his teacher, Russell turned toward philosophy. His works include a critique of Leibniz, analyses of mind and of matter, and a history of Western thought.

Russell became a public figure because of his involvement in social issues. Deeply aware of human loneliness, he was "passionately desirous of finding ways of diminishing this tragic isolation." During World War I he was an antiwar crusader, and he was imprisoned briefly. Again in the 1960s he championed peace. He wrote many books on social issues, winning the Nobel Prize for Literature in 1950.

> **EQUIVALENCES**
>
> A conditional statement and its contrapositive are equivalent. Also, the converse and the inverse are equivalent.

EXAMPLE 2 **Determining Related Conditional Statements**

For the conditional statement $\sim p \rightarrow q$, write each of the following.

(a) the converse   **(b)** the inverse   **(c)** the contrapositive

Solution

(a) The converse of $\sim p \rightarrow q$ is $q \rightarrow \sim p$.

(b) The inverse is $\sim(\sim p) \rightarrow \sim q$, which simplifies to $p \rightarrow \sim q$.

(c) The contrapositive is $\sim q \rightarrow \sim(\sim p)$, which simplifies to $\sim q \rightarrow p$.   ■

Alternative Forms of "If *p*, then *q*"

The conditional statement "If *p*, then *q*" can be stated in several other ways in English. Consider this statement.

> If you take Tylenol, then you will find relief from your symptoms.

It can also be written as follows.

> Taking Tylenol is *sufficient* for relieving your symptoms.

According to this statement, taking Tylenol is enough to relieve your symptoms. Taking other medications or using other treatment techniques *might* also result in symptom relief, but at least we *know* that taking Tylenol will. Thus $p \rightarrow q$ can be written "*p* is sufficient for *q*." Knowing that *p* has occurred is sufficient to guarantee that *q* will also occur.

On the other hand, consider this statement, which has a different structure.

> Fresh ingredients are necessary for making a good pizza.   (*)

This statement claims that fresh ingredients are one condition for making a good pizza. But there may be other conditions (such as a working oven, for example). The statement labeled (*) could be written as

> If you want good pizza, then you need fresh ingredients.

As this example suggests, $p \rightarrow q$ is the same as "*q* is necessary for *p*." In other words, if *q* doesn't happen, then neither will *p*. Notice how this idea is closely related to the idea of equivalence between a conditional statement and its contrapositive.

Kurt Gödel (1906–1978) is widely regarded as the most influential mathematical logician of the twentieth century. He proved by his Incompleteness Theorem that the search for a set of axioms from which all mathematical truths could be proved was futile. In particular, "the vast structure of the *Principia Mathematica* of Whitehead and Russell was inadequate for deciding all mathematical questions."

After the death of his friend **Albert Einstein** (1879–1955), Gödel developed paranoia, and his life ended tragically when, convinced he was being poisoned, he refused to eat, essentially starving himself to death.

COMMON TRANSLATIONS OF $p \rightarrow q$

The conditional $p \rightarrow q$ can be translated in any of the following ways, none of which depends on the truth or falsity of $p \rightarrow q$.

If p, then q.	p is sufficient for q.
If p, q.	q is necessary for p.
p implies q.	All p are q.
p only if q.	q if p.

Example: If you live in Alamogordo, then you live in New Mexico.   Statement

You live in New Mexico if you live in Alamogordo.
You live in Alamogordo only if you live in New Mexico.
Living in New Mexico is necessary for living in Alamogordo.
Living in Alamogordo is sufficient for living in New Mexico.
All residents of Alamogordo are residents of New Mexico.
Being a resident of Alamogordo implies residency in New Mexico.

Common translations

EXAMPLE 3 Rewording Conditional Statements

Rewrite each statement in the form "If p, then q."

(a) You'll get sick if you eat that.

(b) Go to the doctor only if your temperature exceeds 101°F.

(c) Everyone at the game had a great time.

Solution

(a) If you eat that, then you'll get sick.

(b) If you go to the doctor, then your temperature exceeds 101°F.

(c) If you were at the game, then you had a great time.   ∎

EXAMPLE 4 Translating from Words to Symbols

Let p represent "A triangle is equilateral," and let q represent "A triangle has three sides of equal length." Write each of the following in symbols.

(a) A triangle is equilateral if it has three sides of equal length.

(b) A triangle is equilateral only if it has three sides of equal length.

Solution

(a) $q \rightarrow p$ **(b)** $p \rightarrow q$ ∎

Biconditionals

The compound statement ***p if and only if q*** (often abbreviated ***p iff q***) is called a **biconditional.** It is symbolized $p \leftrightarrow q$ and is interpreted as the conjunction of the two conditionals $p \rightarrow q$ and $q \rightarrow p$.

Using symbols, the conjunction of the conditionals $p \rightarrow q$ and $q \rightarrow p$ is written $(q \rightarrow p) \wedge (p \rightarrow q)$ so that, by definition,

$$p \leftrightarrow q \equiv (q \rightarrow p) \wedge (p \rightarrow q).$$ Biconditional

The truth table for the biconditional $p \leftrightarrow q$ can be determined using this definition.

TRUTH TABLE FOR THE BICONDITIONAL   *p if and only if q*

p if and only if q

p	q	$p \leftrightarrow q$
T	T	T
T	F	F
F	T	F
F	F	T

A biconditional is true when both component statements have the same truth value. It is false when they have different truth values.

Principia Mathematica, the title chosen by Whitehead and Russell, was a deliberate reference to *Philosophiae naturalis principia mathematica,* or "mathematical principles of the philosophy of nature," Isaac Newton's epochal work of 1687. Newton's *Principia* pictured a kind of "clockwork universe" that ran via his Law of Gravitation. Newton independently invented the calculus, unaware that Leibniz had published his own formulation of it earlier.

EXAMPLE 5   **Determining Whether Biconditionals Are True or False**

Determine whether each biconditional statement is *true* or *false*.

(a) $6 + 8 = 14$ if and only if $11 + 5 = 16$

(b) $6 = 5$ if and only if $12 \neq 12$

(c) Mars is a moon if and only if Jupiter is a planet.

Solution

(a) Both $6 + 8 = 14$ and $11 + 5 = 16$ are true. By the truth table for the biconditional, this biconditional is true.

(b) Both component statements are false, so by the last line of the truth table for the biconditional, this biconditional statement is true.

(c) Because the first component Mars is a moon is false, and the second is true, this biconditional statement is false. ∎

Summary of Truth Tables

Truth tables have been derived for several important types of compound statements.

SUMMARY OF BASIC TRUTH TABLES

1. $\sim p$, the **negation** of p, has truth value opposite that of p.

2. $p \wedge q$, the **conjunction,** is true only when both p and q are true.

3. $p \vee q$, the **disjunction,** is false only when both p and q are false.

4. $p \rightarrow q$, the **conditional,** is false only when p is true and q is false.

5. $p \leftrightarrow q$, the **biconditional,** is true only when both p and q have the same truth value.

3.4   EXERCISES

*For each given conditional statement (or statement that can be written as a conditional), write (**a**) the converse, (**b**) the inverse, and (**c**) the contrapositive in if . . . then form. In some of the exercises, it may be helpful to first restate the given statement in if . . . then form.*

1. If beauty were a minute, then you would be an hour.

2. If you lead, then I will follow.

3. If it ain't broke, don't fix it.

4. If I had a nickel for each time that happened, I would be rich.

5. Walking in front of a moving car is dangerous to your health.

6. Milk contains calcium.

7. Birds of a feather flock together.

8. A rolling stone gathers no moss.

9. If you build it, he will come.

10. Where there's smoke, there's fire.

11. $p \rightarrow \sim q$ 12. $\sim p \rightarrow q$

13. $\sim p \rightarrow \sim q$ 14. $\sim q \rightarrow \sim p$

15. $p \rightarrow (q \vee r)$ (*Hint:* Use one of De Morgan's laws as necessary.)

16. $(r \vee \sim q) \rightarrow p$ (*Hint:* Use one of De Morgan's laws as necessary.)

17. Discuss the equivalences that exist among a given conditional statement, its converse, its inverse, and its contrapositive.

18. State the contrapositive of "If the square of a natural number is odd, then the natural number is odd." The two statements must have the same truth value. Use several examples and inductive reasoning to decide whether both are true or both are false.

Write each statement in the form "*if p, then q.*"

19. If the Kings go to the playoffs, pigs will fly.

20. If I score 90% or higher on my test, I'll go to a movie.

21. Legs of 3 and 4 imply a hypotenuse of 5.

22. "This is a leap year" implies that next year is not.

23. All whole numbers are rational numbers.

24. No irrational numbers are rational.

25. Doing logic puzzles is sufficient for driving me crazy.

26. Being in Kalamazoo is sufficient for being in Michigan.

27. Two coats of paint are necessary to cover the graffiti.

28. Being an environmentalist is necessary for being elected.

29. Employment will improve only if the economy recovers.

30. The economy will recover only if employment improves.

31. No whole numbers are not integers.

32. No integers are irrational numbers.

33. The Phillies will win the pennant when their pitching improves.

34. The grass will be greener when we're on the other side.

35. A rectangle is a parallelogram with perpendicular adjacent sides.

36. A square is a rectangle with two adjacent sides equal.

37. A triangle with two perpendicular sides is a right triangle.

38. A parallelogram is a four-sided figure with opposite sides parallel.

39. The square of a three-digit number whose units digit is 5 will end in 25.

40. An integer whose units digit is 0 or 5 is divisible by 5.

41. One of the following statements is not equivalent to all the others. Which one is it?
    A. *r* only if *s*.   B. *r* implies *s*.
    C. If *r*, then *s*.   D. *r* is necessary for *s*.

42. Many students have difficulty interpreting *necessary* and *sufficient*. Use the statement "Being in Vancouver is sufficient for being in North America" to explain why "*p* is sufficient for *q*" translates as "if *p*, then *q*."

43. Use the statement "To be an integer, it is necessary that a number be rational" to explain why "*p* is necessary for *q*" translates as "if *q*, then *p*."

44. Explain why the statement "A week has eight days if and only if October has forty days" is true.

Identify each statement as true *or* false.

45. $6 = 9 - 3$ if and only if $8 + 2 = 10$.

46. $3 + 1 \neq 7$ if and only if $8 \neq 8$.

47. $8 + 7 \neq 15$ if and only if $3 \times 5 \neq 8$.

48. $6 \times 2 = 18$ if and only if $9 + 7 \neq 16$.

49. George H. W. Bush was president if and only if George W. Bush was not president.

50. McDonald's sells Whoppers if and only if Apple manufactures Ipods.

Two statements that can both be true about the same object are **consistent.** *For example, "It is green" and "It weighs 60 pounds" are consistent statements. Statements that cannot both be true about the same object are called* **contrary.** *"It is a Nissan" and "It is a Mazda" are contrary. In Exercises 51–55, label each pair of statements as either* contrary *or* consistent.

51. Michael Jackson is alive. Michael Jackson is dead.

52. That book is nonfiction. That same book costs more than $150.

53. This number is a whole number. This same number is irrational.

54. This number is positive. This same number is a natural number.

55. This number is an integer. This same number is a rational number.

56. Refer to the "For Further Thought" on **page 99** at the end of **Section 3.2.** Verify that the logic circuit at the top of the next column (consisting of only NOR gates) is equivalent to the biconditional $A \leftrightarrow B$. Build a truth table for the statement $\sim (p \veebar q)$ (the negation of the Exclusive OR statement). What is the relationship between this and the biconditional?

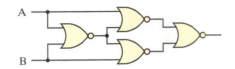

Exercises 57 and 58 refer to the Chapter Opener on **page 83.** *Humphrey the hero deduced from the riddle that there are twin trolls who take shifts guarding the magical doors. One of them tells only truths, the other only lies. When the troll standing guard sees Humphrey, he says,*

> *"If truly Truthful Troll I be, then go thou east and be thou free."*

Humphrey needs to decide which troll he is addressing, call him by name, and tell him which door he would like opened.

There are two things Humphrey must get right: the name of the troll and the proper door. If he misidentifies the troll, no door will be opened. If he correctly names the troll and picks the wrong door, he will be confined to the dungeon behind it.

57. Because the troll either always lies or always tells the truth, Humphrey knows that

(1) if the troll is Truthful Troll, then the conditional statement he uttered is true, and

(2) if the conditional statement he uttered is true, then he is Truthful Troll.

Let p represent "the troll is Truthful Troll" and let q represent "the door to the east leads to freedom." Express the statements in (1) and (2) in symbolic form.

58. The conjunction of the answers from **Exercise 57** is a biconditional that must be true.

(a) Build a truth table for this biconditional.

(b) Use the fact that it *must* be true to solve Humphrey's riddle.

| **3.5** | **ANALYZING ARGUMENTS WITH EULER DIAGRAMS** |

OBJECTIVES

1 Define logical arguments.

2 Use Euler diagrams to analyze arguments with universal quantifiers.

3 Use Euler diagrams to analyze arguments with existential quantifiers.

Logical Arguments

With inductive reasoning we observe patterns to solve problems. Now we study how deductive reasoning may be used to determine whether logical arguments are valid or invalid.

A logical argument is made up of **premises** (assumptions, laws, rules, widely held ideas, or observations) and a **conclusion.** Recall that *deductive* reasoning involves drawing specific conclusions from given general premises. When reasoning from the premises of an argument to obtain a conclusion, we want the argument to be valid.

Leonhard Euler (1707–1783) won the Academy prize and edged out du Châtelet and Voltaire. That was a minor achievement, as was the invention of "Euler circles" (which antedated Venn diagrams). Euler was the most prolific mathematician of his generation despite blindness that forced him to dictate from memory.

> **VALID AND INVALID ARGUMENTS**
>
> An argument is **valid** if the fact that all the premises are true forces the conclusion to be true. An argument that is not valid is **invalid.** It is called a **fallacy.**

"Valid" and "true" do not have the same meaning—an argument can be valid even though the conclusion is false (see Example 4), *or invalid even though the conclusion is true* (see Example 6).

Arguments with Universal Quantifiers

Several techniques can be used to check whether an argument is valid. One such technique is based on **Euler diagrams.**

Leonhard Euler (pronounced "Oiler") was one of the greatest mathematicians who ever lived. He is immortalized in mathematics history with the important irrational number *e*, named in his honor. This number appears throughout mathematics and is discussed in **Chapters 6 and 8.**

▌ EXAMPLE 1 Using an Euler Diagram to Determine Validity

Is the following argument valid?

> No accidents happen on purpose.
> Spilling the beans was an accident.
> ——————————————
> The beans were not spilled on purpose.

Solution

To begin, draw regions to represent the first premise. Because no accidents happen on purpose, the region for "accidents" goes outside the region for "things that happen on purpose," as shown in **Figure 8.**

The second premise, "Spilling the beans was an accident," suggests that "spilling the beans" belongs in the region representing "accidents." Let *x* represent "spilling the beans." **Figure 9** shows that "spilling the beans" is not in the region for "things that happen on purpose." If both premises are true, the conclusion that the beans were not spilled on purpose is also true. The argument is valid. ■

Figure 8

Figure 9

▌ EXAMPLE 2 Using an Euler Diagram to Determine Validity

Is the following argument valid?

> All rainy days are cloudy.
> Today is not cloudy.
> ——————————
> Today is not rainy.

Solution

In **Figure 10**, the region for "rainy days" is drawn entirely inside the region for "cloudy days." Since "Today is *not* cloudy," place an *x* for "today" *outside* the region for "cloudy days." See **Figure 11.** Placing the *x* outside the region for "cloudy days" forces it also to be outside the region for "rainy days." Thus, if the two premises are true, then it is also true that today is not rainy. The argument is valid. ■

Figure 10 (Cloudy days / Rainy days)

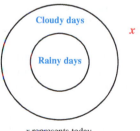

x represents today.
Figure 11 ■

EXAMPLE 3 **Using an Euler Diagram to Determine Validity**

Is the following argument valid?

> All magnolia trees have green leaves.
> That plant has green leaves.
> _____
> That plant is a magnolia tree.

Solution

Figure 12

The region for "magnolia trees" goes entirely inside the region for "things that have green leaves." See **Figure 12.** The *x* that represents "that plant" must go inside the region for "things that have green leaves," but it can go either inside or outside the region for "magnolia trees." Even if the premises are true, we are not forced to accept the conclusion as true. This argument is invalid. It is a fallacy. ■

EXAMPLE 4 **Using an Euler Diagram to Determine Validity**

Is the following argument valid?

> All expensive things are desirable.
> All desirable things make you feel good.
> All things that make you feel good make you live longer.
> _____
> All expensive things make you live longer.

Solution

A diagram for the argument is given in **Figure 13.** If each premise is true, then the conclusion must be true because the region for "expensive things" lies completely within the region for "things that make you live longer." Thus, the argument is valid. (This argument is an example of the fact that a *valid* argument need *not* have a true conclusion.)

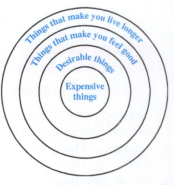

Figure 13   ■

Arguments with Existential Quantifiers

EXAMPLE 5 **Using an Euler Diagram to Determine Validity**

Is the following argument valid?

> Many students drive Hondas.
> I am a student.
> _____
> I drive a Honda.

Solution

Figure 14

Figure 15

The first premise is sketched in **Figure 14,** where many (but not necessarily *all*) students drive Hondas. There are two possibilities for *I,* as shown in **Figure 15.** One possibility is that *I* drive a Honda. The other is that *I* don't. Since the truth of the premises does not force the conclusion to be true, the argument is invalid. ■

EXAMPLE 6 Using an Euler Diagram to Determine Validity

Is the following argument valid?

All fish swim.
All whales swim.
_____
A whale is not a fish.

Solution

The premises lead to two possibilities. **Figure 16** shows the set of fish and the set of whales as intersecting, while **Figure 17** does not. Both diagrams are valid interpretations of the given premises, but only one supports the conclusion.

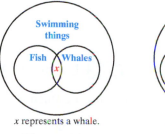

x represents a whale.   *x* represents a whale.

Figure 16   **Figure 17**

Because the truth of the premises does not force the conclusion to be true, the argument is invalid. Even though we know the conclusion to be true, this knowledge is not deduced from the premises. ∎

FOR FURTHER THOUGHT

Common Fallacies

Discussions in everyday conversation, politics, and advertising provide a nearly endless stream of examples of **fallacies**—arguments exhibiting illogical reasoning. There are many general forms of fallacies, and we now present descriptions and examples of some of the more common ones. (Much of this list is adapted from the document Reader Mission Critical, located on the San Jose State University Web site, www.sjsu.edu)

1. **Circular Reasoning** (also called *Begging the Question*) The person making the argument assumes to be true what he is trying to prove.

   Husband: What makes you say that this dress makes you look fat?

   Wife: Because it does.

   Here the wife makes her case by stating what she wants to prove.

2. **False Dilemma** (also called the *Either-Or Fallacy*, or the *Black and White Fallacy*) Presenting two options with the assumption that they are contradictions (that is, the truth of one implies the falsity of the other) when in fact they are not, is the basis of a common fallacy.

Politician: America: Love it or leave it.

This argument implies only two choices. It is possible that someone may love America and yet leave, while someone else may not love America and yet stay.

3. **Loaded Question and Complex Claims** This fallacy involves one person asking a question or making a statement that is constructed in such a way as to obtain an answer in which the responder agrees to something with which he does not actually agree.

Teenager Beth to her father: I hope you enjoyed embarrassing me in front of my friends.

If Beth gets the expected response "No, I didn't enjoy it," the answer allows Beth to interpret that while her father didn't enjoy it, he did indeed embarrass her.

4. **Post Hoc Reasoning** An argument that is based on the false belief that if event A preceded event B, then A must have caused B is called *post hoc reasoning.*

Johnny: I wore my Hawaiian shirt while watching all three playoff games, and my team won all three games. So I am going to wear that shirt every time I watch them.

The fact that Johnny put the same shirt on before each game has nothing to do with the outcomes of the games.

5. **Red Herring** (also called *Smoke Screen,* or *Wild Goose Chase*) This fallacy involves introducing an irrelevant topic to divert attention away from the original topic, allowing the person making the argument to seemingly prevail.

(The following script is from a political advertisement during the 2008 presidential campaign, intended to establish that John McCain lacked understanding of the economy.)

Maybe you're struggling just to pay the mortgage on your home. But recently, John McCain said, "The fundamentals of our economy are strong." Hmm. Then again, that same day, when asked how many houses he owns, McCain lost track. He couldn't remember. Well, it's seven. Seven houses. And here's one house America can't afford to let John McCain move into (showing a picture of the White House).

The advertisement shifted the focus to the number of houses McCain owned, which had nothing to do with the state of the economy, or the ability of "average" citizens to make their mortgage payments.

6. **Shifting the Burden of Proof** A person making a claim usually is required to support that claim. In this fallacy, if the claim is difficult to support, that person turns the burden of proof of that claim over to someone else.

Employee: You accuse me of embezzling money? That's ridiculous.

Employer: Well, until you can prove otherwise, you will just have to accept it as true.

If money has been disappearing, it is up to the employer to prove that this employee is guilty. The burden of proof is on the employer, but he is insinuating that the employee must prove that he is not the one taking the money.

7. **Straw Man** This fallacy involves creating a false image (like a scarecrow, or straw man) of someone else's position in an argument.

Dan Quayle: I have as much experience in the Congress as Jack Kennedy did when he sought the presidency.

Lloyd Bentsen: Senator, I served with Jack Kennedy. I knew Jack Kennedy. Jack Kennedy was a friend of mine. And Senator, you're no Jack Kennedy.

Dan Quayle: That was really uncalled for, Senator.

Lloyd Bentsen: You're the one that was making the comparison, Senator.

While this was the defining moment of the 1988 vice-presidential debate, Bentsen expertly used the straw man fallacy. Quayle did not compare himself or his accomplishments to those of Kennedy, but merely stated that he had spent as much time in Congress as Kennedy had when the latter ran for president.

For Group or Individual Investigation

Use the Internet to investigate the following additional logical fallacies.

Appeal to Authority	Appeal to Common Belief	Common Practice
Two Wrongs		Wishful Thinking
Appeal to Fear	Indirect Consequences	Appeal to Pity
Appeal to Prejudice	Appeal to Loyalty	Appeal to Vanity
Guilt by Association	Appeal to Spite	Hasty Generalization
	Slippery Slope	

3.5 EXERCISES

Decide whether each argument is valid or invalid.

1. All amusement parks have thrill rides.
   Universal Orlando is an amusement park.

   Universal Orlando has thrill rides.

2. All disc jockeys play music.
   Calvin is a disc jockey.

   Calvin plays music.

3. All celebrities have problems.
That man has problems.
That man is a celebrity.

4. All Southerners speak with an accent.
Nick speaks with an accent.
Nick is a Southerner.

5. All dogs love to bury bones.
Puddles does not love to bury bones.
Puddles is not a dog.

6. All vice presidents use cell phones.
Bob does not use a cell phone.
Bob is not a vice president.

7. All residents of Colorado know how to breathe thin air.
Julie knows how to breathe thin air.
Julie lives in Colorado.

8. All drivers must have a photo I.D.
Kay has a photo I.D.
Kay is a driver.

9. Some dinosaurs were plant eaters.
Danny was a plant eater.
Danny was a dinosaur.

10. Some philosophers are absent minded.
Nicole is a philosopher.
Nicole is absent minded.

11. Many nurses belong to unions.
Heather is a nurse.
Heather belongs to a union.

12. Some trucks have sound systems.
Some trucks have gun racks.
Some trucks with sound systems have gun racks.

13. Refer to **Example 3.** If the second premise and the conclusion were interchanged, would the argument then be valid?

14. Refer to **Example 4.** Give a different conclusion from the one given there so that the argument is still valid.

Construct a valid argument based on the Euler diagram shown.

15.   **16.**

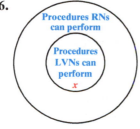

x represents Erin.   *x* represents vaccinations.

As mentioned in the text, an argument can have a true conclusion yet be invalid. In these exercises, each argument has a true conclusion. Identify each argument as valid *or* invalid.

17. All birds fly.
All planes fly.
A bird is not a plane.

18. All actors have cars.
All cars use gas.
All actors have gas.

19. All chickens have beaks.
All hens are chickens.
All hens have beaks.

20. All chickens have beaks.
All birds have beaks.
All chickens are birds.

21. Amarillo is northeast of El Paso.
Amarillo is northeast of Deming.
El Paso is northeast of Deming.

22. Beaverton is north of Salem.
Salem is north of Lebanon.
Beaverton is north of Lebanon.

23. No whole numbers are negative.
−3 is negative.
−3 is not a whole number.

24. A scalene triangle has a longest side.
A scalene triangle has a largest angle.
The largest angle in a scalene triangle is opposite the longest side.

In Exercises 25–30, the premises marked A, B, *and* C *are followed by several possible conclusions. Take each conclusion in turn, and check whether the resulting argument is* valid *or* invalid.

A. *All people who drive contribute to air pollution.*

B. *All people who contribute to air pollution make life a little worse.*

C. *Some people who live in a suburb make life a little worse.*

25. Some people who live in a suburb contribute to air pollution.

26. Some people who live in a suburb drive.

27. Suburban residents never drive.

28. Some people who contribute to air pollution live in a suburb.

29. Some people who make life a little worse live in a suburb.

30. All people who drive make life a little worse.

3.6

ANALYZING ARGUMENTS WITH TRUTH TABLES

OBJECTIVES

1 Use truth tables to determine validity of arguments.

2 Recognize valid and invalid argument forms.

3 Be familiar with the arguments of Lewis Carroll.

Using Truth Tables to Determine Validity

In **Section 3.5** we used Euler diagrams to test the validity of arguments. While Euler diagrams often work well for simple arguments, difficulties can develop with more complex ones, because Euler diagrams must show every possible case. In complex arguments, it is hard to be sure that all cases have been considered.

In deciding whether to use Euler diagrams to test the validity of an argument, look for quantifiers such as "all," "some," or "no." These words often indicate arguments best tested by Euler diagrams. If these words are absent, it may be better to use truth tables to test the validity of an argument.

EXAMPLE 1 Using a Truth Table to Determine Validity

Determine whether the argument is *valid* or *invalid*.

> If there is a problem, then I must fix it.
> There is a problem.
> _____
> I must fix it.

Solution

To test the validity of this argument, we begin by assigning the letters p and q to represent these statements.

p represents "There is a problem."

q represents "I must fix it."

Now we write the two premises and the conclusion in symbols.

> Premise 1: $p \rightarrow q$
> Premise 2: p
> _____
> Conclusion: q

To decide if this argument is valid, we must determine whether the conjunction of both premises implies the conclusion for all possible combinations of truth values for p and q. Therefore, write the conjunction of the premises as the antecedent of a conditional statement, and write the conclusion as the consequent.

$$[(p \rightarrow q) \quad \wedge \quad p] \quad \rightarrow \quad q$$

premise   **and**   **premise**   **implies**   **conclusion**

Finally, construct the truth table for this conditional statement, as shown below.

p	q	$p \rightarrow q$	$(p \rightarrow q) \wedge p$	$[(p \rightarrow q) \wedge p] \rightarrow q$
T	T	T	T	T
T	F	F	F	T
F	T	T	F	T
F	F	T	F	T

Because the final column, shown in color, indicates that the conditional statement that represents the argument is true for all possible truth values of p and q, the statement is a tautology. Thus, the argument is valid. ■

In the 2007 Spanish film ***La Habitacion de Fermat (Fermat's Room),*** four mathematicians are invited to dinner, only to discover that the room in which they are meeting is designed to eventually crush them as walls creep in closer and closer. The only way for them to delay the inevitable is to answer enigmas, questions, puzzles, problems, and riddles that they are receiving on a cell phone.

One of the enigmas deals with a hermetically sealed room that contains a single light bulb. There are three switches outside the room, all of them are off, and only one of these switches controls the bulb. You are allowed to flip any or all of the switches as many times as you wish before you enter the room, but once you enter, you cannot return to the switches outside. How can you determine which one controls the bulb? (The answer is on **page 124.**)

Answer to the Light Bulb question on page 123.

Label the switches 1, 2, and 3. Turn switch 1 on and leave it on for several minutes. Then turn switch 1 off, turn switch 2 on, and then immediately enter the room. If the bulb is on, then you know that switch 2 controls it. If the bulb is off, touch it to see if it is still warm. If it is, then switch 1 controls it. If the bulb is not warm, then switch 3 controls it.

The pattern of the argument in **Example 1**

$$p \to q$$
$$\underline{p}$$
$$q$$

is called **modus ponens,** or the *law of detachment.*

To test the validity of an argument using a truth table, follow the steps in the box.

TESTING THE VALIDITY OF AN ARGUMENT WITH A TRUTH TABLE

Step 1 Assign a letter to represent each component statement in the argument.

Step 2 Express each premise and the conclusion symbolically.

Step 3 Form the symbolic statement of the entire argument by writing the *conjunction* of *all* the premises as the antecedent of a conditional statement, and the conclusion of the argument as the consequent.

Step 4 Complete the truth table for the conditional statement formed in Step 3. If it is a tautology, then the argument is valid; otherwise, it is invalid.

EXAMPLE 2 Using a Truth Table to Determine Validity

Determine whether the argument is *valid* or *invalid.*

If my check arrives in time, I'll register for fall semester.
I've registered for fall semester.
_____
My check arrived in time.

Solution

Let p represent "My check arrives (arrived) in time." Let q represent "I'll register (I've registered) for fall semester." The argument can be written as follows.

$$p \to q$$
$$\underline{q}$$
$$p$$

To test for validity, construct a truth table for the statement $[(p \to q) \land q] \to p$.

p	q	$p \to q$	$(p \to q) \land q$	$[(p \to q) \land q] \to p$
T	T	T	T	T
T	F	F	F	T
F	T	T	T	F
F	F	T	F	T

The final column of the truth table contains an F. The argument is invalid. ∎

If a conditional and its converse were logically equivalent, then an argument of the type found in **Example 2** would be valid. Because a conditional and its converse are *not* equivalent, the argument is an example of what is sometimes called the **fallacy of the converse.**

EXAMPLE 3 Using a Truth Table to Determine Validity

Determine whether the argument is *valid* or *invalid.*

If I can avoid sweets, I can avoid the dentist.
I can't avoid the dentist.
_____
I can't avoid sweets.

Solution

If *p* represents "I can avoid sweets" and *q* represents "I can avoid the dentist," the argument is written as follows.

$$p \rightarrow q$$
$$\frac{\sim q}{\sim p}$$

The symbolic statement of the entire argument is as follows.

$$[(p \rightarrow q) \wedge \sim q] \rightarrow \sim p$$

The truth table for this argument indicates a tautology, and the argument is valid.

p	*q*	*p* → *q*	~*q*	(*p* → *q*) ∧ ~*q*	~*p*	[(*p* → *q*) ∧ ~*q*] → ~*p*
T	T	T	F	F	F	T
T	F	F	T	F	F	T
F	T	T	F	F	T	T
F	F	T	T	T	T	T

The pattern of reasoning of this example is called **modus tollens,** or the *law of contraposition,* or *indirect reasoning.*   ■

With reasoning similar to that used to name the fallacy of the converse, the fallacy

$$p \rightarrow q$$
$$\frac{\sim p}{\sim q}$$

Concluding ~*q* from ~*p* wrongly assumes ~*p* → ~*q*, the *inverse* of the given premise *p* → *q*.

is called the **fallacy of the inverse.** An example of such a fallacy is "If it rains, I get wet. It doesn't rain. Therefore, I don't get wet."

EXAMPLE 4   Using a Truth Table to Determine Validity

Determine whether the argument is *valid* or *invalid.*

> I'll buy a car or I'll take a vacation.
> I won't buy a car.
> ———————————————
> I'll take a vacation.

Solution

If *p* represents "I'll buy a car" and *q* represents "I'll take a vacation," the argument is symbolized as follows.

$$p \vee q$$
$$\frac{\sim p}{q}$$

We must set up a truth table for the statement $[(p \vee q) \wedge \sim p] \rightarrow q$.

p	*q*	*p* ∨ *q*	~*p*	(*p* ∨ *q*) ∧ ~*p*	[(*p* ∨ *q*) ∧ ~*p*] → *q*
T	T	T	F	F	T
T	F	T	F	F	T
F	T	T	T	T	T
F	F	F	T	F	T

The statement is a tautology and the argument is valid. Any argument of this form is valid by the law of **disjunctive syllogism.**   ■

EXAMPLE 5 Using a Truth Table to Determine Validity

Determine whether the argument is *valid* or *invalid*.

> If it squeaks, then I use WD-40.
> If I use WD-40, then I must go to the hardware store.
> If it squeaks, then I must go to the hardware store.

Solution

Let p represent "It squeaks," let q represent "I use WD-40," and let r represent "I must go to the hardware store." The argument takes on the following form.

$$p \to q$$
$$q \to r$$
$$\overline{p \to r}$$

Make a truth table for this statement, which requires eight rows.

$$[(p \to q) \land (q \to r)] \to (p \to r)$$

p	q	r	$p \to q$	$q \to r$	$p \to r$	$(p \to q) \land (q \to r)$	$[(p \to q) \land (q \to r)] \to (p \to r)$
T	T	T	T	T	T	T	T
T	T	F	T	F	F	F	T
T	F	T	F	T	T	F	T
T	F	F	F	T	F	F	T
F	T	T	T	T	T	T	T
F	T	F	T	F	T	F	T
F	F	T	T	T	T	T	T
F	F	F	T	T	T	T	T

This argument is valid because the final statement is a tautology. This pattern of argument is called **reasoning by transitivity,** or the *law of hypothetical syllogism*. ∎

Valid and Invalid Argument Forms

A summary of the valid forms of argument presented so far follows.

VALID ARGUMENT FORMS

Modus Ponens	Modus Tollens	Disjunctive Syllogism	Reasoning by Transitivity
$p \to q$	$p \to q$	$p \lor q$	$p \to q$
p	$\sim q$	$\sim p$	$q \to r$
q	$\sim p$	q	$p \to r$

The following is a summary of invalid forms (or fallacies).

INVALID ARGUMENT FORMS (FALLACIES)

Fallacy of the Converse	Fallacy of the Inverse
$p \to q$	$p \to q$
q	$\sim p$
p	$\sim q$

In a scene near the beginning of the 1974 film *Monty Python and the Holy Grail,* an amazing application of **poor logic** leads to the apparent demise of a supposed witch. Some peasants have forced a young woman to wear a nose made of wood. The convoluted argument they make is this: Witches and wood are both burned, and because witches are made of wood, and wood floats, and ducks also float, if she weighs the same as a duck, then she is made of wood and, therefore, is a witch!

Setting the Table Correctly If an argument has the form

$$p_1$$
$$p_2$$
$$\vdots$$
$$\underline{p_n,}$$
$$c$$

then **Step 3** in the testing process calls for the statement

$$(p_1 \wedge p_2 \wedge \ldots \wedge p_n) \to c.$$

EXAMPLE 6 Using a Truth Table to Determine Validity

Determine whether the argument is *valid* or *invalid*.

> If Eddie goes to town, then Mabel stays at home.
> If Mabel does not stay at home, then Rita will cook.
> Rita will not cook. Therefore, Eddie does not go to town.

Solution

In an argument written in this manner, the premises are given first, and the conclusion is the statement that follows the word "Therefore." Let *p* represent "Eddie goes to town," let *q* represent "Mabel stays at home," and let *r* represent "Rita will cook." Then the argument is symbolized as follows.

$$p \to q$$
$$\sim q \to r$$
$$\underline{\sim r}$$
$$\sim p$$

When an argument contains more than two premises, it is necessary to determine the truth values of the conjunction of *all* of them.

> *If at least one premise in a conjunction of several premises is false, then the entire conjunction is false.*

To test validity, set up a truth table for this statement.

$$[(p \to q) \wedge (\sim q \to r) \wedge \sim r] \to \sim p$$

p	q	r	$p \to q$	$\sim q$	$\sim q \to r$	$\sim r$	$(p \to q) \wedge (\sim q \to r) \wedge \sim r$	$\sim p$	$[(p \to q) \wedge (\sim q \to r) \wedge \sim r] \to \sim p$
T	T	T	T	F	T	F	F	F	T
T	T	F	T	F	T	T	T	F	F
T	F	T	F	T	T	F	F	F	T
T	F	F	F	T	F	T	F	F	T
F	T	T	T	F	T	F	F	T	T
F	T	F	T	F	T	T	T	T	T
F	F	T	T	T	T	F	F	T	T
F	F	F	T	T	F	T	F	T	T

Because the final column does not contain all Ts, the statement is not a tautology. The argument is invalid. ■

Arguments of Lewis Carroll

Consider the following verse, which has been around for many years.

> *For want of a nail, the shoe was lost.*
> *For want of a shoe, the horse was lost.*
> *For want of a horse, the rider was lost.*
> *For want of a rider, the battle was lost.*
> *For want of a battle, the war was lost.*
> *Therefore, for want of a nail, the war was lost.*

Each line of the verse may be written as an *if . . . then* statement. For example, the first line may be restated as "If a nail is lost, then the shoe is lost." The conclusion, "For want of a nail, the war was lost," follows from the premises, because repeated use of the law of transitivity applies. Arguments such as the one used by Lewis Carroll in the next example often take a similar form.

Alice's Adventures in Wonderland is the most famous work of **Charles Dodgson** (1832–1898), better known as **Lewis Carroll,** who was a mathematician and logician. He popularized recreational mathematics with this story and its sequel, *Through the Looking-Glass.* More than a century later, Raymond Smullyan continues this genre in his book *Alice in Puzzle-land* and many others.

EXAMPLE 7   Supplying a Conclusion to Ensure Validity

Supply a conclusion that yields a valid argument for the following premises.

> Babies are illogical.
>
> Nobody is despised who can manage a crocodile.
>
> Illogical persons are despised.

Solution

First, write each premise in the form *if . . . then. . . .*

> If you are a baby, then you are illogical.
>
> If you can manage a crocodile, then you are not despised.
>
> If you are illogical, then you are despised.

Let p represent "you are a baby," let q represent "you are logical," let r represent "you can manage a crocodile," and let s represent "you are despised." The statements can be written symbolically.

$$p \rightarrow \sim q$$
$$r \rightarrow \sim s$$
$$\sim q \rightarrow s$$

Begin with any letter that appears only once. Here p appears only once. Using the contrapositive of $r \rightarrow \sim s$, which is $s \rightarrow \sim r$, rearrange the statements as follows.

$$p \rightarrow \sim q$$
$$\sim q \rightarrow s$$
$$s \rightarrow \sim r$$

From the three statements, repeated use of reasoning by transitivity gives the conclusion

$$p \rightarrow \sim r, \text{ which leads to a valid argument.}$$

In words, the conclusion is "If you are a baby, then you cannot manage a crocodile," or, as Lewis Carroll would have written it, "Babies cannot manage crocodiles." ∎

WHEN Will I Ever USE This ?

Suppose you are a pediatric nurse administering flu vaccination for a 6-year-old patient. The flowchart on the next page is a dosing algorithm provided by the CDC for children 6 months to 8 years of age.

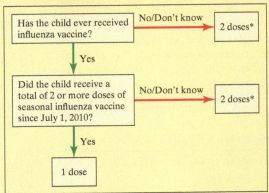

Influenza Dosing Algorithm

* Doses should be administered at least 4 weeks apart.
Source: cdc.gov

The patient was vaccinated for the first time last year, so you place the patient on a 2-dose regimen. Use a truth table to determine the validity of your action.

We let

p be "The child has received flu vaccine in the past,"

q be "The child has received 2 or more doses of flu vaccine since July 1, 2010,"

and r be "The child needs 2 doses this season."

Then the algorithm and your treatment decision can be expressed symbolically by the following argument.

$$(p \wedge q) \leftrightarrow \sim r$$
$$\underline{\sim q}$$
$$r$$

Make a truth table for the argument. Since there are three component statements, we require eight rows.

p	q	r	[((p	$\wedge$	q)	$\leftrightarrow$	$\sim r$)	$\wedge$	$\sim q$]	$\rightarrow$	r
T	T	T								T	T
T	T	F	T	T	T	T	T	F	F	T	F
T	F	T								T	T
T	F	F	T	F	F	F	T	F	T	T	F
F	T	T								T	T
F	T	F	F	F	T	F	T	F	F	T	F
F	F	T								T	T
F	F	F	F	F	F	F	T	F	T	T	F
			①	②	①	④	③	⑥	⑤	⑧	⑦

The truth table shows that your argument is a tautology, so your action was valid according to the CDC.

Note that in the odd rows of the table, we used the fact that the consequent *r* was true to conclude that the conditional was true, thus avoiding the effort required to "build" those rows piece by piece. In the even rows, we used the shortcut method introduced in **Example 10** of **Section 3.2**.

Each argument either is valid by one of the forms of valid arguments discussed in this section, or is a fallacy by one of the forms of invalid arguments discussed. (See the summary boxes.) Decide whether the argument is valid or a fallacy, and give the form that applies.

1. If Rascal Flatts comes to town, then I will go to the concert.

If I go to the concert, then I'll call in sick for work.

If Rascal Flatts comes to town, then I'll call in sick for work.

2. If you use binoculars, then you get a glimpse of the bald eagle.

If you get a glimpse of the bald eagle, then you'll be amazed.

If you use binoculars, then you'll be amazed.

3. If Marina works hard enough, she will get a promotion.

Marina works hard enough.

She will get a promotion.

4. If Isaiah's ankle heals on time, he'll play this season.

His ankle heals on time.

He'll play this season.

5. If he doesn't have to get up at 3:00 A.M., he's ecstatic.

He's ecstatic.

He doesn't have to get up at 3:00 A.M.

6. "A mathematician is a device for turning coffee into theorems." (quote from Paul Erdos)

You turn coffee into theorems.

You are a mathematician.

7. If Clayton pitches, the Dodgers win.

The Dodgers do not win.

Clayton does not pitch.

8. If Josh plays, the opponent gets shut out.

The opponent does not get shut out.

Josh does not play.

9. "If you're going through hell, keep going." (quote from Winston Churchill)

You're not going through hell.

Don't keep going.

10. "If you can't get rid of the skeleton in your closet, you'd best teach it to dance." (quote from George Bernard Shaw)

You can get rid of the skeleton in your closet.

You'd best not teach it to dance.

11. She uses e-commerce or she pays by credit card.

She does not pay by credit card.

She uses e-commerce.

12. Mia kicks or Drew passes.

Drew does not pass.

Mia kicks.

Use a truth table to determine whether the argument is valid or invalid.

13. $p \vee q$
p
$\sim q$

14. $p \wedge \sim q$
p
$\sim q$

15. $\sim p \rightarrow \sim q$
q
p

16. $p \vee \sim q$
p
$\sim q$

17. $p \rightarrow q$
$q \rightarrow p$
$p \wedge q$

18. $\sim p \rightarrow q$
p
$\sim q$

19. $p \rightarrow \sim q$
q
$\sim p$

20. $p \rightarrow \sim q$
$\sim p$
$\sim q$

21. $(p \rightarrow q) \wedge (q \rightarrow p)$
p
$p \vee q$

22. $(p \wedge q) \vee (p \vee q)$
q
p

23. $(\sim p \vee q) \wedge (\sim p \rightarrow q)$
p
$\sim q$

24. $(r \wedge p) \rightarrow (r \vee q)$
$q \wedge p$
$r \vee p$

25. $(\sim p \wedge r) \rightarrow (p \vee q)$
$\sim r \rightarrow p$
$q \rightarrow r$

26. $(p \rightarrow \sim q) \vee (q \rightarrow \sim r)$
$p \vee \sim r$
$r \rightarrow p$

27. Earlier we showed how to analyze arguments using Euler diagrams. Refer to **Example 5** in this section, restate each premise and the conclusion using a quantifier, and then draw an Euler diagram to illustrate the relationship.

28. Explain in a few sentences how to determine the statement for which a truth table will be constructed so that the arguments that follow in **Exercises 29–38** can be analyzed for validity.

Determine whether each argument is valid or invalid.

29. Joey loves to watch movies. If Terry likes to jog, then Joey does not love to watch movies. If Terry does not like to jog, then Carrie drives a school bus. Therefore, Carrie drives a school bus.

30. If Hurricane Gustave hit that grove of trees, then the trees are devastated. People plant trees when disasters strike and the trees are not devastated. Therefore, if people plant trees when disasters strike, then Hurricane Gustave did not hit that grove of trees.

31. If Yoda is my favorite *Star Wars* character, then I hate Darth Vader. I hate Luke Skywalker or Darth Vader. I don't hate Luke Skywalker. Therefore, Yoda is not my favorite character.

32. Carrie Underwood sings or Joe Jonas is not a teen idol. If Joe Jonas is not a teen idol, then Jennifer Hudson does not win a Grammy. Jennifer Hudson wins a Grammy. Therefore, Carrie Underwood does not sing.

33. The Cowboys will make the playoffs if and only if Troy comes back to play. Jerry doesn't coach the Cowboys or Troy comes back to play. Jerry does coach the Cowboys. Therefore, the Cowboys will not be in the playoffs.

34. If I've got you under my skin, then you are deep in the heart of me. If you are deep in the heart of me, then you are not really a part of me. You are deep in the heart of me or you are really a part of me. Therefore, if I've got you under my skin, then you are really a part of me.

35. If Dr. Hardy is a department chairman, then he lives in Atlanta. He lives in Atlanta and his first name is Larry. Therefore, if his first name is not Larry, then he is not a department chairman.

36. If I were your woman and you were my man, then I'd never stop loving you. I've stopped loving you. Therefore, I am not your woman or you are not my man.

37. All men are created equal. All people who are created equal are women. Therefore, all men are women.

38. All men are mortal. Socrates is a man. Therefore, Socrates is mortal.

39. A recent DirecTV commercial had the following script: "When the cable company keeps you on hold, you feel trapped. When you feel trapped, you need to feel free. When you need to feel free, you try hang-gliding. When you try hang-gliding, you crash into things. When you crash into things, the grid goes down. When the grid goes down, crime goes up, and when crime goes up, your dad gets punched over a can of soup . . ."

   (a) Use reasoning by transitivity and all the component statements to draw a valid conclusion.

   (b) If we added the line, "Your dad does not get punched over a can of soup," what valid conclusion could be drawn?

40. Molly made the following observation: "If I want to determine whether an argument leading to the statement

$$[(p \rightarrow q) \wedge \sim q] \rightarrow \sim p$$

is valid, I only need to consider the lines of the truth table that lead to T for the column that is headed $(p \rightarrow q) \wedge \sim q$." Molly was very perceptive. Can you explain why her observation was correct?

In the arguments used by Lewis Carroll, it is helpful to restate a premise in if . . . then *form in order to more easily identify a valid conclusion. The following premises come from Lewis Carroll. Write each premise in* if . . . then *form.*

41. All my poultry are ducks.

42. None of your sons can do logic.

43. Guinea pigs are hopelessly ignorant of music.

44. No teetotalers are pawnbrokers.

45. No teachable kitten has green eyes.

46. Opium-eaters have no self-command.

47. I have not filed any of them that I can read.

48. All of them written on blue paper are filed.

Exercises 49–54 involve premises from Lewis Carroll. Write each premise in symbols, and then, in the final part, give a conclusion that yields a valid argument.

49. Let *p* be "it is a duck," *q* be "it is my poultry," *r* be "one is an officer," and *s* be "one is willing to waltz."

   (a) No ducks are willing to waltz.

   (b) No officers ever decline to waltz.

   (c) All my poultry are ducks.

   (d) Give a conclusion that yields a valid argument.

50. Let *p* be "one is able to do logic," *q* be "one is fit to serve on a jury," *r* be "one is sane," and *s* be "he is your son."

   (a) Everyone who is sane can do logic.

   (b) No lunatics are fit to serve on a jury.

   (c) None of your sons can do logic.

   (d) Give a conclusion that yields a valid argument.

51. Let *p* be "one is honest," *q* be "one is a pawnbroker," *r* be "one is a promise-breaker," *s* be "one is trustworthy," *t* be "one is very communicative," and *u* be "one is a wine-drinker."

   (a) Promise-breakers are untrustworthy.

   (b) Wine-drinkers are very communicative.

   (c) A person who keeps a promise is honest.

   (d) No teetotalers are pawnbrokers. (*Hint:* Assume "teetotaler" is the opposite of "wine-drinker.")

   (e) One can always trust a very communicative person.

   (f) Give a conclusion that yields a valid argument.

52. Let *p* be "it is a guinea pig," *q* be "it is hopelessly ignorant of music," *r* be "it keeps silent while the *Moonlight Sonata* is being played," and *s* be "it appreciates Beethoven."

   (a) Nobody who really appreciates Beethoven fails to keep silent while the *Moonlight Sonata* is being played.

   (b) Guinea pigs are hopelessly ignorant of music.

   (c) No one who is hopelessly ignorant of music ever keeps silent while the *Moonlight Sonata* is being played.

   (d) Give a conclusion that yields a valid argument.

53. Let *p* be "it begins with 'Dear Sir'," *q* be "it is crossed," *r* be "it is dated," *s* be "it is filed," *t* be "it is in black ink,"

u be "it is in the third person," *v* be "I can read it," *w* be "it is on blue paper," *x* be "it is on one sheet," and *y* be "it is written by Brown."

   (a) All the dated letters are written on blue paper.

   (b) None of them are in black ink, except those that are written in the third person.

   (c) I have not filed any of them that I can read.

   (d) None of them that are written on one sheet are undated.

   (e) All of them that are not crossed are in black ink.

   (f) All of them written by Brown begin with "Dear Sir."

   (g) All of them written on blue paper are filed.

   (h) None of them written on more than one sheet are crossed.

   (i) None of them that begin with "Dear Sir" are written in the third person.

   (j) Give a conclusion that yields a valid argument.

54. Let *p* be "he is going to a party," *q* be "he brushes his hair," *r* be "he has self-command," *s* be "he looks fascinating," *t* be "he is an opium-eater," *u* be "he is tidy," and *v* be "he wears white kid gloves."

   (a) No one who is going to a party ever fails to brush his hair.

   (b) No one looks fascinating if he is untidy.

   (c) Opium-eaters have no self-command.

   (d) Everyone who has brushed his hair looks fascinating.

   (e) No one wears white kid gloves unless he is going to a party. (*Hint:* "*a* unless *b*" ≡ $\sim b \rightarrow a$.)

   (f) A man is always untidy if he has no self-command.

   (g) Give a conclusion that yields a valid argument.

CHAPTER 3   SUMMARY

KEY TERMS

3.1
symbolic logic
truth value
statement
compound
   statement
component
   statements
connectives
negation
quantifiers

3.2
conjunction
truth table
disjunction
equivalent statements

3.3
conditional statement
antecedent
consequent
tautology

3.4
converse
inverse
contrapositive
biconditional

3.5
argument
premises
conclusion
valid

fallacy
Euler diagram

3.6
modus ponens
modus tollens
disjunctive syllogism
fallacy of the converse
fallacy of the inverse
reasoning by
   transitivity

NEW SYMBOLS

∨ disjunction
∧ conjunction
~ negation

→ implication
↔ biconditional
≡ equivalence

TEST YOUR WORD POWER

See how well you have learned the vocabulary in this chapter.

1. A **statement** is
   A. a sentence that asks a question, the answer to which may be true or false.
   B. a directive giving specific instructions.
   C. a sentence declaring something that is either true or false, but not both at the same time.
   D. a paradoxical sentence with no truth value.

2. A **disjunction** (inclusive) is
   A. a compound statement that is true only if both of its component statements are true.
   B. a compound statement that is true if one or both of its component statements is/are true.
   C. a compound statement that is false if either of its component statements is false.
   D. a compound statement that is true if exactly one of its component statements is true.

3. A **conditional** statement is
   A. a statement that may be true or false, depending on some condition.
   B. an idea that can be stated only under certain conditions.
   C. a statement using the connective *if . . . then*.
   D. a statement the antecedent of which is implied by the consequent.

4. The **inverse** of a conditional statement is
   A. the result when the antecedent and consequent are negated.
   B. the result when the antecedent and consequent are interchanged.
   C. the result when the antecedent and consequent are interchanged and negated.
   D. logically equivalent to the conditional.

5. A **fallacy** is
   A. an argument with a false conclusion.
   B. an argument whose conclusion is not supported by the premises.
   C. a valid argument.
   D. an argument containing at least one false premise.

6. **Fallacy of the converse** is
   A. the reason the converse of a conditional is not equivalent to the conditional.
   B. an invalid argument form that assumes the converse of a premise.
   C. an invalid argument form that denies the converse of a premise.
   D. an invalid argument form that assumes the inverse of a premise.

ANSWERS
1. C 2. B 3. C 4. A 5. B 6. B

QUICK REVIEW

Concepts	Examples
3.1 Statements and Quantifiers A **statement** is a declarative sentence that is either true or false (not both simultaneously). A **compound statement** is made up of two or more **component statements** joined by **connectives** (*not, and, or, if . . . then*). **Quantifiers** indicate how many members in a group being considered exhibit a particular property or characteristic. Universal quantifiers indicate *all* members, and existential quantifiers indicate *at least one* member. The **negation** of a statement has the opposite truth value of that statement in all cases.	Consider the following statement. "If it rains this month, then we'll have a green spring." It is a compound statement made up of the following two component statements joined by the connective *if . . . then*. "It rains this month" and "We'll have a green spring," The statement "All five of those birds can fly" contains a universal quantifier. Its negation is "At least one of those five birds cannot fly," which contains an existential quantifier.

Concepts   *Examples*

3.2 Truth Tables and Equivalent Statements

Given two component statements p and q, their **conjunction** is symbolized $p \wedge q$ and is true only when both component statements are true.

p	q	$p \wedge q$	$p \vee q$
T	T	T	T
T	F	F	T
F	T	F	T
F	F	F	F

Their **disjunction**, symbolized $p \vee q$, is false only when both component statements are false.

If p represents "$7 < 10$" and q represents "$4 < 3$," then the second row of the truth table above shows that $p \wedge q$ is false and $p \vee q$ is true.

The truth value of a compound statement is found by substituting T or F for each component statement, and then working from inside parentheses out, determining truth values for larger parts of the overall statement, until the entire statement has been evaluated.

    When this process is carried out for all possible combinations of truth values for the component statements, a **truth table** results.

The truth table for the statement $\sim p \vee (q \wedge p)$ is shown below, with circled numbers indicating the order in which columns were determined.

p	q	$\sim p$	$\vee$	$(q$	$\wedge$	$p)$
T	T	F	T	T	T	T
T	F	F	F	F	F	T
F	T	T	T	T	F	F
F	F	T	T	F	F	F
		①	③	①	②	①

Equivalent statements have the same truth value for all combinations of truth values for the component statements. To determine whether two statements are equivalent, construct truth tables for both and see if the final truth values agree in all rows.

The statements $\sim(p \wedge q)$ and $\sim p \vee \sim q$ are equivalent, as shown in the table.

p	q	$\sim$	$(p$	$\wedge$	$q)$	$\sim p$	$\vee$	$\sim q$
T	T	F	T	T	T	F	F	F
T	F	T	T	F	F	F	T	T
F	T	T	F	F	T	T	T	F
F	F	T	F	F	F	T	T	T

De Morgan's laws can be used to quickly find negations of disjunctions and conjunctions.

$$\sim(p \vee q) \equiv \sim p \wedge \sim q$$
$$\sim(p \wedge q) \equiv \sim p \vee \sim q$$

To find the negation of the statement "I love chess and I had breakfast," let p represent "I love chess" and let q represent "I had breakfast." Then the above statement becomes $p \wedge q$. Its negation $\sim(p \wedge q) \equiv \sim p \vee \sim q$ translates to

"I don't love chess or I didn't have breakfast."

3.3 The Conditional and Circuits

A **conditional statement** uses the *if . . . then* connective and is symbolized $p \rightarrow q$, where p is the **antecedent** and q is the **consequent.**

If p represents "You are mighty" and q represents "I am flighty," then the conditional statement $p \rightarrow q$ is expressed as

"*If* you are mighty, *then* I am flighty."

The conditional is false if the antecedent is true and if the consequent is false. Otherwise, the conditional is true. This is because q is only *required* to be true on the *condition* that p is true, but q may "voluntarily" be true even if p is false. That is, p is sufficient for q, but not necessary.

The statement $(6 < 1) \rightarrow (3 = 7)$ is true because the antecedent is false.

The statement "If you are reading this book, then it is the year 1937" is false, because the antecedent is true and the consequent is false.

Concepts	Examples

Concepts

The conditional $p \rightarrow q$ is equivalent to the disjunction $\sim p \lor q$, and its negation is $p \land \sim q$, as shown by a comparison of their truth tables.

Examples

p	q	$p \rightarrow q$	$\sim p \lor q$	$p \land \sim q$
T	T	T T T	F T T	T F F
T	F	T F F	F F F	T T T
F	T	F T T	T T T	F F F
F	F	F T F	T T F	F F T

The statement "All mice love cheese" can be stated, "*If* it's a mouse, *then* it loves cheese." This is equivalent to saying, "It's not a mouse or it loves cheese." The negation of this statement is "It's a mouse and it does not love cheese."

Electrical circuits are analogous to logical statements, with *parallel* circuits corresponding to disjunctions, and *series* circuits corresponding to conjunctions. Each switch (modeled by an arrow) represents a component statement. When a switch is closed, it allows current to pass through. The circuit represents a true statement when current flows from one end of the circuit to the other.

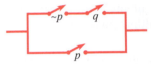

This corresponds to the logical statement $p \lor (\sim p \land q)$. Current will flow from one end to the other if p is true *or* if p is false and q is true. This statement is equivalent to

$$(p \lor \sim p) \land (p \lor q), \quad \text{which simplifies to} \quad p \lor q.$$

3.4 The Conditional and Related Statements

Given a conditional statement $p \rightarrow q$, its **converse, inverse,** and **contrapositive** are defined as follows.

$$\text{Converse: } q \rightarrow p$$
$$\text{Inverse: } \sim p \rightarrow \sim q$$
$$\text{Contrapositive: } \sim q \rightarrow \sim p$$

The conditional $p \rightarrow q$ can be translated in many ways:

If p, then q.	p is sufficient for q.
If p, q.	q is necessary for p.
p implies q.	All p are q.
p only if q.	q if p.

For the **biconditional** statement "p if and only if q,"

$$p \leftrightarrow q \equiv (p \rightarrow q) \land (q \rightarrow p).$$

It is true only when p and q have the same truth value.

Consider the statement "If it's a pie, it tastes good."

Converse: "If it tastes good, it's a pie.

Inverse: "If it's not a pie, it doesn't taste good."

Contrapositive: "If it doesn't taste good, it's not a pie."

Statement	If . . . then form
You'll be sorry if I go.	If I go, then you'll be sorry.
Today is Tuesday only if yesterday was Monday.	If today is Tuesday, then yesterday was Monday.
All nurses wear comfortable shoes.	If you are a nurse, then you wear comfortable shoes.

The statement "$5 < 9$ if and only if $3 > 7$" is false because the component statements have opposite truth values. The first is true, while the second is false.

Summary of Basic Truth Tables

1. $\sim p$, the **negation** of p, has truth value opposite that of p.
2. $p \land q$, the **conjunction,** is true only when both p and q are true.
3. $p \lor q$, the **disjunction,** is false only when both p and q are false.
4. $p \rightarrow q$, the **conditional,** is false only when p is true and q is false.
5. $p \leftrightarrow q$, the **biconditional,** is true only when both p and q have the same truth value.

p	$\sim p$
T	F
F	T

p	q	$p \land q$	$p \lor q$	$p \rightarrow q$	$p \leftrightarrow q$
T	T	T T T	T T T	T T T	T T T
T	F	T F F	T T F	T F F	T F F
F	T	F F T	F T T	F T T	F F T
F	F	F F F	F F F	F T F	F T F

Concepts	Examples

3.5 **Analyzing Arguments with Euler Diagrams**

A logical **argument** consists of premises and a conclusion. An argument is considered **valid** if the truth of the premises forces the conclusion to be true. Otherwise, it is **invalid.**

Euler diagrams can be used to determine whether an argument is valid or invalid.

To draw a Euler diagram, follow these steps:

1. Use the first premise to draw regions. (Arguments with multiple premises may involve multiple regions.)

2. Place an x in the diagram to represent the subject of the argument.

Consider this argument. Notice the universal quantifier "all."

All dogs are animals.
Dotty is a dog.
Dotty is an animal.

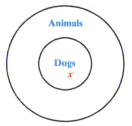

x represents Dotty.

We see from the Euler diagram that the truth of the premises forces the conclusion, that Dotty is an animal, to be true. Thus the argument is valid.

Consider the following argument. Notice the existential quantifier "some."

Some animals are warmblooded.
Albie is an animal.
Albie is warmblooded.

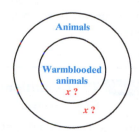

x represents Albie.

We see from the Euler diagram that the location of x is uncertain, so the truth of the premises does not force the conclusion to be true. Thus, the argument is invalid.

3.6 **Analyzing Arguments with Truth Tables**

An argument with premises $p_1, p_2, \ldots, p_n$ and conclusion c can be tested for validity by constructing a truth table for the statement

$$(p_1 \wedge p_2 \wedge \ldots \wedge p_n) \rightarrow c.$$

If all rows yield T for this statement (that is, if it is a tautology), then the argument is valid. Otherwise, the argument is invalid.

Consider this argument. "If I eat ice cream, I regret it later. I didn't regret it later. Therefore, I didn't eat ice cream."

Let p represent "I eat ice cream" and q represent "I regret it later." Then the argument can be expressed as the compound statement

$$[(p \rightarrow q) \wedge \sim q] \rightarrow \sim p.$$

Test its validity using a truth table, which shows that the statement is a tautology. The argument is valid (by **modus tollens**).

Valid Argument Forms

Modus Ponens	Modus Tollens	Disjunctive Syllogism	Reasoning by Transitivity
$p \rightarrow q$	$p \rightarrow q$	$p \vee q$	$p \rightarrow q$
p	$\sim q$	$\sim p$	$q \rightarrow r$
q	$\sim p$	q	$p \rightarrow r$

p	q	$[(p$	$\rightarrow$	$q)$	$\wedge$	$\sim q]$	$\rightarrow$	$\sim p$
T	T				F	F	T	F
T	F	T	F	F	F	T	T	F
F	T						T	T
F	F						T	T

Concepts	Examples

Invalid Argument Forms

Fallacy of the Converse	Fallacy of the Inverse
$p \rightarrow q$	$p \rightarrow q$
q	$\sim p$
p	$\sim q$

Consider this argument.

> If I drink coffee, I get jittery.
> I didn't drink coffee.
> _____
> I don't get jittery.

Let p represent "I drink coffee" and q represent "I get jittery."

Test the argument, using a truth table for the statement

$$[(p \rightarrow q) \wedge \sim p] \rightarrow \sim q.$$

p	q	$[(p$	$\rightarrow$	$q)$	$\wedge$	$\sim p]$	$\rightarrow$	$\sim q$
T	T		T		F	F	**T**	F
T	F						**T**	T
F	T		T		T	T	**F**	F
F	F						**T**	T

This argument is invalid by **fallacy of the inverse.**

CHAPTER 3   TEST

Write a negation for each statement.

1. $6 - 3 = 3$

2. All men are created equal.

3. Some members of the class went on the field trip.

4. If I fall in love, it will be forever.

5. She applied and did not get a student loan.

Let p represent "You will love me" and let q represent "I will love you." Write each statement in symbols.

6. If you won't love me, then I will love you.

7. I will love you if you will love me.

8. I won't love you if and only if you won't love me.

Using the same statements as for Exercises 6–8, write each of the following in words.

9. $\sim p \wedge q$ **10.** $\sim (p \vee \sim q)$

In each of the following, assume that p is true and that q and r are false. Find the truth value of each statement.

11. $\sim q \wedge \sim r$ **12.** $r \vee (p \wedge \sim q)$

13. $r \rightarrow (s \vee r)$ (The truth value of the statement s is unknown.)

14. $p \leftrightarrow (p \rightarrow q)$

15. Explain in your own words why, if p is a statement, the biconditional $p \leftrightarrow \sim p$ must be false.

16. State the necessary conditions for each of the following.

(a) a conditional statement to be false

(b) a conjunction to be true

(c) a disjunction to be false

(d) a biconditional to be true

Construct a truth table for each of the following.

17. $p \wedge (\sim p \vee q)$

18. $\sim (p \wedge q) \rightarrow (\sim p \vee \sim q)$

Decide whether each statement is true *or* false.

19. Some negative integers are whole numbers.

20. All irrational numbers are real numbers.

Write each conditional statement in if . . . then form.

21. All integers are rational numbers.

22. Being a rhombus is sufficient for a polygon to be a quadrilateral.

23. Being divisible by 2 is necessary for a number to be divisible by 4.

24. She digs dinosaur bones only if she is a paleontologist.

For each statement in Exercises 25 and 26, write (**a**) *the converse,* (**b**) *the inverse, and* (**c**) *the contrapositive.*

25. If a picture paints a thousand words, the graph will help me understand it.

26. $\sim p \rightarrow (q \wedge r)$ (Use one of De Morgan's laws as necessary.)

27. Use an Euler diagram to determine whether the argument is *valid* or *invalid*.

All members of that athletic club save money.

Don is a member of that athletic club.

Don saves money.

28. Match each argument in parts (a)–(d) in the next column with the law that justifies its validity, or the fallacy of which it is an example, in choices A–F.

   A. Modus ponens

   B. Modus tollens

   C. Reasoning by transitivity

   D. Disjunctive syllogism

   E. Fallacy of the converse

   F. Fallacy of the inverse

(**a**) If he eats liver, then he'll eat anything.

   He eats liver.

   He'll eat anything.

(**b**) If you use your seat belt, you will be safer.

   You don't use your seat belt.

   You won't be safer.

(**c**) If I hear *Mr. Bojangles,* I think of her.

   If I think of her, I smile.

   If I hear *Mr. Bojangles,* I smile.

(**d**) She sings or she dances.

   She does not sing.

   She dances.

Use a truth table to determine whether each argument is valid or invalid.

29. If I write a check, it will bounce. If the bank guarantees it, then it does not bounce. The bank guarantees it. Therefore, I don't write a check.

30. $\sim p \rightarrow \sim q$

   $q \rightarrow p$

   $p \vee q$

Numeration Systems

4

Network administrators oversee the operation and maintenance of computer networks for businesses and organizations. In order to ensure that computers can communicate with other hosts on the network and the Internet, each computer must be assigned an IP address. Often, for reasons of security or efficiency, an administrator may divide the network into subnetworks. Tasks like these require an understanding of some of the numeration systems covered in this chapter.

<table>
<tr><td>**4.1**</td><td>**HISTORICAL NUMERATION SYSTEMS**</td></tr>
</table>

OBJECTIVES

1　Understand the basics of numeration.

2　Understand ancient Egyptian numeration.

3　Understand ancient Roman numeration.

4　Understand classical Chinese numeration.

Basics of Numeration

The various ways of symbolizing and working with the counting numbers are called **numeration systems.** The symbols representing the numbers are called **numerals.**

Numeration systems have developed over many millennia of human history. Ancient documents provide insight into methods used by the early Sumerian peoples, the Egyptians, the Babylonians, the Greeks, the Romans, the Chinese, the Hindus, and the Mayan people, as well as others.

Keeping accounts by matching may have developed as humans established permanent settlements and began to grow crops and raise livestock. People might have kept track of the number of sheep in a flock by matching pebbles with the sheep, for example. The pebbles could then be kept as a record of the number of sheep.

A more efficient method is to keep a **tally stick.** With a tally stick, one notch or **tally** is made on a stick for each sheep. Tally marks provide a crude and inefficient numeration system. For example, the numeral for the number thirteen might be

$$||\,|\,|\,|\,|\,|\,|\,|\,|\,|\,|\,|\,, \leftarrow \text{13 tally marks}$$

which requires the recording of 13 symbols, and later interpretation requires careful counting of symbols.

Even today, tally marks are used, especially when keeping track of things that occur one or a few at a time, over space or time. To facilitate the counting of the tally, we often use a sort of "grouping" technique as we go.

$$\cancel{||||}\ \cancel{||||}\ |\,|\,| \leftarrow \text{Numeral (tally) for 13}$$

A long evolution of numeration systems throughout recorded history would take us from tally marks to our own modern system, the **Hindu-Arabic system,** which utilizes the set of symbols

$$\{1, 2, 3, 4, 5, 6, 7, 8, 9, 0\}.$$

Ancient Egyptian Numeration

An essential feature common to all more advanced numeration systems is **grouping,** which allows for less repetition of symbols, making numerals easier to interpret. Most historical systems, including our own, have used groups of ten, reflecting the common practice of learning to count by using the fingers. The size of the groupings (again, usually ten) is called the **base** of the number system.

The ancient Egyptian system is an example of a **simple grouping system.** It utilized ten as its base, and its various symbols are shown in **Table 1** on the next page. The symbol for 1 (|) is repeated, in a tally scheme, for 2, 3, and so on up to 9. A new symbol is introduced for 10 (∩), and that symbol is repeated for 20, 30, and so on, up to 90. This pattern enabled the Egyptians to express numbers up to 9,999,999 with just the seven symbols shown in the table.

The numbers denoted by the seven Egyptian symbols are all *powers* of the base ten.

$$10^0 = 1, \quad 10^1 = 10, \quad 10^2 = 100, \quad 10^3 = 1000, \quad 10^4 = 10,000,$$

$$10^5 = 100,000, \quad 10^6 = 1,000,000$$

These expressions, called *exponential expressions,* were first defined in **Section 1.1.** In the expression 10^4, for example, 10 is the *base* and 4 is the *exponent.* Recall that the exponent indicates the number of repeated factors of the base to be multiplied.

Tally sticks like this one were used by the English in about 1400 A.D. to keep track of financial transactions. Each notch stands for one pound sterling.

Much of our knowledge of **Egyptian mathematics** comes from the **Rhind papyrus,** from about 3800 years ago. A small portion of this papyrus, showing methods for finding the area of a triangle, is reproduced here.

Table 1 Early Egyptian Symbols

Number	Symbol	Description
1	I	Stroke
10	∩	Heel bone
100	9	Scroll
1000	⌇	Lotus flower
10,000	⌐	Pointing finger
100,000	⌒	Burbot fish
1,000,000	⚊	Astonished person

EXAMPLE 1 **Interpreting an Egyptian Numeral**

Write the number below in Hindu-Arabic form.

$$\text{⌒⌒} \text{⌇⌇⌇⌇} \text{999} \text{∩∩∩∩III} \\ \text{∩∩∩IIII}$$

Solution

Refer to **Table 1** for the values of the Egyptian symbols. Each ⌒ represents 100,000. Therefore, two ⌒s represent $2 \cdot 100{,}000$, or 200,000. Proceed as shown.

two	⌒	$2 \cdot 100{,}000 =$	200,000
four	⌇	$4 \cdot 1000 =$	4000
three	9	$3 \cdot 100 =$	300
nine	∩	$9 \cdot 10 =$	90
seven	I	$7 \cdot 1 =$	7

$$\overline{204{,}397} \leftarrow \text{Answer} \quad ■$$

EXAMPLE 2 **Creating an Egyptian Numeral**

Write 376,248 in Egyptian form.

Solution

$$3 \quad 7 \quad 6, \quad 2 \quad 4 \quad 8$$
$$\downarrow \quad \downarrow \quad \downarrow \quad \downarrow \quad \downarrow \quad \downarrow$$

⌒⌒⌒ ⌐⌐⌐⌐ ⌇⌇⌇ 99 ∩∩IIII
⌒ ⌐⌐⌐⌐ ⌇⌇⌇ ∩∩IIII   Refer to Table 1 as needed.   ■

An Egyptian tomb painting shows scribes tallying the count of a grain harvest. **Egyptian mathematics** was oriented more to practicality than was Greek or Babylonian mathematics, although the Egyptians did have a formula for finding the volume of a certain portion of a pyramid.

The position or order of the symbols makes no difference in a simple grouping system. Each of the numerals 99∩∩∩∩IIII, IIII∩∩∩99, and II∩∩99∩II would be interpreted as 234. In **Examples 1 and 2,** like symbols are grouped together, and groups of greater-valued symbols are positioned to the left.

A simple grouping system is well suited to addition and subtraction.

⌇⌇ 99 ∩∩∩ II   We use a + sign for convenience and draw a line
                  under the numbers being added, although the Egyp-
+ ⌇ 999 ∩ IIIIII tians did not do this.

Sum: ⌇⌇⌇ 999 ∩∩ IIII   Two Is plus six Is is equal to eight Is, and so on.
     ⌇⌇⌇ 99 ∩∩ IIII

Archaeological investigation has provided much of what we know about the numeration systems of ancient peoples.

Regrouping, or "carrying," is needed when more than nine of the same symbol result.

Subtraction is done in much the same way, as shown in the next example.

EXAMPLE 3 Subtracting Egyptian Numerals

Work each subtraction problem.

(a)

(b)

Solution

(a)

As with addition, work from right to left and subtract.

(b) To subtract four Is from two Is, "borrow" one heel bone, which is equivalent to ten Is. Finish the problem after writing ten additional Is on the right.

one ∩ = ten Is

A procedure such as those described above is called an **algorithm:** a rule or method for working a problem. The Egyptians used an interesting algorithm for multiplication that requires only an ability to add and to double numbers, as shown in **Example 4.** For convenience, this example uses our symbols rather than theirs.

EXAMPLE 4 Using the Egyptian Multiplication Algorithm

A rectangular room in an archaeological excavation measures 19 cubits by 70 cubits. (A cubit, based on the length of the forearm, from the elbow to the tip of the middle finger, was approximately 18 inches.) Find the area of the room.

Solution

Multiply the width and length to find the area of a rectangle. Build two columns of numbers as shown at the top of the next page. Start the first column with 1, the second with 70. Each column is built downward by doubling the number above. Keep going until the first column contains numbers that can be added to equal 19. Then add the corresponding numbers from the second column.

$$\begin{array}{ll} & \rightarrow \quad 1 \qquad 70 \ \leftarrow \\ & \rightarrow \quad 2 \qquad 140 \ \leftarrow \\ 1 + 2 + 16 = 19 & \quad 4 \qquad 280 \qquad 70 + 140 + 1120 = 1330 \\ & \quad 8 \qquad 560 \\ & \rightarrow \quad 16 \quad 1120 \ \leftarrow \end{array}$$

Thus $19 \cdot 70 = 1330$, and the area of the given room is 1330 square cubits. ■

Ancient Roman Numeration

Roman numerals are still used today, mainly for decorative purposes, on clock faces, for heading numbers in outlines, chapter numbers in books, copyright dates of movies, and so on. The base is again 10, with distinct symbols for 1, 10, 100, and 1000. The Romans, however, deviated from pure simple grouping in several ways. For the symbols and some examples, see **Tables 2 and 3,** respectively.

Table 2 Roman Symbols

Number	Symbol
1	I
5	V
10	X
50	L
100	C
500	D
1000	M

Table 3

Selected Roman Numerals

Number	Numeral
6	VI
12	XII
19	XIX
30	XXX
49	XLIX
85	LXXXV
25,040	$\overline{\text{XXV}}$ XL
35,000	$\overline{\text{XXXV}}$
5,105,004	$\overline{\overline{\text{V}}}\,\overline{\text{CV}}$ IV
7,000,000	$\overline{\overline{\text{VII}}}$

SPECIAL FEATURES OF THE ROMAN SYSTEM

1. In addition to symbols for 1, 10, 100, and 1000, "extra" symbols denote 5, 50, and 500. This allows less symbol repetition within a numeral. It is like a secondary base-five grouping functioning within the base-ten simple grouping.

2. A *subtractive feature* was introduced, whereby a smaller-valued symbol, placed immediately to the left of one of larger value, meant to subtract. Thus IV = 4, while VI = 6. Only certain combinations were used in this way:
   (a) I preceded only V or X.
   (b) X preceded only L or C.
   (c) C preceded only D or M.

3. A *multiplicative feature,* rather than more symbols, allowed for larger numbers:
   (a) A bar over a numeral meant to multiply by 1000.
   (b) A double bar meant to multiply by 1000^2—that is, by 1,000,000.

Adding and subtracting with Roman numerals is very similar to the Egyptian method, except that the subtractive feature of the Roman system sometimes makes the processes more involved. With Roman numerals we cannot add IV and VII to get the sum VVIII by simply combining like symbols. (Even XIII would be incorrect.) The safest method is to rewrite IV as IIII, then add IIII and VII, getting VIIIIII. We convert this to VVI, and then to XI by regrouping. Subtraction, which is similar, is shown in the following example.

EXAMPLE 5 Subtracting Roman Numerals

Janus, a Roman official, has 26 servants. If, on a given Saturday, he has excused 14 of them to attend a Lucky Lyres concert at the Forum, how many are still at home to serve the banquet?

Solution

To find the answer, we subtract XIV from XXVI. Set up the problem in terms of simple grouping numerals (that is, XIV is rewritten as XIIII).

$$
\begin{array}{l}
\text{Problem:} \quad \begin{array}{r} \text{XXVI} \\ -\ \text{XIV} \\ \hline \end{array} \quad
\begin{array}{l}\text{Problem restated without}\\ \text{subtractive notation:}\end{array} \quad
\begin{array}{r} \text{XXVI} \\ -\ \text{XIIII} \\ \hline \end{array}
\end{array}
$$

$$
\text{Regrouped:} \quad \begin{array}{r} \text{XXIIIIII} \\ -\ \text{XIIII} \\ \hline \text{XII} \end{array} \leftarrow \text{Answer}
$$

Since four **I**s cannot be subtracted from one **I,** we have "borrowed" in the top numeral, writing XXVI as XXIIIIII. The subtraction can then be carried out. Janus has 12 servants home for the banquet. ■

Computation, in early forms, was often aided by mechanical devices just as it is today. The Roman merchants, in particular, did their figuring on a counting board, or **counter,** on which lines or grooves represented 1s, 10s, 100s, etc., and on which the spaces between the lines represented 5s, 50s, 500s, and so on. Discs or beads (called *calculi,* the word for "pebbles") were positioned on the board to denote numbers, and *calculations* were carried out by moving the discs around and simplifying.

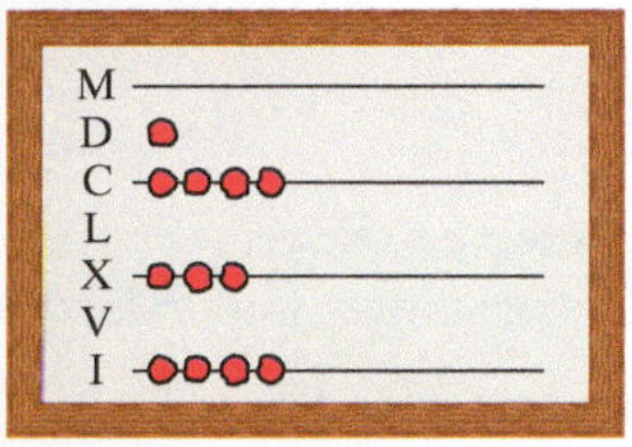

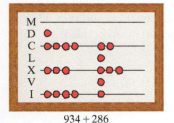

934

Figure 1

EXAMPLE 6 Adding on a Roman Counting Board

A Roman merchant wants to calculate the sum $934 + 286$. Use counting boards to carry out the following steps.

(a) Represent the first number, 934.

(b) Represent the second number, 286, beside the first.

(c) Represent the sum, in simplified form.

Solution

(a) See **Figure 1**.   **(b)** See **Figure 2**.

(c) See **Figure 3**. The simplified answer is MCCXX, or 1220. In the process of simplification, five discs on the bottom line were replaced by a single disc in the V space. This made two Vs that were replaced by an additional disc on the X line. Five of those on the X line were then replaced by one in the L space, and this process continued until the disc on the M line finally appeared. ■

$934 + 286$

Figure 2

Classical Chinese Numeration

The preceding examples show that simple grouping, although an improvement over tallying, still requires considerable repetition of symbols. To denote 90, for example, the ancient Egyptian system must utilize nine ∩s: ∩∩∩∩∩/∩∩∩∩. If an additional symbol (a "multiplier") was introduced to represent nine, say "9," then 90 could be denoted 9 ∩. All possible numbers of repetitions of powers of the base could be handled by introducing a separate multiplier symbol for each counting number less than the base.

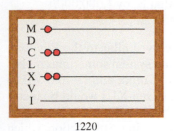

1220

Figure 3

Table 4
Chinese Symbols

Number	Symbol
1	一
2	二
3	三
4	四
5	五
6	六
7	七
8	八
9	九
10	十
100	百
1000	千
0	零

Just such a system was developed long ago in China. We show the predominant Chinese version, which used the symbols shown in **Table 4**. We call this type of system a **multiplicative grouping system.** In general, a numeral in such a system would contain pairs of symbols, each pair containing a multiplier (with some counting-number value less than the base) and then a power of the base. The Chinese numerals are read from top to bottom rather than from left to right.

If the Chinese system were *pure* multiplicative grouping, the number 2018 would be denoted as shown in **Figure 4**. But three special features of the system show that they had started to move beyond multiplicative grouping toward something more efficient.

二 2
•
千 1000
+
零 0
•
百 100
+
一 1
•
十 10
+
八 8
•
一 1

2018 in pure multiplicative grouping

Figure 4

SPECIAL FEATURES OF THE CHINESE SYSTEM

1. A single symbol, rather than a pair, denotes the number of 1s. The multiplier (1, 2, 3, 4, . . . , or 9) is written, but the power of the base (10^0) is omitted. See **Figure 5** (and also **Examples 7(a), (b), and (c)**).
2. In the 10s pair, if the multiplier is 1 it is omitted. See **Figure 6** (and **Example 8(a)**).
3. When a particular power of the base is totally missing, the omission is denoted with the zero symbol. See **Figure 7** (and **Examples 7(b) and 8(b)**). If two or more consecutive powers are missing, just one zero symbol denotes the total omission. (See **Example 7(c)**.) The omission of 1s and 10s and any other powers occurring at the extreme bottom of a numeral need not be denoted at all. (See **Example 7(d)**.)

二 2
•
千 1000
+
零 0
•
百 100
+
一 1
•
十 10
+
八 8

2018 with feature 1

Figure 5

二 2
•
千 1000
+
零 0
•
百 100
+
十 10
+
八 8

2018 with features 1 and 2

Figure 6

二 2
•
千 1000
+
零 0
+
十 10
+
八 8

2018 with features 1, 2, and 3

Figure 7

Note that, for clarification in the following examples, we have emphasized the grouping into pairs by spacing and by colored braces. These features were *not* part of the actual numerals in practice.

EXAMPLE 7 Interpreting Chinese Numerals

Interpret each Chinese numeral.

(a) 三 千 一 百 六 十 四 **(b)** 七 百 零 三 **(c)** 五 千 零 九 **(d)** 四 千 二 百

Solution

(a)

三 千 } $3 \cdot 1000 = 3000$
一 百 } $1 \cdot 100 = 100$
六 十 } $6 \cdot 10 = 60$
四 $4(\cdot 1) = 4$
Total: 3164

(b)

七 百 } $7 \cdot 100 = 700$
零 $0(\cdot 10) = 00$
三 $3(\cdot 1) = 3$
Total: 703

This illustration is of a **quipu.** In *Ethnomathematics: A Multicultural View of Mathematical Ideas*, Marcia Ascher writes:

A quipu is an assemblage of colored knotted cotton cords. Cotton cordage and cloth were of unparalleled importance in Inca culture. The colors of the cords, the way the cords are connected, the relative placement of the cords, the spaces between the cords, the types of knots on the individual cords, and the relative placement of the knots are all part of the logical-numerical recording.

(c)

$$5 \cdot 1000 = 5000$$
$$0(\cdot 100) = 000$$
$$0(\cdot 10) = 00$$
$$9(\cdot 1) = 9$$
Total: 5009

(d)

$$4 \cdot 1000 = 4000$$
$$2 \cdot 100 = 200$$
Total: 4200

EXAMPLE 8 Creating Chinese Numerals

Write a Chinese numeral for each number.

(a) 614   **(b)** 5090

Solution

(a) The number 614 is made up of six 100s, one 10, and four 1s, as depicted at the right.

$$6 \cdot 100:$$
$$(1 \cdot)10:$$
$$4(\cdot 1):$$

(b) The number 5090 consists of five 1000s, no 100s, and nine 10s (no 1s).

$$5 \cdot 1000:$$
$$0(\cdot 100):$$
$$9 \cdot 10:$$

4.1   EXERCISES

Convert each Egyptian numeral to Hindu-Arabic form.

1. 𓏺𓎆𓎆𓍢𓍢𓍢||||

2. 𓎆𓎆𓈎𓈎𓈎𓈎𓍢||

3. 𓆼𓆼𓆼𓆼 𓅆𓅆𓅆 𓏺 𓈎𓈎𓈎𓈎 𓍢𓍢 |||||
         𓆼𓆼𓆼 𓅆𓅆𓅆 𓈎𓈎𓈎𓈎 ||||

4. 𓆼𓆼𓆼 𓎆𓎆𓎆𓎆𓎆 𓈎𓈎𓍢𓍢𓍢|

Convert each Hindu-Arabic numeral to Egyptian form.

5. 23,135   **6.** 427

7. 8,657,000   **8.** 306,090

Chapter 1 of the book of Numbers in the Bible describes a census of the draft-eligible men of Israel after Moses led them out of Egypt into the Desert of Sinai, about 1450 B.C. Write an Egyptian numeral for the number of available men from each tribe listed.

9. 59,300 from the tribe of Simeon

10. 46,500 from the tribe of Reuben

11. 74,600 from the tribe of Judah

12. 45,650 from the tribe of Gad

13. 62,700 from the tribe of Dan

14. 32,200 from the tribe of Manasseh

Convert each Roman numeral to Hindu-Arabic form.

15. CLXXIII   **16.** MDXCVII

17. $\overline{\text{XIV}}$ **18.** $\overline{\overline{\text{V}}}\,\overline{\text{CXXI}}\text{CD}$

Convert each Hindu-Arabic numeral to Roman form.

19. 2861   **20.** 749

21. 25,619   **22.** 6,402,524

Convert each Chinese numeral to Hindu-Arabic form.

23.   **24.**   **25.**   **26.**

Convert each Hindu-Arabic numeral to Chinese form.

27. 965

28. 63

29. 7012

30. 2416

Though Chinese art forms began before written history, their highest development was achieved during four particular dynasties. Write traditional Chinese numerals for the beginning and ending dates of each dynasty listed.

31. Ming (1368 to 1644)

32. Sung (960 to 1279)

33. T'ang (618 to 907)

34. Han (202 B.C. to A.D. 220)

Work each addition or subtraction problem, using regrouping as necessary. Convert each answer to Hindu-Arabic form.

35. (Egyptian numeral addition)

36. (Egyptian numeral addition)

37. (Egyptian numeral addition)

38. (Egyptian numeral addition)

39. (Egyptian numeral subtraction)

40. (Egyptian numeral subtraction)

41. (Egyptian numeral subtraction)

42. (Egyptian numeral subtraction)

43. MCDXII
  + DCIX

44. $\overline{\text{XXIII}}$CXIX
  + $\overline{\text{XIV}}$CDXII

45. MCCCXXII
  − CDXIX

46. $\overline{\text{XII}}$CCCVI
  − MMCXXXII

Use the Egyptian algorithm to find each product.

47. 32 · 47

48. 29 · 75

49. 64 · 127

50. 52 · 131

In Exercises 51 and 52, convert all numbers to Egyptian numerals. Multiply using the Egyptian algorithm, and add using the Egyptian symbols. Give the final answer using a Hindu-Arabic numeral.

51. *Value of a Biblical Treasure* The book of Ezra in the Bible describes the return of the exiles to Jerusalem. To rebuild the temple, the King of Persia gave them the following items: thirty golden basins, a thousand silver basins, four hundred ten silver bowls, and thirty golden bowls. Find the total value of this treasure if each gold basin is worth 3000 shekels, each silver basin is worth 500 shekels, each silver bowl is worth 50 shekels, and each golden bowl is worth 400 shekels.

52. *Total Bill for King Solomon* King Solomon told the King of Tyre (now Lebanon) that he needed the best cedar for his temple, and that he would "pay you for your men whatever sum you fix." Find the total bill to Solomon if the King of Tyre used the following numbers of men: 5500 tree cutters at 2 shekels per week each, for a total of 7 weeks; 4600 sawers of wood at 3 shekels per week each, for a total of 32 weeks; and 900 sailors at 1 shekel per week each, for a total of 16 weeks.

Explain why each step would be an improvement in the development of numeration systems.

53. progressing from carrying groups of pebbles to making tally marks on a stick

54. progressing from tallying to simple grouping

55. utilizing a subtractive technique within simple grouping, as the Romans did

56. progressing from simple grouping to multiplicative grouping

The ancient Egyptian system described in this section was simple grouping, used a base of ten, and contained seven distinct symbols. The largest number expressible in that system is 9,999,999. Identify the largest number expressible in each of the following simple grouping systems. (In Exercises 61–64, d can be any counting number.)

57. base ten, five distinct symbols

58. base ten, ten distinct symbols

59. base five, five distinct symbols

60. base five, ten distinct symbols

61. base ten, d distinct symbols

62. base five, d distinct symbols

63. base seven, d distinct symbols

64. base b, d distinct symbols (where b is any counting number 2 or greater)

65. The Chinese system presented in the text has symbols for 1 through 9 and also for 10, 100, and 1000. What is the greatest number expressible in that system?

66. The Chinese system did eventually adopt two additional symbols, for 10,000 and 100,000. What greatest number could then be expressed?

4.2   MORE HISTORICAL NUMERATION SYSTEMS

OBJECTIVES

1 Understand the basic features of positional numeration systems.

2 Understand Hindu-Arabic numeration.

3 Understand Babylonian numeration.

4 Understand Mayan numeration.

5 Understand Greek numeration.

Basics of Positional Numeration

A simple grouping system relies on repetition of symbols to denote the number of each power of the base. A multiplicative grouping system uses multipliers in place of repetition, which is more efficient. The ultimate in efficiency is attained with a **positional system** in which only multipliers are used. The various powers of the base require no separate symbols, because the power associated with each multiplier can be understood from the position that the multiplier occupies in the numeral.

If the Chinese system had evolved into a positional system, then the numeral for 7482 could have been written

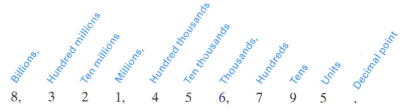

In the positional version on the left, the lowest symbol is understood to represent two 1s (10^0), the next one up denotes eight 10s (10^1), then four 100s (10^2), then seven 1000s (10^3). Each symbol in a numeral now has both a *face value,* associated with that particular symbol (the multiplier value), and a *place value* (a power of the base), associated with the place, or position, occupied by the symbol.

> **POSITIONAL NUMERATION**
>
> In a positional numeral, each symbol (called a **digit**) conveys two things:
>
> 1. **face value**—the inherent value of the symbol
>
> 2. **place value**—the power of the base that is associated with the position that the digit occupies in the numeral.

Hindu-Arabic Numeration

The place values in a Hindu-Arabic numeral, from right to left, are 1, 10, 100, 1000, and so on. The three 4s in the number 46,424 all have the same face value but different place values. The first 4, on the left, denotes four 10,000s, the next one denotes four 100s, and the one on the right denotes four 1s. Place values (in base ten) are named as shown here.

Billions,	Hundred millions	Ten millions	Millions,	Hundred thousands	Ten thousands	Thousands,	Hundreds	Tens	Units	Decimal point
8,	3	2	1,	4	5	6,	7	9	5	.

This numeral is read as eight billion, three hundred twenty-one million, four hundred fifty-six thousand, seven hundred ninety-five.

To work successfully, a positional system must have a symbol for zero to serve as a **placeholder** in case one or more powers of the base are not needed. Because of this requirement, some early numeration systems took a long time to evolve into a positional form, or never did. Although the traditional Chinese system does utilize a zero symbol, it never did incorporate all the features of a positional system, but remained essentially a multiplicative grouping system.

The one numeration system that did achieve the maximum efficiency of positional form is our own system, the **Hindu-Arabic** system. Its symbols have been traced to the Hindus of 200 B.C. They were picked up by the Arabs and eventually transmitted to Spain, where a late tenth-century version appeared like this:

$$I\ Z\ Z\ \chi\ Y\ 6\ 7\ 8\ 9$$

The earliest stages of the system evolved under the influence of navigational, trade, engineering, and military requirements. And in early modern times, the advance of astronomy and other sciences led to a structure well suited to fast and accurate computation.

The purely positional form that the system finally assumed was introduced to the West by Leonardo Fibonacci of Pisa (1170–1250) early in the thirteenth century, but widespread acceptance of standardized symbols and form was not achieved until the invention of printing during the fifteenth century. Since that time, no better system of numeration has been devised, and the positional base-ten Hindu-Arabic system is commonly used around the world today.

The Hindu-Arabic system and notation will be investigated further in **Sections 4.3 and 4.4.** The systems we consider next, the Babylonian and the Mayan, achieved the main ideas of positional numeration without fully developing them.

Babylonian Numeration

The Babylonians used a base of 60 in their system. Because of this, in theory they would then need distinct symbols for numbers from 1 through 59 (just as we have symbols for 1 through 9). However, the Babylonian method of writing on clay with wedge-shaped sticks gave rise to only *two* symbols, as shown in **Table 5**. The number 47 would be written

Table 5

Babylonian Symbols

Number	Symbol
1	▼
10	‹

$$\text{‹‹‹‹▼▼▼▼▼▼▼}\qquad \text{or}\qquad \begin{array}{l}\text{‹‹▼▼▼▼}\\ \text{‹‹ ▼▼▼}\end{array}. \qquad \text{The number 47}$$

Since the Babylonian system had base sixty, the "digit" on the right in a multi-digit number represented the number of 1s, with the second "digit" from the right giving the number of 60s. The third digit would give the number of 3600s $(60 \cdot 60 = 3600)$, and so on.

SPECIAL FEATURES OF THE BABYLONIAN SYSTEM

1. Rather than using distinct symbols for each number less than the base (sixty), the Babylonians expressed face values in base-ten simple grouping, using only the two symbols

   ‹ for 10   and   ▼ for 1.

   The system is, therefore, base-ten simple grouping *within* base-sixty positional.

2. The earliest Babylonian system lacked a placeholder symbol (zero), so missing powers of the base were difficult to express. Blank spaces within a numeral would be subject to misinterpretation.

EXAMPLE 1   Converting Babylonian Numerals to Hindu-Arabic

Convert each Babylonian numeral to Hindu-Arabic form.

(a) ⟨⟨⟨⟨⟨▼▼▼   (b) ⟨⟨⟨▼▼▼▼ ⟨⟨▼▼ / ⟨⟨ ▼▼▼▼ ⟨⟨▼▼   (c) ⟨⟨▼▼▼▼▼⟨▼⟨⟨⟨▼▼▼▼▼▼

Solution

(a) Here we have five 10s and three 1s.

$$5 \cdot 10 = 50$$
$$3 \cdot 1 = 3$$
$$\overline{53} \leftarrow \text{Answer}$$

(b) This "two-digit" Babylonian number represents twenty-two 1s and fifty-eight 60s.

$$22 \cdot 1 = 22$$
$$58 \cdot 60 = 3480$$
$$\overline{3502} \leftarrow \text{Answer}$$

(c) Here we have a three-digit number.

$$36 \cdot 1 = 36$$
$$11 \cdot 60 = 660$$
$$25 \cdot 3600 = 90{,}000$$
$$\overline{90{,}696} \leftarrow \text{Answer}$$

EXAMPLE 2   Converting Hindu-Arabic Numerals to Babylonian

Convert each Hindu-Arabic numeral to Babylonian form.

(a) 733   (b) 75,904   (c) 43,233

Solution

(a) To write 733 in Babylonian, we will need some 60s and some 1s. Divide 60 into 733. The quotient is 12, with a remainder of 13. Thus we need twelve 60s and thirteen 1s.

⟨▼▼⟨▼▼▼ ← 733

(b) For 75,904, we need some 3600s, as well as some 60s and some 1s. Divide 75,904 by 3600. The answer is 21, with a remainder of 304. Divide 304 by 60. The quotient is 5, with a remainder of 4.

⟨⟨▼ ▼▼▼▼▼ ▼▼▼▼ ← 75,904

(c) Divide 43,233 by 3600. The answer is 12, with a remainder of 33. We need no 60s here. In a system such as ours we would use a 0 to show that no 60s are needed. Since the early Babylonians had no such symbol, they merely left a space.

⟨▼▼ ⟨⟨⟨▼▼▼ ← 43,233

Example 2(c) illustrates the problem presented by the lack of a symbol for zero. In our system we know that 202 is not the same as 2002 or 20,002. The lack of a zero symbol was a major difficulty with the very early Babylonian system. A symbol for zero was introduced about 300 B.C.

Mayan Numeration

The Mayan Indians of Central America and Mexico also used what is basically a positional system. Like the Babylonians, the Mayans did not use base ten—they used base twenty, with a twist. In a true base-twenty system, the digits would represent 1s, 20s, $20 \cdot 20 = 400$s, $20 \cdot 400 = 8000$s, and so on. The Mayans used 1s, 20s, $18 \cdot 20 = 360$s, $20 \cdot 360 = 7200$s, and so on. It is possible that they multiplied 20 by 18 (instead of 20) because $18 \cdot 20$ is close to the number of days in a year, convenient for astronomy. The symbols of the Mayan system are shown in **Table 6**.

Table 6

Mayan Symbols

Number	Symbol
0	⬭
1	·
5	—

The Mayans were one of the first civilizations to invent a placeholder. They had a zero symbol many hundreds of years before it reached western Europe. Mayan numerals are written from top to bottom, just as in the classical Chinese system.

SPECIAL FEATURES OF THE MAYAN SYSTEM

1. Rather than using distinct symbols for each number less than the base (twenty), the Mayans expressed face values in base-five simple grouping, using only the two symbols — for 5 and · for 1. The system is, therefore, base-five simple grouping *within* base-twenty positional.

2. Place values in base-twenty would normally be

$$1, \quad 20, \quad 20^2 = 400, \quad 20^3 = 8000, \quad 20^4 = 160{,}000, \quad \text{and so on.}$$

   However, the Mayans multiplied by 18 rather than 20 in just one case, so the place values are

$$1, \quad 20, \quad 20 \cdot 18 = 360, \quad 360 \cdot 20 = 7200, \quad 7200 \cdot 20 = 144{,}000, \quad \text{and so on.}$$

EXAMPLE 3 Converting Mayan Numerals to Hindu-Arabic

Convert each Mayan numeral to Hindu-Arabic form.

(a) **(b)**

Solution

(a) The top group of symbols represents two 360s, the middle group represents nine 20s, and the bottom group represents eleven 1s.

$$2 \cdot 360 = 720$$
$$9 \cdot 20 = 180$$
$$11 \cdot 1 = \underline{11}$$
$$911 \leftarrow \text{Answer}$$

(b)

$$6 \cdot 360 = 2160$$
$$0 \cdot 20 = 0$$
$$13 \cdot 1 = \underline{13}$$
$$2173 \leftarrow \text{Answer}$$

EXAMPLE 4 Converting Hindu-Arabic Numerals to Mayan

Convert each Hindu-Arabic numeral to Mayan form.

(a) 277   **(b)** 1238

Solution

(a) The number 277 requires thirteen 20s (divide 277 by 20) and seventeen 1s.

$\leftarrow 277$

(b) Divide 1238 by 360. The quotient is 3, with remainder 158. Divide 158 by 20. The quotient is 7, with remainder 18. Thus we need three 360s, seven 20s, and eighteen 1s.

$\leftarrow 1238$

Table 7

Greek Symbols

Number	Symbol
1	α
2	β
3	γ
4	δ
5	ϵ
6	ς
7	ζ
8	η
9	θ
10	ι
20	κ
30	λ
40	μ
50	ν
60	ξ
70	o
80	π
90	φ
100	ρ
200	σ
300	τ
400	υ
500	ϕ
600	χ
700	ψ
800	ω
900	λ

Greek Numeration

The classical Greeks of Ionia assigned values to the 24 letters of their ordinary alphabet, together with three obsolete Phoenician letters (the digamma ς for 6, the koppa φ for 90, and the sampi λ for 900). See **Table 7**. This scheme, usually called a **ciphered system,** makes all counting numbers less than 1000 easily represented. It avoids repetitions of symbols but requires vast multiplication tables for 27 distinct symbols. Computation would be very burdensome. The base is 10, but the system is quite different from simple grouping, multiplicative grouping, or positional.

EXAMPLE 5 Converting Greek Numerals to Hindu-Arabic

Convert each Greek numeral to Hindu-Arabic form.

(a) $\lambda\alpha$ (b) $\tau\xi\epsilon$ (c) $\lambda\varphi\theta$ (d) $\chi\delta$

Solution

(a) 31   (b) 365   (c) 999   (d) 604

For numbers larger than 999, the Greeks introduced two additional techniques.

SPECIAL FEATURES OF THE GREEK SYSTEM

1. Multiples of 1000 (up to 9000) are indicated with a small stroke next to a units symbol. For example, 9000 would be denoted $\prime\theta$.

2. Multiples of 10,000 are indicated by the letter M (from the word *myriad*, meaning ten thousand) with the multiple (a units symbol) shown above the M. The number 50,000 would be denoted $\overset{\epsilon}{M}$.

EXAMPLE 6 Converting Hindu-Arabic Numerals to Greek

Convert each Hindu-Arabic numeral to Greek form.

(a) 3000   (b) 40,000   (c) 7694   (d) 88,888

Solution

(a) $\prime\gamma$ (b) $\overset{\delta}{M}$ (c) $\prime\zeta\chi\varphi\delta$ (d) $\overset{\eta}{M}\prime\eta\omega\pi\eta$

4.2 EXERCISES

Identify each numeral in Exercises 1–20 as Babylonian, Mayan, or Greek. Give the equivalent in the Hindu-Arabic system.

1. ≝

2. ⟨⟨▼▼

3. ⟨⟨⟨▼▼

4. ≝

5. $\sigma\lambda\delta$

6. $\omega o\beta$

7. ≝

8. ≝

9. ⟨⟨▼⟨⟨▼▼

10. ⟨▼▼▼▼⟨⟨⟨▼▼

11. ⟨⟨▼▼ ⟨⟨▼ ⟨⟨▼

12. ⟨⟨⟨▼▼▼ ⟨⟨ ▼▼ ⟨⟨⟨▼

13. ≝

14. ≝

15. ≝

16. ≝

17. ⟨⟨▼▼⟨⟨▼⟨▼▼▼▼

18. ⟨⟨ ▼▼ ⟨⟨ ▼▼▼▼ ⟨⟨ ⟨⟨▼▼ ⟨ ▼▼▼▼ ⟨⟨▼▼▼

19. $\overset{\alpha}{M}\prime\epsilon\rho\mu\theta$

20. $\overset{\eta}{M}\omega\eta$

Write each number as a Babylonian numeral.

21. 21　　**22.** 32　　**23.** 293

24. 412　　**25.** 1514　　**26.** 3280

27. 5190　　**28.** 43,205　　**29.** ⠂⠂⠄⠄⠄ ⬭

30. ⠂ ⬭ ⠄　　**31.** $\chi\kappa\epsilon$　　**32.** $\overset{\beta}{M}$

Write each number as a Mayan numeral.

33. 12　　**34.** 32　　**35.** 151

36. 208　　**37.** 4694　　**38.** 4328

39. ⟨𝅘𝅥𝅯⟨⟨𝅘𝅥𝅯⟨𝅘 𝅘⟨⟨𝅘𝅥𝅯⟨𝅘

40. ⟨𝅘𝅥𝅯 𝅘𝅥𝅯⟨𝅘𝅥𝅯 𝅘 𝅘⟨𝅘

41. $\tau\nu\theta$　　**42.** $\tau\pi\alpha$

Write each number as a Greek numeral.

43. 39　　**44.** 51　　**45.** 92

46. 106　　**47.** 412　　**48.** 381

49. 2769　　**50.** 9814　　**51.** 54,726

52. 80,102　　**53.** ⟨𝅘⟨⟨𝅘　　**54.** 𝅘⟨𝅘⟨𝅘𝅘

55. ⠂⠂⠄⠄⠄ ⠄⠄⠄

56. ⬭ ⠄⠄ ⠄⠄⠄

4.3　ARITHMETIC IN THE HINDU-ARABIC SYSTEM

OBJECTIVES

1 Express Hindu-Arabic numerals in expanded form.
2 Explore historical calculation devices.

Expanded Form

The historical development of numeration culminated in positional systems. The most successful of these is the Hindu-Arabic system, which has base ten and, therefore, has place values that are powers of 10. Exponential expressions, or powers, are the basis of expanded form in a positional system.

EXAMPLE 1　Evaluating Powers

Find each power.

(a) 10^3　　**(b)** 7^2　　**(c)** 2^8

Solution

(a) $10^3 = 10 \cdot 10 \cdot 10 = 1000$
(10^3 is read "10 cubed," or "10 to the third power.")

(b) $7^2 = 7 \cdot 7 = 49$
(7^2 is read "7 squared," or "7 to the second power.")

(c) $2^8 = 2 \cdot 2 \cdot 2 \cdot 2 \cdot 2 \cdot 2 \cdot 2 \cdot 2 = 256$
(2^8 is read "2 to the eighth power.")

■

To simplify work with exponents, it is agreed that

$$a^0 = 1, \quad \text{for any nonzero number } a.$$

Thus, $7^0 = 1$, $52^0 = 1$, and so on. At the same time,

$$a^1 = a, \quad \text{for any number } a.$$

For example, $8^1 = 8$, and $25^1 = 25$. The exponent 1 is usually omitted.

By using exponents, numbers can be written in **expanded form** in which the value of the digit in each position is made clear. For example,

$$924 = 900 + 20 + 4$$
$$= (9 \cdot \mathbf{100}) + (2 \cdot \mathbf{10}) + (4 \cdot \mathbf{1})$$
$$= (9 \cdot \mathbf{10^2}) + (2 \cdot \mathbf{10^1}) + (4 \cdot \mathbf{10^0}). \quad 100 = 10^2, 10 = 10^1, \text{ and } 1 = 10^0$$

This Iranian stamp should remind us that counting on fingers (and toes) is an age-old practice. In fact, our word **digit**, referring to the numerals 0–9, comes from a Latin word for "finger" (or "toe"). Aristotle first noted the relationships between fingers and base ten in Greek numeration. Anthropologists go along with the notion. Some cultures, however, have used two, three, or four as number bases, for example, counting on the joints of the fingers or the spaces between them.

There is much evidence that early humans (in various cultures) used their fingers to represent numbers. As calculations became more complicated, **finger reckoning,** as illustrated above, became popular. The Romans became adept at this sort of calculating, carrying it to 10,000 or perhaps higher.

EXAMPLE 2   Writing Numbers in Expanded Form

Write each number in expanded form.

(a) 1906   **(b)** 46,424

Solution

(a) $1906 = (1 \cdot 10^3) + (9 \cdot 10^2) + (0 \cdot 10^1) + (6 \cdot 10^0)$

Because $0 \cdot 10^1 = 0$, this term could be omitted, but the form is clearer with it included.

(b) $46{,}424 = (4 \cdot 10^4) + (6 \cdot 10^3) + (4 \cdot 10^2) + (2 \cdot 10^1) + (4 \cdot 10^0)$ ∎

EXAMPLE 3   Simplifying Expanded Numbers

Simplify each expansion.

(a) $(3 \cdot 10^5) + (2 \cdot 10^4) + (6 \cdot 10^3) + (8 \cdot 10^2) + (7 \cdot 10^1) + (9 \cdot 10^0)$

(b) $(2 \cdot 10^1) + (8 \cdot 10^0)$

Solution

(a) $(3 \cdot 10^5) + (2 \cdot 10^4) + (6 \cdot 10^3) + (8 \cdot 10^2) + (7 \cdot 10^1) + (9 \cdot 10^0) = 326{,}879$

(b) $(2 \cdot 10^1) + (8 \cdot 10^0) = 28$ ∎

Expanded notation and the **distributive property** can be used to see why standard algorithms for addition and subtraction really work.

DISTRIBUTIVE PROPERTY

For all real numbers a, b, and c,

$$(b \cdot a) + (c \cdot a) = (b + c) \cdot a.$$

Example: $(3 \cdot 10^4) + (2 \cdot 10^4) = (3 + 2) \cdot 10^4$
$$= 5 \cdot 10^4$$

EXAMPLE 4   Adding Expanded Forms

Use expanded notation to add 23 and 64.

Solution

$$23 = (2 \cdot 10^1) + (3 \cdot 10^0)$$
$$+\ 64 = (6 \cdot 10^1) + (4 \cdot 10^0)$$
$$(8 \cdot 10^1) + (7 \cdot 10^0) = 87 \quad \text{Sum}$$ ∎

Finger Counting The first digits many people used for counting were their fingers. In Africa the Zulu used the method shown here to count to ten. They started on the left hand with palm up and fist closed. The Zulu finger positions for 1–5 are shown above on the left. The Zulu finger positions for 6–10 are shown on the right.

EXAMPLE 5   Subtracting Expanded Forms

Use expanded notation to subtract 254 from 695.

Solution

$$695 = (6 \cdot 10^2) + (9 \cdot 10^1) + (5 \cdot 10^0)$$
$$-\ 254 = (2 \cdot 10^2) + (5 \cdot 10^1) + (4 \cdot 10^0)$$
$$(4 \cdot 10^2) + (4 \cdot 10^1) + (1 \cdot 10^0) = 441 \quad \text{Difference}$$ ∎

The **Carmen de Algorismo** (opening verses shown here) by Alexander de Villa Dei, thirteenth century, popularized the new art of "algorismus":

. . . from these twice five figures 0 9 8 7 6 5 4 3 2 1 of the Indians we benefit . . .

The *Carmen* related that Algor, an Indian king, invented the art. But actually, "algorism" (or "algorithm") comes in a roundabout way from the name Muhammad ibn Musa al-Khorârizmi, an Arabian mathematician of the ninth century, whose arithmetic book was translated into Latin. Furthermore, this Muhammad's book on equations, *Hisab al-jabr w'almuqâbalah*, yielded the term "algebra" in a similar way.

EXAMPLE 6 Carrying in Expanded Form

Use expanded notation to add 75 and 48.

Solution

$$75 = (7 \cdot 10^1) + (5 \cdot 10^0)$$
$$+\,48 = (4 \cdot 10^1) + (8 \cdot 10^0)$$
$$\overline{\qquad\quad (11 \cdot 10^1) + (13 \cdot 10^0)}$$

The units position (10^0) has room for only one digit, so we modify $13 \cdot 10^0$.

$$13 \cdot 10^0 = (10 \cdot 10^0) + (3 \cdot 10^0) \qquad \text{Distributive property}$$
$$= (1 \cdot 10^1) + (3 \cdot 10^0) \qquad 10 \cdot 10^0 = 1 \cdot 10^1$$

The 1 from 13 moved to the left (carried) from the units position to the tens position.

$$\overbrace{\qquad\qquad}^{13\,\cdot\,10^0}$$
$$(11 \cdot 10^1) + (1 \cdot 10^1) + (3 \cdot 10^0)$$
$$= (12 \cdot 10^1) + (3 \cdot 10^0) \qquad \text{Distributive property}$$
$$= (10 \cdot 10^1) + (2 \cdot 10^1) + (3 \cdot 10^0) \qquad \text{Modify } 12 \cdot 10^1.$$
$$= (1 \cdot 10^2) + (2 \cdot 10^1) + (3 \cdot 10^0) \qquad 10 \cdot 10^1 = 1 \cdot 10^2$$
$$= 123 \qquad \text{Sum} \qquad ∎$$

EXAMPLE 7 Borrowing in Expanded Form

Use expanded notation to subtract 186 from 364.

Solution

$$364 = (3 \cdot 10^2) + (6 \cdot 10^1) + (4 \cdot 10^0)$$
$$-186 = (1 \cdot 10^2) + (8 \cdot 10^1) + (6 \cdot 10^0)$$

We cannot subtract 6 from 4. The units position borrows from the tens position.

$$(3 \cdot 10^2) + (6 \cdot 10^1) + (4 \cdot 10^0)$$
$$= (3 \cdot 10^2) + (5 \cdot 10^1) + (1 \cdot 10^1) + (4 \cdot 10^0) \qquad \text{Distributive property}$$
$$= (3 \cdot 10^2) + (5 \cdot 10^1) + (10 \cdot 10^0) + (4 \cdot 10^0) \qquad 1 \cdot 10^1 = 10 \cdot 10^0$$
$$= (3 \cdot 10^2) + (5 \cdot 10^1) + (14 \cdot 10^0) \qquad \text{Distributive property}$$

We cannot take 8 from 5 in the tens position, so we borrow from the hundreds.

$$(3 \cdot 10^2) + (5 \cdot 10^1) + (14 \cdot 10^0)$$
$$= (2 \cdot 10^2) + (1 \cdot 10^2) + (5 \cdot 10^1) + (14 \cdot 10^0) \qquad \text{Distributive property}$$
$$= (2 \cdot 10^2) + (10 \cdot 10^1) + (5 \cdot 10^1) + (14 \cdot 10^0) \qquad 1 \cdot 10^2 = 10 \cdot 10^1$$
$$= (2 \cdot 10^2) + (15 \cdot 10^1) + (14 \cdot 10^0) \qquad \text{Distributive property}$$

Now we can complete the subtraction.

$$(2 \cdot 10^2) + (15 \cdot 10^1) + (14 \cdot 10^0)$$
$$-\,(1 \cdot 10^2) + (8 \cdot 10^1) + (6 \cdot 10^0)$$
$$\overline{(1 \cdot 10^2) + (7 \cdot 10^1) + (8 \cdot 10^0) = 178} \qquad \text{Difference} \qquad ∎$$

Smartphones perform mathematical calculations and many other functions as well.

Examples **4–7** used expanded notation and the distributive property to clarify our usual addition and subtraction methods. In practice, our actual work for these four problems would appear as follows.

$$
\begin{array}{cccc}
 & & & \overset{1}{} & \overset{2\ 15}{} \\
23 & 695 & 75 & 3\,\cancel{6}^{1}4 \\
+\ 64 & -\ 254 & +\ 48 & -\ 1\ 8\ 6 \\
\hline
87 & 441 & 123 & 1\ 7\ 8
\end{array}
$$

The procedures seen in this section also work for positional systems with bases other than ten.

Historical Calculation Devices

Because our numeration system is based on powers of ten, it is often called the **decimal system,** from the Latin word *decem,* meaning ten.* Over the years, many methods have been devised for speeding calculations in the decimal system.

One of the oldest calculation methods is the **abacus,** a device made with a series of rods with sliding beads and a dividing bar. Reading from right to left, the rods have values of 1, 10, 100, 1000, and so on. The bead above the bar has five times the value of those below. Beads moved *toward* the bar are in the "active" position, and those toward the frame are ignored. In our illustrations of *abaci* (plural form of *abacus*), such as in **Figure 8,** the activated beads are shown in black.

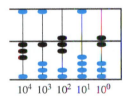

$10^4 \ 10^3 \ 10^2 \ 10^1 \ 10^0$

Figure 8

EXAMPLE 8   **Reading an Abacus**

What number is shown on the abacus in **Figure 8**?

Solution

Find the number as follows.

> Beads above the bar have five times the value.

$(3 \cdot 10{,}000) + (1 \cdot 1000) + [(1 \cdot 500) + (2 \cdot 100)] + 0 \cdot 10 + [(1 \cdot 5) + (1 \cdot 1)]$

$\quad = 30{,}000 + 1000 + (500 + 200) + 0 + (5 + 1)$

$\quad = 31{,}706$ ∎

The **speed and accuracy of the abacus** are well known, according to www.ucmasusa.com. In a contest held between a Japanese **soroban** (the Japanese version of the abacus) expert and a highly skilled desk-calculator operator, the abacus won on addition, subtraction, division, and combinations of these operations. The electronic calculator won only on multiplication.

As paper became more readily available, people gradually switched from devices like the abacus (though these still are commonly used in some areas) to paper-and-pencil methods of calculation. One early scheme, used in India and Persia, was the **lattice method,** which arranged products of single digits into a diagonalized lattice.

_____

*December was the tenth month in an old form of the calendar. It is interesting to note that *decem* became *dix* in the French language; a ten-dollar bill, called a "dixie," was in use in New Orleans before the Civil War. "Dixie Land" was a nickname for that city before Dixie came to refer to all the Southern states, as in Daniel D. Emmett's song, written in 1859.

John Napier's most significant mathematical contribution, developed over a period of at least 20 years, was the concept of **logarithms,** which, among other things, allow multiplication and division to be accomplished with addition and subtraction. It was a great computational advantage given the state of mathematics at the time (1614).

Napier, a supporter of John Knox and James I, published a widely read anti-Catholic work that analyzed the biblical book of Revelation. He concluded that the Pope was the Antichrist and that the Creator would end the world between 1688 and 1700. Napier is one of many who, over the years, have miscalculated the end of the world.

EXAMPLE 9 Using the Lattice Method for Products

Find the product 38 · 794 by the lattice method.

Solution

Step 1   Write the problem, with one number at the side and one across the top.

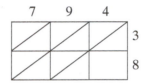

Step 2   Within the lattice, write the products of all pairs of digits from the top and side.

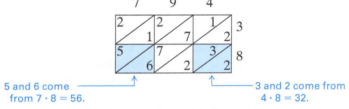

5 and 6 come from 7 · 8 = 56.   3 and 2 come from 4 · 8 = 32.

Step 3   Starting at the right of the lattice, add diagonally, carrying as necessary.

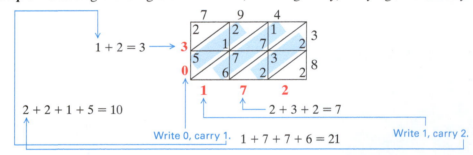

$1 + 2 = 3$

$2 + 2 + 1 + 5 = 10$ $2 + 3 + 2 = 7$

Write 0, carry 1. $1 + 7 + 7 + 6 = 21$ Write 1, carry 2.

Step 4   Read the answer around the left side and bottom: 38 · 794 = **30,172.** ■

The Scottish mathematician John Napier (1550–1617) introduced a significant calculating tool called **Napier's rods,** or **Napier's bones.** Napier's invention, based on the lattice method of multiplication, is widely acknowledged as a very early forerunner of modern computers. It consisted of a set of strips, several for each digit 0 through 9, on which multiples of each digit appeared in a sort of lattice column. See **Figure 9**.

An additional strip, the *index,* is laid beside any of the others to indicate the multiplier at each level. **Figure 10** shows that 7 × 2806 = **19,642.**

Napier's rods were an early step toward modern computers.

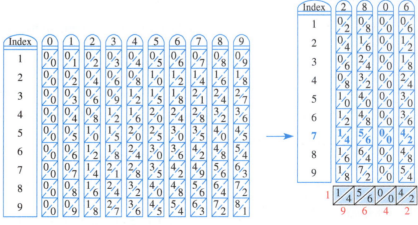

Figure 9   **Figure 10**

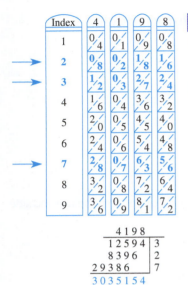

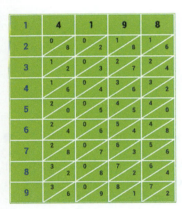

Free on the Android Market is an **app for Napier's rods** (or "Napier's abacus"). The app prompts the user for a number and then generates a picture showing the rods for those digits next to the index.

For a way to include a little magic with your calculations, check out
http://digicc.com/fido.

EXAMPLE 10   Multiplying with Napier's Rods

Use Napier's rods to find the product of 723 and 4198.

Solution

We line up the rods for 4, 1, 9, and 8 next to the index, as in **Figure 11**. The product $3 \cdot 4198$ is found as described in **Example 9** and written at the bottom of the figure. Then $2 \cdot 4198$ is found similarly and written below, shifted one place to the left. (Why?) Finally, the product $7 \cdot 4198$ is written shifted two places to the left.

The final answer is found by addition.

$$723 \cdot 4198 = 3{,}035{,}154$$ ∎

Another paper-and-pencil method of multiplication is the **Russian peasant method,** which is similar to the Egyptian method of doubling explained in **Section 4.1.** To multiply 37 and 42 by the Russian peasant method, make two columns headed by 37 and 42. Form the first column by dividing 37 by 2 again and again, ignoring any remainders. Stop when 1 is obtained. Form the second column by doubling each number down the column.

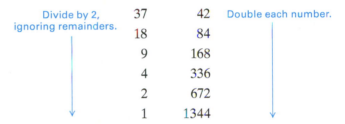

Now add up only the second-column numbers that correspond to odd numbers in the first column. Omit those corresponding to even numbers in the first column.

	37	42	
	18	84	
Identify odd →	9	168	← Add these numbers.
	4	336	
	2	672	
→	1	1344	←

$$37 \cdot 42 = 42 + 168 + 1344 = 1554 \leftarrow \text{Answer}$$

Most people use standard algorithms for adding and subtracting, carrying or borrowing when appropriate, as illustrated following **Example 7.** An interesting alternative is the **nines complement method** for subtracting. To use this method, we first agree that the nines complement of a digit n is $9 - n$. For example, the nines complement of 0 is 9, of 1 is 8, of 2 is 7, and so on, down to the nines complement of 9, which is 0.

To carry out the nines complement method, complete the following steps:

Step 1   Align the digits as in the standard subtraction algorithm.

Step 2   Add leading zeros, if necessary, in the subtrahend so that both numbers have the same number of digits.

Step 3   Replace each digit in the subtrahend with its nines complement, and then add.

Step 4   Finally, delete the leading digit (1), and add 1 to the remaining part of the sum.

EXAMPLE 11   Using the Nines Complement Method

Use the nines complement method to subtract 2803 – 647.

Solution

	Step 1	Step 2	Step 3	Step 4
	2803	2803	2803	2155
	– 647	– 0647	+ 9352	+ 1
			12,155	2156 Difference

FOR FURTHER THOUGHT

Calculating on the Abacus

The abacus has been (and still is) used to perform rapid calculations. Add 526 and 362 as shown.

Start with 526 on the abacus.

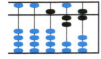

To add 362, start by "activating" an additional 2 on the 1s rod.

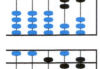

Next, activate an additional 6 on the 10s rod.

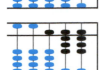

Finally, activate an additional 3 on the 100s rod.

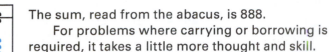

The sum, read from the abacus, is 888.

For problems where carrying or borrowing is required, it takes a little more thought and skill.

For Group or Individual Investigation

1. Use an abacus to add: 13,728 + 61,455. Explain each step of your procedure.

2. Use an abacus to subtract: 6512 – 4816. Explain each step of your procedure.

4.3   EXERCISES

Write each number in expanded form.

1. 73   2. 265   3. 8335   4. 12,398

5. three thousand, six hundred twenty-four

6. fifty-two thousand, one hundred eighteen

7. fourteen million, two hundred six thousand, forty

8. two hundred twelve million, eleven thousand, nine hundred sixteen

Simplify each expansion.

9. $(7 \cdot 10^1) + (5 \cdot 10^0)$

10. $(8 \cdot 10^2) + (2 \cdot 10^1) + (0 \cdot 10^0)$

11. $(4 \cdot 10^3) + (3 \cdot 10^2) + (8 \cdot 10^1) + (0 \cdot 10^0)$

12. $(5 \cdot 10^5) + (0 \cdot 10^4) + (3 \cdot 10^3) + (5 \cdot 10^2) + (6 \cdot 10^1) + (8 \cdot 10^0)$

13. $(7 \cdot 10^7) + (4 \cdot 10^5) + (1 \cdot 10^3) + (9 \cdot 10^0)$

14. $(3 \cdot 10^8) + (8 \cdot 10^6) + (2 \cdot 10^4) + (3 \cdot 10^0)$

In each of the following, add in expanded notation.

15. 37 + 42   16. 732 + 417

In each of the following, subtract in expanded notation.

17. 85 – 32   18. 835 – 534

Perform each addition using expanded notation.

19. 64 + 45   20. 663 + 272

21. 434 + 299   22. 6755 + 4827

Perform each subtraction using expanded notation.

23. 54 – 48   24. 383 – 78

25. 855 – 649   26. 816 – 335

Perform each subtraction using the nines complement method.

27. 283 – 41   28. 536 – 425

29. 50,000 – 199   30. 40,002 – 4846

Identify the number represented on each abacus.

31.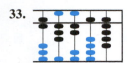

32.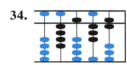

33.

34.

Sketch an abacus to show each number.

35. 38

36. 183

37. 2547

38. 70,163

Use the lattice method to find each product.

39. 65 · 29

40. 32 · 741

41. 525 · 73

42. 912 · 483

*Refer to **Example 10** where Napier's rods were used. Then complete Exercises 43 and 44.*

43. Find the product of 723 and 4198 by completing the lattice process shown here.

44. Explain how Napier's rods could have been used in **Example 10** to set up one complete lattice product rather than adding three individual (shifted) lattice products. Illustrate with a sketch.

*Use Napier's rods (**Figure 9**) to find each product.*

45. 8 · 62

46. 32 · 73

47. 26 · 8354

48. 526 · 4863

Use the Russian peasant method to find each product.

49. 5 · 92

50. 41 · 53

51. 62 · 529

52. 145 · 63

The Hindu-Arabic system is positional and uses ten as the base. Describe any advantages or disadvantages that may have resulted in each case.

53. Suppose the base had been larger, say twelve or twenty.

54. Suppose the base had been smaller, maybe five or eight.

| **4.4** | **CONVERSION BETWEEN NUMBER BASES** |

OBJECTIVES

1 Convert numerals from one base to another.

2 Use different bases in the context of computer mathematics.

General Base Conversions

In this section we consider bases other than ten, but we use the familiar Hindu-Arabic symbols. We indicate bases other than ten with a spelled-out subscript, as in the numeral 43_{five}. *Whenever a number appears without a subscript, it is assumed that the intended base is ten.* Be careful how you read (or verbalize) numerals here. The numeral 43_{five} is read "four three base five." (Do *not* read it as "forty-three," as that terminology implies base ten and names a totally different number.)

Table 8 gives powers of some numbers used as alternative bases.

Table 8 Selected Powers of Some Alternative Number Bases

	Fourth Power	**Third Power**	**Second Power**	**First Power**	**Zero Power**
Base two	16	8	4	2	1
Base five	625	125	25	5	1
Base seven	2401	343	49	7	1
Base eight	4096	512	64	8	1
Base sixteen	65,536	4096	256	16	1

For example, the base-two row of **Table 8** indicates that

$$2^4 = 16, \quad 2^3 = 8, \quad 2^2 = 4, \quad 2^1 = 2, \quad \text{and} \quad 2^0 = 1.$$

Table 9
Base-Five Numerals

Base Ten	Base Five	Base Ten	Base Five
0	0	16	31
1	1	17	32
2	2	18	33
3	3	19	34
4	4	20	40
5	10	21	41
6	11	22	42
7	12	23	43
8	13	24	44
9	14	25	100
10	20	26	101
11	21	27	102
12	22	28	103
13	23	29	104
14	24	30	110
15	30	31	111

We begin with the base-five system, which requires just five distinct symbols: 0, 1, 2, 3, and 4. **Table 9** compares base-five and decimal (base-ten) numerals for the whole numbers 0 through 31. Notice that because only the symbols 0, 1, 2, 3, and 4 are used in base five, we must use two digits in base five when we get to 5_{ten}.

5_{ten} is expressed as one 5 and no 1s—that is, as 10_{five}.

6_{ten} becomes 11_{five} (one 5 and one 1).

While base five uses fewer distinct symbols than base ten (an apparent advantage because there are fewer symbols to learn), it often requires more digits than base ten to denote the same number (a disadvantage because more symbols must be written).

You will find that in any base, if you denote the base "b," then the base itself will be 10_b, just as occurred in base five. For example,

$7_{ten} = 10_{seven}$, $16_{ten} = 10_{sixteen}$, and so on.

EXAMPLE 1   Converting from Base Five to Base Ten

Convert 1342_{five} to decimal form.

Solution

Referring to the powers of five in **Table 8**, we see that this number has one 125, three 25s, four 5s, and two 1s.

$$1342_{five} = (1 \cdot 125) + (3 \cdot 25) + (4 \cdot 5) + (2 \cdot 1)$$
$$= 125 + 75 + 20 + 2$$
$$= 222$$

A shortcut for converting from base five to decimal form, which is *particularly useful when you use a calculator,* can be derived as follows.

$$1342_{five} = (1 \cdot 5^3) + (3 \cdot 5^2) + (4 \cdot 5) + 2$$
$$= ((1 \cdot 5^2) + (3 \cdot 5) + 4) \cdot 5 + 2 \quad \text{Factor 5 out of the three quantities in parentheses.}$$
$$= (((1 \cdot 5) + 3) \cdot 5 + 4) \cdot 5 + 2 \quad \text{Factor 5 out of the two "inner" quantities.}$$

The inner parentheses around $1 \cdot 5$ are not needed because the product is done automatically before the 3 is added. Therefore, we can write

$$1342_{five} = ((1 \cdot 5 + 3) \cdot 5 + 4) \cdot 5 + 2 = 222.$$

This series of products and sums is easily done as an uninterrupted sequence of operations on a calculator, with no intermediate results written down. The same method works for converting to base ten from any other base.

CALCULATOR SHORTCUT FOR BASE CONVERSION

To convert from another base to decimal form, follow these steps.

Step 1 Start with the first digit on the left and multiply by the base.

Step 2 Then add the next digit, multiply again by the base, and so on.

Step 3 Add the last digit on the right. ***Do not multiply it by the base.***

With some calculators, only the digits, the multiplications, and the additions need to be entered in order. With others, you may need to press the $=$ key following each addition of a digit. If you handle grouped expressions on your calculator by actually entering parentheses, then enter the expression just as illustrated above and in the following example. (The number of left parentheses to start with will be two fewer than the number of digits in the original numeral.)

EXAMPLE 2   Using the Calculator Shortcut

Use the calculator shortcut to convert 244314_{five} to decimal form.

Solution

$$244314_{\text{five}} = ((((2 \cdot 5 + 4) \cdot 5 + 4) \cdot 5 + 3) \cdot 5 + 1) \cdot 5 + 4$$

$$= 9334$$

> Note the four left parentheses for a 6-digit numeral.

EXAMPLE 3   Converting from Base Ten to Base Five

Convert 497 from decimal form to base five.

Solution

The base-five place values, starting from the right, are

$$1, \quad 5, \quad 25, \quad 125, \quad 625, \quad \text{and so on.}$$

Because 497 is between 125 and 625, it will require no 625s, but some 125s, as well as possibly some 25s, 5s, and 1s.

- Dividing 497 by 125 yields a quotient of 3, which is the proper number of 125s.

- The remainder, 122, is divided by 25 (the next place value) to find the proper number of 25s. The quotient is 4, with remainder 22, so we need four 25s.

- Dividing 22 by 5 yields 4, with remainder 2, so we need four 5s.

- Dividing 2 by 1 yields 2 (with remainder 0), so we need two 1s.

Thus 497 consists of three 125s, four 25s, four 5s, and two 1s, so $497 = 3442_{\text{five}}$. More concisely, this process can be written as follows.

$$497 \div 125 = 3 \qquad \text{Remainder 122}$$
$$122 \div 25 = 4 \qquad \text{Remainder 22}$$
$$22 \div 5 = 4 \qquad \text{Remainder 2}$$
$$2 \div 1 = 2 \qquad \text{Remainder 0}$$
$$497 = 3442_{\text{five}}$$

Check:
$$3442_{\text{five}} = (3 \cdot 125) + (4 \cdot 25) + (4 \cdot 5) + (2 \cdot 1)$$
$$= 375 + 100 + 20 + 2$$
$$= 497 \checkmark$$

The symbol here is the ancient Chinese "**yin-yang**," in which the black and the white enfold each other, each containing a part of the other. A kind of duality is conveyed between destructive (yin) and beneficial (yang) aspects.

Leibniz (1646–1716) studied Chinese ideograms in search of a universal symbolic language and promoted East–West cultural contact.

Niels Bohr (1885–1962), famous Danish Nobel laureate in physics (atomic theory), adopted the yin-yang symbol in his coat of arms to depict his principle of *complementarity*, which he believed was fundamental to reality at the deepest levels. Bohr also pushed for East–West cooperation.

In its 1992 edition, *The World Book Dictionary* first judged "yin-yang" to have been used enough to become a permanent part of our ever-changing language, assigning to it the definition "made up of opposites."

The calculator shortcut for converting from another base to decimal form involved repeated *multiplications* by the other base. (See **Example 2.**) A shortcut for converting from decimal form to another base makes use of repeated *divisions* by the other base. Just divide the original decimal numeral, and the resulting quotients in turn, by the desired base until the quotient 0 appears.

Woven fabric is a binary system of threads going lengthwise (warp threads—tan in the diagram above) and threads going crosswise (weft or woof). At any point in a fabric, either warp or weft is on top, and the variation creates the pattern.

Nineteenth-century looms for weaving operated using punched cards "programmed" for a pattern. The looms were set up with hooked needles, the hooks holding the warp. Where there were holes in cards, the needles moved, the warp lifted, and the weft passed under. Where no holes were, the warp did not lift, and the weft was on top. The system parallels the on–off system in calculators and computers. In fact, these looms were models in the development of modern calculating machinery.

Joseph Marie Jacquard (1752–1834) is credited with improving the mechanical loom so that mass production of fabric was feasible.

EXAMPLE 4 **Using a Shortcut to Convert from Base Ten**

Repeat **Example 3** using the shortcut just described.

Solution

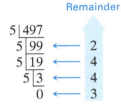

$$
\begin{array}{r|r}
5 & 497 \\
5 & 99 \leftarrow 2 \\
5 & 19 \leftarrow 4 \\
5 & 3 \leftarrow 4 \\
& 0 \leftarrow 3
\end{array}
$$

Read the answer from the remainder column, reading from the bottom up.

$$497 = 3442_{\text{five}}$$

To see why this shortcut works, notice the following:

- The first division shows that four hundred ninety-seven 1s are equivalent to ninety-nine 5s and two 1s. (The two 1s are set aside and account for the last digit of the answer.)
- The second division shows that ninety-nine 5s are equivalent to nineteen 25s and four 5s. (The four 5s account for the next digit of the answer.)
- The third division shows that nineteen 25s are equivalent to three 125s and four 25s. (The four 25s account for the next digit of the answer.)
- The fourth (and final) division shows that the three 125s are equivalent to no 625s and three 125s. The remainders, as they are obtained *from top to bottom*, give the number of 1s, then 5s, then 25s, then 125s.

The methods for converting between bases ten and five, including the shortcuts, can be adapted for conversions between base ten and any other base.

EXAMPLE 5 **Converting from Base Seven to Base Ten**

Convert 6343_{seven} to decimal form, by expanding in powers, and by using the calculator shortcut.

Solution

$$
\begin{aligned}
6343_{\text{seven}} &= (6 \cdot 7^3) + (3 \cdot 7^2) + (4 \cdot 7^1) + (3 \cdot 7^0) \\
&= (6 \cdot 343) + (3 \cdot 49) + (4 \cdot 7) + (3 \cdot 1) \\
&= 2236
\end{aligned}
$$

Calculator shortcut: $6343_{\text{seven}} = ((6 \cdot 7 + 3) \cdot 7 + 4) \cdot 7 + 3 = 2236.$

EXAMPLE 6 **Converting from Base Ten to Base Seven**

Convert 7508 to base seven.

Solution

Divide 7508 by 7, then divide the resulting quotient by 7, until a quotient of 0 results.

$$
\begin{array}{r|r}
7 & 7508 \\
7 & 1072 \leftarrow 4 \\
7 & 153 \leftarrow 1 \\
7 & 21 \leftarrow 6 \\
7 & 3 \leftarrow 0 \\
& 0 \leftarrow 3
\end{array}
$$

From the remainders, reading bottom to top, $7508 = 30614_{\text{seven}}$.

To handle conversions between arbitrary bases (where neither is ten), go from the given base to base ten and then to the desired base.

EXAMPLE 7 Converting between Two Bases Other Than Ten

Convert 3164_{seven} to base five.

Solution

Convert to decimal form.

$$3164_{\text{seven}} = (3 \cdot 7^3) + (1 \cdot 7^2) + (6 \cdot 7^1) + (4 \cdot 7^0)$$
$$= (3 \cdot 343) + (1 \cdot 49) + (6 \cdot 7) + (4 \cdot 1)$$
$$= 1029 + 49 + 42 + 4$$
$$= 1124$$

Convert this decimal result to base five.

Remainder

$$\begin{array}{r} 5\,|\,\underline{1124} \\ 5\,|\,\underline{224} \longleftarrow 4 \\ 5\,|\,\underline{44} \longleftarrow 4 \\ 5\,|\,\underline{8} \longleftarrow 4 \\ 5\,|\,\underline{1} \longleftarrow 3 \\ 0 \longleftarrow 1 \end{array}$$

From the remainders, $3164_{\text{seven}} = 13444_{\text{five}}$. ∎

Computer Mathematics

There are three alternative base systems that are most useful in computer applications—**binary** (base two), **octal** (base eight), and **hexadecimal** (base sixteen).

Computers and handheld calculators use the binary system for their internal calculations because that system consists of only two symbols, 0 and 1. All numbers can then be represented by electronic "switches," where "on" indicates 1 and "off" indicates 0. The octal and hexadecimal systems have been used extensively by programmers who work with internal computer codes and for communication between the CPU (central processing unit) and a printer or other output device.

The binary system is extreme in that it has only two available symbols (0 and 1). Thus, representing numbers in binary form requires more digits than in any other base. **Table 10** shows the whole numbers up to 31 expressed in binary form.

Table 10

Base-Two Numerals

Base Ten	Base Two	Base Ten	Base Two
0	0	16	10000
1	1	17	10001
2	10	18	10010
3	11	19	10011
4	100	20	10100
5	101	21	10101
6	110	22	10110
7	111	23	10111
8	1000	24	11000
9	1001	25	11001
10	1010	26	11010
11	1011	27	11011
12	1100	28	11100
13	1101	29	11101
14	1110	30	11110
15	1111	31	11111

EXAMPLE 8 Converting from Binary to Decimal

Convert 110101_{two} to decimal form.

Solution

$$110101_{\text{two}} = (1 \cdot 2^5) + (1 \cdot 2^4) + (0 \cdot 2^3) + (1 \cdot 2^2) + (0 \cdot 2^1) + (1 \cdot 2^0)$$
$$= (1 \cdot 32) + (1 \cdot 16) + (0 \cdot 8) + (1 \cdot 4) + (0 \cdot 2) + (1 \cdot 1)$$
$$= 32 + 16 + 0 + 4 + 0 + 1$$
$$= 53$$

Calculator shortcut: $110101_{\text{two}} = ((((1 \cdot 2 + 1) \cdot 2 + 0) \cdot 2 + 1) \cdot 2 + 0) \cdot 2 + 1$
$$= 53$$

> Note the four left parentheses for a 6-digit numeral.

∎

EXAMPLE 9 Converting from Decimal to Octal

Convert 9583 to octal form.

Solution

Divide repeatedly by 8, writing the remainders at the side.

$$
\begin{array}{r|r}
8 & 9583 \\
8 & 1197 \leftarrow \quad 7 \\
8 & 149 \leftarrow \quad 5 \\
8 & 18 \leftarrow \quad 5 \\
8 & 2 \leftarrow \quad 2 \\
 & 0 \leftarrow \quad 2
\end{array}
$$

Remainder

From the remainders, $9583 = 22557_{\text{eight}}$.   ∎

The hexadecimal system, having base sixteen (and 16 being greater than 10), presents a new problem. Because distinct symbols are needed for all whole numbers from 0 up to one less than the base, base sixteen requires more symbols than are normally used in our decimal system. Computer programmers commonly use the letters **A, B, C, D, E, and F** as hexadecimal digits for **the numbers ten through fifteen, respectively.**

EXAMPLE 10 Converting from Hexadecimal to Decimal

Convert $FA5_{\text{sixteen}}$ to decimal form.

Solution

$$
\begin{aligned}
\mathbf{FA}5_{\text{sixteen}} &= (\mathbf{15} \cdot 16^2) + (\mathbf{10} \cdot 16^1) + (5 \cdot 16^0) \\
&= 3840 + 160 + 5 \\
&= 4005
\end{aligned}
$$

F and A represent 15 and 10, respectively.   ∎

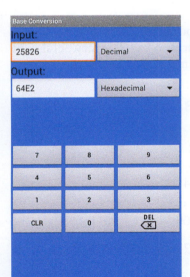

EXAMPLE 11 Converting from Decimal to Hexadecimal

Convert 748 from decimal form to hexadecimal form.

Solution

Use repeated division by 16.

$$
\begin{array}{r|r}
16 & 748 \\
16 & 46 \leftarrow \quad 12 \leftarrow \quad C \\
16 & 2 \leftarrow \quad 14 \leftarrow \quad E \\
 & 0 \leftarrow \quad 2 \leftarrow \quad 2
\end{array}
$$

Remainder   Hexadecimal notation

From the remainders at the right, $748 = 2EC_{\text{sixteen}}$.   ∎

The decimal whole numbers 0 through 17 are shown in **Table 11** on the next page, along with their equivalents in the common computer-oriented bases (two, eight, and sixteen). Conversions among binary, octal, and hexadecimal can generally be accomplished by the shortcuts illustrated in the remaining examples.

The binary system is the natural one for internal computer workings because of its compatibility with the two-state electronic switches. It is very cumbersome, however, for human use, because so many digits occur even in the numerals for relatively small numbers. The octal and hexadecimal systems are the choices of computer programmers mainly because of their close relationship with the binary system. ***Both eight and sixteen are powers of two.***

Table 11 Some Decimal Equivalents in the Common
Computer-Oriented Bases

Decimal (Base Ten)	Hexadecimal (Base Sixteen)	Octal (Base Eight)	Binary (Base Two)
0	0	0	0
1	1	1	1
2	2	2	10
3	3	3	11
4	4	4	100
5	5	5	101
6	6	6	110
7	7	7	111
8	8	10	1000
9	9	11	1001
10	A	12	1010
11	B	13	1011
12	C	14	1100
13	D	15	1101
14	E	16	1110
15	F	17	1111
16	10	20	10000
17	11	21	10001

Table 12

Octal	Binary
0	000
1	001
2	010
3	011
4	100
5	101
6	110
7	111

When conversions involve one base that is a power of the other, there is a quick conversion shortcut available. For example, because $8 = 2^3$, every octal digit (0 through 7) can be expressed as a 3-digit binary numeral. See **Table 12**.

EXAMPLE 12 Converting from Octal to Binary

Convert 473_{eight} to binary form.

Solution

Replace each octal digit with its 3-digit binary equivalent. (Leading zeros can be omitted only when they occur in the leftmost group.) Combine the equivalents into a single binary numeral.

$$
\begin{array}{ccc}
4 & 7 & 3_{eight} \\
\downarrow & \downarrow & \downarrow \\
100 & 111 & 011_{two}
\end{array}
$$

$$473_{eight} = 100111011_{two}$$ ■

EXAMPLE 13 Converting from Binary to Octal

Convert 10011110_{two} to octal form.

Solution

Start at the right and break the digits into groups of three. Then convert the groups to their octal equivalents.

Finally, $10011110_{two} = 236_{eight}.$

$$
\begin{array}{ccc}
10 & 011 & 110_{two} \\
\downarrow & \downarrow & \downarrow \\
2 & 3 & 6_{eight}
\end{array}
$$ ■

The following message was seen on the front of a T-shirt. "There are only 10 types of people in the world: Those who understand **binary** and those who don't." Do YOU understand this message?

Because $16 = 2^4$, every hexadecimal digit can be equated to a 4-digit binary numeral (see **Table 13** below), and conversions between binary and hexadecimal forms can be done in a manner similar to that used in **Examples 12 and 13.**

EXAMPLE 14 **Converting from Hexadecimal to Binary**

Convert $8B4F_{sixteen}$ to binary form.

Solution

Each hexadecimal digit yields a 4-digit binary equivalent.

$$
\begin{array}{cccc}
8 & B & 4 & F_{sixteen} \\
\downarrow & \downarrow & \downarrow & \downarrow \\
1000 & 1011 & 0100 & 1111_{two}
\end{array}
$$

Combining these groups of digits, $8B4F_{sixteen} = 1000101101001111_{two}$. ■

Several games and tricks are based on the binary system. For example, **Table 14** can be used to find the age of a person 31 years old or younger. The person need only tell you the columns that contain his or her age. For example, suppose Kathy says that her age appears in columns B, C, and D. To find her age, add the numbers from the top row of these columns:

Kathy is $2 + 4 + 8 = 14$ years old.

Do you see how this trick works? (See **Exercises 65–68.**)

Table 13

Hexadecimal	Binary
0	0000
1	0001
2	0010
3	0011
4	0100
5	0101
6	0110
7	0111
8	1000
9	1001
A	1010
B	1011
C	1100
D	1101
E	1110
F	1111

Table 14

A	B	C	D	E
1	2	4	8	16
3	3	5	9	17
5	6	6	10	18
7	7	7	11	19
9	10	12	12	20
11	11	13	13	21
13	14	14	14	22
15	15	15	15	23
17	18	20	24	24
19	19	21	25	25
21	22	22	26	26
23	23	23	27	27
25	26	28	28	28
27	27	29	29	29
29	30	30	30	30
31	31	31	31	31

In the world of computers, each binary digit is called a **bit,** and eight bits are called a **byte.** Computers receive, process, and transmit sequences of 1s and 0s (binary code). When one computer sends information to another computer, it *encapsulates* the information with binary strings that indicate, among other things, file type (such as text, photo, or video) and the *address* of the target computer. This Internet Protocol (IP) address is expressed in **dotted decimal notation,** which features four decimal numerals, each ranging from 0 to 255, separated by decimal points.

Converting an IP Address

Convert the IP address 192.168.1.108 to binary form.

Solution

Each decimal numeral ranging from 0 to 255 can be expressed in eight bits.

$$192 = 128 + 64 = 2^7 + 2^6 \qquad\qquad\quad = 11000000_{\text{two}}$$

$$168 = 128 + 32 + 8 = 2^7 + 2^5 + 2^3 \qquad = 10101000_{\text{two}}$$

$$1 = 2^0 \qquad\qquad\qquad\qquad\qquad\qquad = 00000001_{\text{two}}$$

$$108 = 64 + 32 + 8 + 4 = 2^6 + 2^5 + 2^3 + 2^2 = 01101100_{\text{two}}$$

Thus the binary representation for this IP address is as follows.

$$11000000.10101000.00000001.01101100$$

■

WHEN Will I Ever USE This ?

Suppose you are a network administrator for a company that has a network address of 192.168.3.0. The *subnet mask* for your network is 255.255.255.0, which means that the first twenty-four bits of any IP address in your network are used to designate the network address. The last eight bits are available to be assigned to individual computers within your network.

$$\underbrace{11000000.10101000.00000011.}_{\text{Network field}}\ \underbrace{00000000}_{\text{Host field}}$$

You have been directed by the owner of the company to set up subnetworks within the network—one for management and one for employees. This will enable you to keep certain information (such as personnel files and financial data) on company servers from being viewed by employees who have no business seeing them. The management subnet will only need up to 20 computers (hosts) connected, but the employee subnet will need as many as 50.

1. What new subnet mask should be used to set up your subnetworks?

2. What IP addresses will be available for assignment to computers on each subnetwork?

Because the largest subnet will need to host as many as 50 computers, you will need at least six bits allocated to hosts. (Five bits would not be enough because that would allow for only $2^5 = 32$ host addresses.)

Since the current network address allows eight bits for hosts within the network, you decide to borrow the two highest-order bits to use for subnets, leaving the last six for hosts.

$$11000000.10101000.00000011.\underbrace{00}_{\substack{\text{Subnet}\\\text{field}}}\underbrace{000000}_{\substack{\text{Host}\\\text{field}}}$$

Network field   Subnet field   Host field

A subnet mask has all 1s in the network and subnet fields, and all 0s in the host field. Thus the new subnet mask is

$$11111111.11111111.11111111.\underbrace{11}_{\substack{\text{Subnet}\\\text{field}}}\underbrace{000000}_{\substack{\text{Host}\\\text{field}}},$$

Network field   Subnet field   Host field

which is expressed in dotted decimal notation as follows.

$$255.255.255.\underset{11000000}{192}$$

Note that this configuration will allow for four subnets, each having 62 addresses available for hosts, as shown in **Table 15**. (The host addresses come in blocks of 64, but the first is reserved for the subnetwork address, and the last, with all 1s in the host field, is reserved as a "broadcast" address for use when messages need to be sent to all hosts on the subnetwork.)

Table 15

Subnet	Subnetwork Address	Range of Host Addresses	Broadcast Address
00	192.168.3.0	192.168.3.1–192.168.3.62	192.168.3.63
01	192.168.3.64	192.168.3.65–192.168.3.126	192.168.3.127
10	192.168.3.128	192.168.3.129–192.168.3.190	192.168.3.191
11	192.168.3.192	192.168.3.193–192.168.3.254	192.168.3.255

You decide to assign the 01 subnet (with address 192.168.3.64) to management, and the 10 subnet (with address 192.168.3.128) to employees. The topology of your company's newly configured network is shown in **Figure 12**.

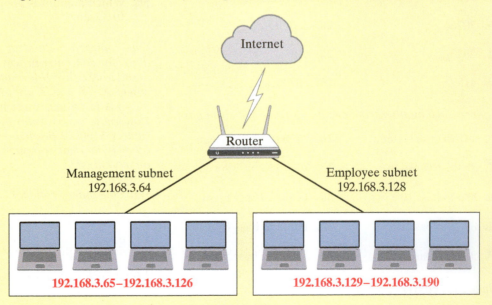

Figure 12

The router can now be programmed to filter traffic between the management and employee subnetworks.

4.4 EXERCISES

List the first twenty counting numbers in each base.

1. seven (Only digits 0 through 6 are used in base seven.)

2. eight (Only digits 0 through 7 are used.)

3. nine (Only digits 0 through 8 are used.)

4. sixteen (The digits 0, 1, 2, . . . , 9, A, B, C, D, E, F are used in base sixteen.)

Write (in the same base) the counting numbers just before and just after the given number. (Do not convert to base ten.)

5. 14_{five}

6. 555_{six}

7. $B6F_{\text{sixteen}}$

8. 10111_{two}

Determine the number of distinct symbols needed in each of the following positional systems.

9. base three

10. base seven

11. base eleven

12. base sixteen

Determine, in each base, the least and greatest four-digit numbers and their decimal equivalents.

13. three

14. sixteen

Convert each number to decimal form by expanding in powers and by using the calculator shortcut.

15. $3BC_{\text{sixteen}}$

16. 34432_{five}

17. 2366_{seven}

18. 101101110_{two}

19. 70266_{eight}

20. $ABCD_{\text{sixteen}}$

21. 2023_{four}

22. 6185_{nine}

23. 41533_{six}

24. 88703_{nine}

Convert each number from decimal form to the given base.

25. 147 to base sixteen

26. 2730 to base sixteen

27. 36401 to base five

28. 70893 to base seven

29. 586 to base two

30. 12888 to base eight

31. 8407 to base three

32. 11028 to base four

33. 9346 to base six

34. 99999 to base nine

Make each conversion as indicated.

35. 43_{five} to base seven

36. 27_{eight} to base five

37. 6748_{nine} to base four

38. $C02_{\text{sixteen}}$ to base seven

Convert each number from octal form to binary form.

39. 367_{eight}

40. 2406_{eight}

Convert each number from binary form to octal form.

41. 100110111_{two}

42. 11010111101_{two}

Make each conversion as indicated.

43. DC_{sixteen} to binary

44. $F111_{\text{sixteen}}$ to binary

45. 101101_{two} to hexadecimal

46. $101111011101000_{\text{two}}$ to hexadecimal

Identify the greatest number from each list.

47. $42_{\text{seven}}, 37_{\text{eight}}, 1D_{\text{sixteen}}$

48. $1101110_{\text{two}}, 414_{\text{five}}, 6F_{\text{sixteen}}$

*There is a theory that twelve would be a better base than ten for general use. This is mainly because twelve has more divisors (1, 2, 3, 4, 6, 12) than ten (1, 2, 5, 10), which makes fractions easier in base twelve. The base-twelve system is called the **duodecimal system**. In the decimal system we speak of a one, a ten, and a hundred (and so on); in the duodecimal system we say a one, a dozen (twelve), and a gross (twelve squared, or one hundred forty-four).*

49. Adam's clients ordered 9 gross, 10 dozen, and 11 copies of *The King* during 2014. How many copies was that in base ten?

50. Which amount is greater: 3 gross, 6 dozen or 2 gross, 19 dozen?

*One common method of converting symbols into binary digits for computer processing is called **ASCII** (American Standard Code of Information Interchange). The uppercase letters A through Z are assigned the numbers 65 through 90, so A has binary code 1000001 and Z has code 1011010. Lowercase letters a through z have codes 97 through 122 (that is, 1100001 through 1111010). ASCII codes, as well as other numerical computer output, normally appear without commas.*

Write the binary code for each letter.

51. C

52. X

53. k

54. q

Break each code into groups of seven digits and write as letters.

55. 1001000100010110011001010000

56. 1000011100100010101011000011001011

Translate each word into an ASCII string of binary digits. (Be sure to distinguish uppercase and lowercase letters.)

57. New

58. Orleans

*Exercises 59–62 refer to the **When Will I Ever Use This?** feature on **page 168.***

59. Explain why an IP address like 192.168.3.30 is a 32-bit address, and convert it to binary.

60. What subnet mask would you use to split a network with address 192.168.7.0 into 8 subnetworks, each having room for 30 hosts?

If the IP address and subnet mask for a computer are known, then the network address can be found by performing a bitwise logical AND on the IP address and subnet mask. That is, convert the dotted decimal address and mask into binary form, AND them together bit-by-bit, and convert the result back to dotted decimal form.

61. Suppose your computer's IP address is 192.172.3.67 and the subnet mask is 255.255.255.192. What is the network address for the network your computer is on?

62. If a computer's IP address is 192.168.3.135, and the subnet mask is 255.255.255.224, what is the network address?

63. Explain why the octal and hexadecimal systems are convenient for people who code for computers.

64. There are thirty-seven counting numbers whose base eight numerals contain two digits but whose base three numerals contain four digits. Find the least and greatest of these numbers.

*Refer to **Table 14** for Exercises 65–68.*

65. After observing the binary forms of the numbers 1–31, identify a common property of all **Table 14** numbers in each of the following columns.

 (a) Column A

 (b) Column B

 (c) Column C

 (d) Column D

 (e) Column E

66. Explain how the "trick" of **Table 14** works.

67. How many columns would be needed for **Table 14** to include all ages up to 63?

68. How many columns would be needed for **Table 14** to include all ages up to 127?

In our decimal system, we distinguish odd and even numbers by looking at their ones (or units) digits. If the ones digit is even (0, 2, 4, 6, or 8), the number is even. If the ones digit is odd (1, 3, 5, 7, or 9), the number is odd. For Exercises 69–76, determine whether this same criterion works for numbers expressed in the given bases.

69. two **70.** three **71.** four

72. five **73.** six **74.** seven

75. eight **76.** nine

77. Consider all even bases. If the above criterion works for all, explain why. If not, find a criterion that does work for all even bases.

78. Consider all odd bases. If the above criterion works for all, explain why. If not, find a criterion that does work for all odd bases.

Determine whether the given base-five numeral represents one that is divisible by five.

79. 3204_{five} **80.** 200_{five}

81. 2310_{five} **82.** 342_{five}

83. If you want to multiply a decimal numeral by 10, what do you do?

84. If you want to multiply a binary numeral by 2, what do you do?

85. Explain why your answers to **Exercises 83 and 84** are correct.

86. If a binary numeral ends in 0, then it is even. How would you divide such a number in half?

*Recall that conversions between binary and octal are simplified because eight is a power of 2: $8 = 2^3$. (See **Examples 12 and 13.**) The same is true of conversions between binary and hexadecimal, because $16 = 2^4$. (See **Example 14.**) Direct conversion between octal and hexadecimal does not work the same way, because 16 is not a power of 8. Explain how to carry out each conversion without using base ten, and give an example.*

87. hexadecimal to octal **88.** octal to hexadecimal

Devise a method (similar to the one for conversions between binary, octal, and hexadecimal) for converting between base three and base nine, and use it to carry out each conversion.

89. 6504_{nine} to base three

90. 81170_{nine} to base three

91. 212201221_{three} to base nine

92. 200121021_{three} to base nine

The colors seen on a computer screen consist of combinations of the primary colors red (R), green (G), and blue (B). The intensity of each of the three ranges from 0 to 255. This allows

$$255^3 = 16,581,375 \text{ distinct colors}$$

(at least theoretically), where a particular color's designation may require up to 9 digits.

For example, the (R, G, B) designation for black is (0, 0, 0), white is (255, 255, 255), and gray is (128, 128, 128).

But the colors are coded in **HTML** (HyperText Markup Language) in hexadecimal, so each primary color intensity then requires no more than two digits. $(255 = \text{FF}_{sixteen}.)$ White is coded FFFFFF, and black is 000000.

Give similar HTML codes for the following colors.

93. gray

94. yellow (255, 255, 0)

95. hot pink (255, 105, 180)

96. the color (171, 205, 239)

CHAPTER 4   SUMMARY

KEY TERMS

4.1
numeral
numeration system
tally stick
tally
Hindu-Arabic system
grouping
base
simple grouping system
algorithm

multiplicative grouping
   system

4.2
positional system
digit
face value
place value
placeholder
ciphered system

4.3
expanded form
distributive property
decimal system
abacus
lattice method
Napier's rods
Russian peasant method
nines complement
   method

4.4
binary
octal
hexadecimal
bit
byte
dotted decimal
   notation

TEST YOUR WORD POWER

See how well you have learned the vocabulary in this chapter.

1. A **simple grouping system** is
   A. a numeration system that uses group position to determine values of numerals.
   B. a numeration system in which the order of the symbols in a numeral makes no difference.
   C. a numeration system that uses grouping symbols (such as parentheses) to simplify numerals.
   D. a system that constructs numerals by grouping different numbers together.

2. A **positional system** is
   A. a numeration system that features a different symbol for each power of the base.
   B. a numeration system that works differently depending on the context in which it is used.
   C. a numeration system that is extremely efficient because only multipliers are used.
   D. a system that uses position to determine the face values of the digits.

3. **Napier's rods (or bones)** is the name of
   A. a calculating device based on the lattice method for multiplication.

   B. a device containing sliding beads and a dividing bar used to speed calculations.
   C. a device based on the Russian peasant method for multiplication.
   D. an adding machine designed around the nines complement method of subtraction.

4. The **hexadecimal** numeration system is
   A. a numeration system needing twelve distinct symbols.
   B. a base-sixteen numeration system using symbols 0–9, A, B, C, D, E, and F.
   C. a base-six numeration system with multipliers corresponding to the vertices of a hexagon.
   D. a system having fewer symbols than the octal numeration system.

ANSWERS
1. B   **2.** C   **3.** A   **4.** B

QUICK REVIEW

Concepts	*Examples*

4.1 Historical Numeration Systems

Numeration systems use symbols called **numerals** to represent numbers. Ancient Egyptian numeration is an example of a **simple grouping system.** It is a base-ten system that uses a different symbol to represent each power of 10 (from 10^0 to 10^6).

Add $123,223 + 3104$ using ancient Egyptian numeration.

The sum can be translated to our Hindu-Arabic system as

126,327.

Early Egyptian Symbols

Number	Symbol
1		
10	∩
100	9
1000	⌁
10,000	⌁
100,000	⌁
1,000,000	⌁

The ancient Roman numeration system is also base ten, but it has symbols for 5, 50, and 500 as well. It uses a subtractive feature (involving writing a smaller-valued symbol immediately to the left of a larger-valued symbol) to reduce repetition of symbols, and a multiplicative feature in which a bar over a symbol means to multiply it by 1000, and a double-bar indicates multiplication by 1,000,000.

Express the difference $20,236 - 10,125$ in Roman numerals.

$$\overline{XX}CCXXXVI$$
$$- \quad \overline{X}CXXV$$
$$\overline{X}CXI$$

Translating the result to Hindu-Arabic numeration gives

10,111.

Classical Chinese numeration is a base-ten **multiplicative grouping system,** with some modifications. Pairs of symbols are written, each including a multiplier (with value less than the base) and a symbol representing a power of the base. The modifications involve exceptions, which include omitting the symbol for 10^0 when displaying the number of 1s, omitting the multiplier for the 10s if the multiplier is 1, and using a single zero symbol in place of one, two, or more consecutive "missing" powers of 10.

Express the numeral 2014 in the classical Chinese numeration system.

Read from top to bottom.

4.2 More Historical Numeration Systems

Positional systems rely on the position of each symbol to determine its place value (that is, for which power of the base it is a multiplier).

In our Hindu-Arabic (base-ten) numeration system, the numeral

3024

means "three thousand twenty-four." The digit 3 is in the *thousands* place, the placeholder 0 is in the *hundreds* place, the digit 2 is in the *tens* place, and 4 is in the *ones* place.

Concepts	Examples

The Babylonian numeration system features only two symbols.

Babylonian Symbols

Number	Symbol
1	▼
10	◄

These symbols are grouped to express the numbers 1–59. The groups are then positioned to represent powers of 60, the base of the system.

The Mayan numeration system uses three symbols.

Mayan Symbols

Number	Symbol
0	◯
1	·
5	—

Most notable is the presence of a symbol for zero. The **place values** in the Mayan system are 1, 20, 360, 7200, 144,000, and so on. Each place value is 20 times the previous place value, with the exception that $360 = 18 \cdot 20$. Thus it is essentially a base-twenty positional system using simple groupings of symbols to form multipliers with face values ranging from 0 to 19.

Greek numeration is a **ciphered system,** using 27 distinct symbols. These symbols can be used to express numbers less than 1000 easily. Multiples of 1000 and 10,000 require the use of special features.

The Babylonian numeral

$$\underbrace{\text{《《▼▼▼}}_{23 \cdot 3600} \quad \underbrace{\text{《▼▼}}_{12 \cdot 60} \quad \underbrace{\text{▼▼▼▼}}_{4 \cdot 1}$$

represents

$$23 \cdot 3600 + 12 \cdot 60 + 4 \cdot 1 = 83{,}524. \quad \text{Hindu-Arabic form}$$

To convert the numeral 83,524 to a Babylonian numeral, divide 83,524 by 3600 (since 3600 is the largest power of 60 that divides into 83,524) to get a quotient of 23 and a remainder of 724. Divide 724 by 60 to get a quotient of 12 with a remainder of 4. Then express these multiples of powers of 60 using the symbols as shown above.

The Mayan numeral

$$\begin{aligned}\text{···} &\longleftarrow 8 \cdot 360\\ \text{◯} &\longleftarrow 0 \cdot 20\\ \text{≟} &\longleftarrow 12 \cdot 1\end{aligned}$$

represents

$$8 \cdot 360 + 0 \cdot 20 + 12 \cdot 1 = 2892. \quad \text{Hindu-Arabic form}$$

To convert the Hindu-Arabic numeral 948 to a Mayan numeral, divide by 360, and divide the remainder by 20 to see that

$$948 = 2 \cdot 360 + 11 \cdot 20 + 8.$$

In Mayan, this is written as follows.

Since α represents 1, ϕ represents 500, μ represents 40, and β represents 2, the Greek numeral $\overset{\alpha}{\text{M}}\phi\mu\beta$ represents

$$10{,}000 + 500 + 40 + 2 = 10{,}542.$$

The Hindu-Arabic numeral 837 is expressed in Greek numeration as follows.

$$\omega\lambda\zeta$$

4.3 Arithmetic in the Hindu-Arabic System

Since the Hindu-Arabic system is a positional base-ten numeration system, the digits in a numeral are multipliers for powers of 10. Writing a numeral in **expanded form** makes this clear.

Adding numerals using expanded form makes clear the need to "carry" when the sum includes more than nine of any power of 10. Similarly, expanded notation makes clear the need to "borrow" when subtracting a larger number of a power of 10 from a smaller number.

Write the numeral 4132 in expanded form.

$$4 \cdot 10^3 + 1 \cdot 10^2 + 3 \cdot 10^1 + 2 \cdot 10^0$$

Add 67 and 56 using expanded notation.

$$\begin{aligned} 67 &= (6 \cdot 10^1) & + (7 \cdot 10^0)\\ + 56 &= (5 \cdot 10^1) & + (6 \cdot 10^0)\\ \hline & (11 \cdot 10^1) & + (13 \cdot 10^0) \end{aligned}$$

We must rewrite both products in this sum.

Concepts	Examples
	$11 \cdot 10^1 = (10 + 1) \cdot 10^1 = 1 \cdot 10^2 + 1 \cdot 10^1$, and $13 \cdot 10^0 = (10 + 3) \cdot 10^0 = 1 \cdot 10^1 + 3 \cdot 10^0$, so the sum $11 \cdot 10^1 + 13 \cdot 10$ becomes $$(1 \cdot 10^2 + 1 \cdot 10^1) + (1 \cdot 10^1 + 3 \cdot 10^0)$$ $$= 1 \cdot 10^2 + (1 \cdot 10^1 + 1 \cdot 10^1) + 3 \cdot 10^0$$ $$= 1 \cdot 10^2 + 2 \cdot 10^1 + 3 \cdot 10^0 = 123.$$ $$= 123.$$

Over time, several calculation devices have been invented to make calculation with Hindu-Arabic numerals more efficient. Two examples are the **abacus** and **Napier's rods.** The abacus consists of a series of rods (corresponding to powers of 10), beads (for the multipliers), and a dividing bar which separates 5s (above the rod in the figure at right) from 1s (below the rod in the figure at right). Napier's rods contain strips of numbers that facilitate use of the **lattice method** for multiplication.

The abacus shows the number 11,707.

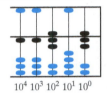

$10^4 \quad 10^3 \quad 10^2 \quad 10^1 \quad 10^0$

Napier's rods show the multiplication of $6 \cdot 2305$.

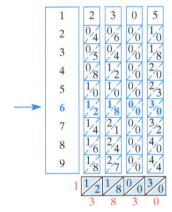

The product is 13,830.

The **nines complement method** of subtraction involves replacing the digits in the subtrahend with their nines complements, adding the resulting number to the minuend, and then deleting the leading 1 and adding 1 to the result.

Calculate the difference $625 - 48$ using the nines complement method.

$$
\begin{array}{rrr}
625 & 625 & 576 \\
-048 & 951 & 1 \\
\hline
 & 1576 & 577 \leftarrow \text{Answer}
\end{array}
$$

4.4 Conversion between Number Bases

Any base-n positional numeration system needs n distinct symbols for expressing numbers.

The hexadecimal (base-sixteen) system uses 0–9 and A, B, C, D, E, and F as digits.

Converting from Another Base to Decimal Form

Convert the base-five numeral 1214_{five} to decimal form.

Step 1 Start with the first digit on the left and multiply by the base.

$$1214_{\text{five}} = ((1 \cdot 5 + 2) \cdot 5 + 1) \cdot 5 + 4 = 184$$

Step 2 Then add the next digit, multiply again by the base, and so on.

Step 3 Add the last digit on the right. ***Do not multiply it by the base.***

Concepts	Examples

We can convert from decimal to any other base n by dividing the decimal numeral by the other base repeatedly, recording the remainders, and dividing quotients until the quotient 0 appears. Then write the remainders in reverse order.

Convert 395 to base four as shown.

Remainder

```
4 ⌐395
 4 ⌐98  ←  3
  4 ⌐24  ←  2
   4 ⌐6  ←  0
    4 ⌐1  ←  2
     4 ⌐0  ←  1
```

Thus $395 = 12023_{four}$.

Computers operate on the basis of the binary system, using 1s and 0s corresponding to switch states of "on" and "off", respectively. Because 8 and 16 are powers of 2, the octal and hexadecimal systems are also helpful in condensing data into shorter strings of symbols.

Octal	Binary
0	000
1	001
2	010
3	011
4	100
5	101
6	110
7	111

Because $8 = 2^3$, each digit in the octal system can be expressed as three digits ("bits") in the binary system.

Convert the octal numeral 375_{eight} to binary. Replace each digit with its three-bit representation.

$$375_{eight} = \underset{3}{\underbrace{011}}\,\underset{7}{\underbrace{111}}\,\underset{5}{\underbrace{101}}_{two}$$

Decimal-to-binary conversion is helpful when working with IP addresses and networking configurations.

Express the IP address 192.168.5.35 in binary form.

11000000.10101000.00000101.00100011

CHAPTER 4   TEST

In each case, identify the numeration system, and give the Hindu-Arabic equivalent.

1. ⌐ 99 ∩∩∩ ||
    Ⴟ 999 ||

2. $\overline{\text{XCDLXXIV}}$

3.
≡
百
λ
十
五

4. ⟨⟨▼▼ ⟨▼▼⟨⟨
    ⟨⟨▼

5.
··
⊖
⁚

6. $\overset{\epsilon}{\text{M}},\gamma\phi\kappa\delta$

Perform each operation using the alternative algorithm specified.

7. $23 \cdot 45$ (Russian peasant or Egyptian method)

8. $246 \cdot 97$ (Lattice method)

9. $21{,}425 - 8198$ (Nines complement method)

Convert each number to base ten.

10. 243_{five}

11. 100101_{two}

12. $BEEF_{sixteen}$

Convert as indicated.

13. 49 to binary

14. 2930 to base five

15. 10101110_{two} to octal

16. 7215_{eight} to hexadecimal

17. 5041_{six} to decimal

18. $BAD_{sixteen}$ to binary

Briefly explain each of the following.

19. the advantage of multiplicative grouping over simple grouping

20. the advantage of positional over multiplicative grouping

21. the advantage, in a positional numeration system, of a smaller base over a larger base

22. the advantage, in a positional numeration system, of a larger base over a smaller base

23. Explain a quick method to convert a base-nine numeral to base three.

24. Illustrate your method from **Exercise 23** by converting 765_{nine} to base three.

Number Theory

5

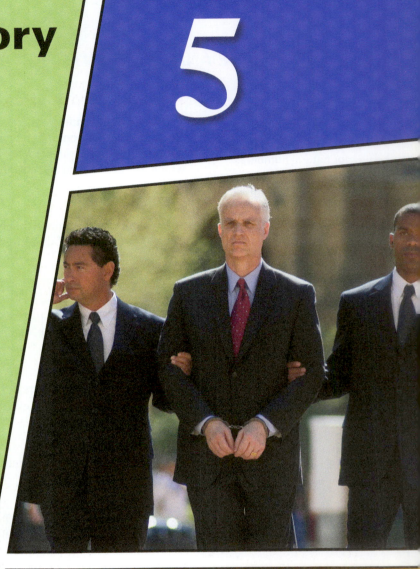

The modern criminal justice system, like most other areas of government, the military, commerce, and personal life, is highly dependent on computer technology and the mathematical theory and algorithms underlying it. Our money, safety, and identity are protected by public key **cryptography,** which includes the processes of **encrypting** (coding) and **decrypting** (decoding) crucial information. Cryptography involves two kinds of problems:

1. Given two prime numbers, find their product.
2. Given the product, find its two prime factors.

You will learn about prime numbers in **Section 5.1.**

Cryptography systems utilize extremely large prime numbers, which are the subject of **Section 5.2.** The success of encryption depends on the fact that, given today's state of computer hardware and software, and given large enough primes, **Problem 1** above can be done, but **Problem 2** *cannot* be done, even with the most powerful computers available.

177

5.1 PRIME AND COMPOSITE NUMBERS

OBJECTIVES

1 Identify prime and composite numbers.

2 Apply divisibility tests for natural numbers.

3 Apply the fundamental theorem of arithmetic.

Do not confuse $b \mid a$ with b/a. The expression $b \mid a$ denotes the **statement** "b divides a." For example, $3 \mid 12$ is a true statement, while $5 \mid 14$ is a false statement. On the other hand, b/a denotes the **operation** "b divided by a." For example, $28/4$ yields the result 7.

The ideas of **even** and **odd natural numbers** are based on the concept of divisibility. A natural number is even if it is divisible by 2 and odd if it is not. Every even number can be written in the form $2k$ (for some natural number k), while every odd number can be written in the form $2k - 1$. Here is another way to say the same thing: 2 divides every even number but fails to divide every odd number. (If a is even, then $2 \mid a$, whereas if a is odd, then $2 \nmid a$.)

For any natural number a, it is true that $a \mid a$ and also that $1 \mid a$.

Primes, Composites, and Divisibility

The famous German mathematician Carl Friedrich Gauss once remarked, "Mathematics is the Queen of Science, and number theory is the Queen of Mathematics." **Number theory** is devoted to the study of the properties of the **natural numbers,** which are also called the **counting numbers** and the **positive integers.**

$$N = \{1, 2, 3, \dots\}$$

A key concept of number theory is the idea of *divisibility*. One counting number is *divisible* by another if dividing the first by the second leaves a remainder 0.

DIVISIBILITY

The natural number a is **divisible** by the natural number b if there exists a natural number k such that $a = bk$. If b divides a, then we write $b \mid a$.

Notice that if b divides a, then the quotient a/b or $\frac{a}{b}$ is a natural number. For example, 4 divides 20 because there exists a natural number k such that

$$20 = 4k.$$

The value of k here is 5, because

$$20 = 4 \cdot 5.$$

The natural number 20 is not divisible by 7, since there is no natural number k satisfying $20 = 7k$. Alternatively, "20 divided by 7 gives quotient 2 with remainder 6," and since there is a nonzero remainder, divisibility does not hold. We write $7 \nmid 20$ to indicate that 7 does *not* divide 20.

If the natural number a is divisible by the natural number b, then b is a **factor** (or **divisor**) of a, and a is a **multiple** of b. For example, 5 is a factor of 30, and 30 is a multiple of 5. Also, 6 is a factor of 30, and 30 is a multiple of 6. The number 30 equals $6 \cdot 5$. This product $6 \cdot 5$ is called a **factorization** of 30. Other factorizations of 30 include

$$3 \cdot 10, \quad 2 \cdot 15, \quad 1 \cdot 30, \quad \text{and} \quad 2 \cdot 3 \cdot 5.$$

EXAMPLE 1 Checking Divisibility

Decide whether the first number is divisible by the second.

(a) 45; 9   **(b)** 60; 7   **(c)** 19; 19   **(d)** 26; 1

Solution

(a) Is there a natural number k that satisfies $45 = 9k$? The answer is yes, because $45 = 9 \cdot 5$, and 5 is a natural number. Therefore, 9 divides 45, written $9 \mid 45$.

(b) Because the quotient $60 \div 7$ is not a natural number, 60 is not divisible by 7, written $7 \nmid 60$.

(c) The quotient $19 \div 19$ is the natural number 1, so 19 is divisible by 19. (*In fact, any natural number is divisible by itself.*)

(d) The quotient $26 \div 1$ is the natural number 26, so 26 is divisible by 1. (*In fact, any natural number is divisible by 1.*)

EXAMPLE 2   Finding Factors

Find all the natural number factors of each number.

(a) 36   **(b)** 50   **(c)** 11

Solution

(a) To find the factors of 36, try to divide 36 by 1, 2, 3, 4, 5, 6, and so on. This gives the following natural number factors of 36: 1, 2, 3, 4, 6, 9, 12, 18, and 36.

(b) The factors of 50 are 1, 2, 5, 10, 25, and 50.

(c) The only natural number factors of 11 are 11 and 1.   ■

How to Use Up Lots of Chalk In 1903, the mathematician F. N. Cole presented before a meeting of the American Mathematical Society his discovery of a factorization of the number

$$2^{67} - 1.$$

He walked up to the chalkboard, raised 2 to the 67th power, and then subtracted 1. Then he moved over to another part of the board and multiplied out

$$193{,}707{,}721 \times 761{,}838{,}257{,}287.$$

The two calculations agreed, and Cole received a standing ovation for a presentation that did not include a single word.

PRIME AND COMPOSITE NUMBERS

A natural number greater than 1 that has only itself and 1 as factors is called a **prime number.** A natural number greater than 1 that is not prime is called **composite.**

Mathematicians agree that the natural number 1 is neither prime nor composite.

ALTERNATIVE DEFINITION OF A PRIME NUMBER

A **prime number** is a natural number that has *exactly* two different natural number factors (which clarifies that 1 is not a prime).

There is a systematic method for identifying prime numbers in a list of numbers: 2, 3, . . . , n. The method, known as the **Sieve of Eratosthenes,** is named after the Greek geographer, poet, astronomer, and mathematician (about 276–192 B.C.).

To construct such a sieve, list all the natural numbers from 2 through some given natural number n, such as 100. The number 2 is prime, but all other multiples of 2 (4, 6, 8, 10, and so on) are composite. Circle the prime 2, and cross out all other multiples of 2. The next number not crossed out and not circled is 3, the next prime. Circle the 3, and cross out all other multiples of 3 (6, 9, 12, 15, and so on) that are not already crossed out. Circle the next prime, 5, and cross out all other multiples of 5 not already crossed out. Continue this process for all primes less than or equal to the square root of the last number in the list. For this list, we may stop with 7, because the next prime, 11, is greater than the square root of 100, which is 10. At this stage, simply circle all remaining numbers that are not crossed out.

Table 1 shows the Sieve of Eratosthenes for 2, 3, 4, . . . , 100.

Table 1   Sieve of Eratosthenes

The 25 primes between 1 and 100 are circled.

	②	③	4̶	⑤	6̶	⑦	8̶	9̶	1̶0̶	⑪	1̶2̶	⑬	1̶4̶
1̶5̶	1̶6̶	⑰	1̶8̶	⑲	2̶0̶	2̶1̶	2̶2̶	㉓	2̶4̶	2̶5̶	2̶6̶	2̶7̶	2̶8̶
㉙	3̶0̶	㉛	3̶2̶	3̶3̶	3̶4̶	3̶5̶	3̶6̶	㊲	3̶8̶	3̶9̶	4̶0̶	㊶	4̶2̶
㊸	4̶4̶	4̶5̶	4̶6̶	㊼	4̶8̶	4̶9̶	5̶0̶	5̶1̶	5̶2̶	㊾ (53)	5̶4̶	5̶5̶	5̶6̶
5̶7̶	5̶8̶	㉟ (59)	6̶0̶	(61)	6̶2̶	6̶3̶	6̶4̶	6̶5̶	6̶6̶	(67)	6̶8̶	6̶9̶	7̶0̶
(71)	7̶2̶	(73)	7̶4̶	7̶5̶	7̶6̶	7̶7̶	7̶8̶	(79)	8̶0̶	8̶1̶	8̶2̶	(83)	8̶4̶
8̶5̶	8̶6̶	8̶7̶	8̶8̶	(89)	9̶0̶	9̶1̶	9̶2̶	9̶3̶	9̶4̶	9̶5̶	9̶6̶	(97)	9̶8̶
9̶9̶	1̶0̶0̶												

In the October 1, 1994 issue of *Science News,* Ivars Peterson gave a fascinating account of the discovery of a 75-year-old **factoring machine** ("Cranking Out Primes: Tracking Down a Long-lost Factoring Machine"). In 1989, Jeffrey Shallit of the University of Waterloo in Ontario came across an article in an obscure 1920 French journal, in which the author, Eugene Olivier Carissan, reported his invention of the factoring apparatus. Shallit and two colleagues embarked on a search for the machine. They contacted all telephone subscribers in France named Carissan and received a reply from Eugene Carissan's daughter. The machine was still in existence and in working condition, stored in a drawer at an astronomical observatory in Floirac, near Bordeaux.

Peterson explains in the article how the apparatus works. Using the machine, Carissan took just ten minutes to prove that 708,158,977 is a prime number, and he was able to factor a 13-digit number. While this cannot compare to what technology can accomplish today, it was a significant achievement for Carissan's day.

EXAMPLE 3 Identifying Prime and Composite Numbers

Decide whether each number is prime or composite.

(a) 89   **(b)** 83,572   **(c)** 629

Solution

(a) Because 89 is circled in **Table 1,** it is prime. If 89 had a smaller prime factor, 89 would have been crossed out as a multiple of that factor.

(b) The number 83,572 is even, so it is divisible by 2. It is composite.

There is only one even prime, the number 2 itself.

(c) For 629 to be composite, there must be a number other than 629 and 1 that divides into it with remainder 0. Start by trying 2, and then 3. Neither works. There is no need to try 4. (If 4 divides with remainder 0 into a number, then 2 will also.) Try 5. There is no need to try 6 or any succeeding even number. (Why?) Try 7. Try 11. (Why not try 9?) Try 13. Keep trying numbers until one works, or until a number is tried whose square exceeds the given number, 629. Try 17.

$$629 \div 17 = 37$$

The number 629 is composite, since

$$629 = 17 \cdot 37.$$ ∎

An aid in determining whether a natural number is divisible by another natural number is called a **divisibility test. Table 2** shows tests for divisibility by the natural numbers 2 through 12 (except for 7 and 11, which are covered in the exercises).

Table 2 Divisibility Tests for Natural Numbers

Divisible by	Test	Example
2	Number ends in 0, 2, 4, 6, or 8. (The last digit is even.)	9,489,994 ends in 4; it is divisible by 2.
3	Sum of the digits is divisible by 3.	897,432 is divisible by 3, since $8 + 9 + 7 + 4 + 3 + 2 = 33$ is divisible by 3.
4	Last two digits form a number divisible by 4.	7,693,432 is divisible by 4, since 32 is divisible by 4.
5	Number ends in 0 or 5.	890 and 7635 are divisible by 5.
6	Number is divisible by both 2 and 3.	27,342 is divisible by 6 since it is divisible by both 2 and 3.
8	Last three digits form a number divisible by 8.	1,437,816 is divisible by 8, since 816 is divisible by 8.
9	Sum of the digits is divisible by 9.	428,376,105 is divisible by 9 since the sum of the digits is 36, which is divisible by 9.
10	The last digit is 0.	897,463,940 is divisible by 10.
12	Number is divisible by both 4 and 3.	376,984,032 is divisible by 12.

The following program, written by Charles W. Gantner and provided courtesy of Texas Instruments, can be used on the TI-83 Plus calculator to list all primes less than or equal to a given natural number N.

```
PROGRAM: PRIMES
: Disp "INPUT N ≥ 2"
: Disp "TO GET"
: Disp "PRIMES ≤ N"
: Input N
: 2 → T
: Disp T
: 1 → A
: Lbl 1
: A + 2 → A
: 3 → B
: If A > N
: Stop
: Lbl 2
: If B ≤ √(A)
: Goto 3
: Disp A
: Pause
: Goto 1
: Lbl 3
: If A/B ≤ int (A/B)
: Goto 1
: B + 2 → B
: Goto 2
```

```
prgmPRIMES

INPUT N≥2
TO GET
PRIMES ≤ N
?6
```

```
TO GET
PRIMES ≤ N
?6
                    2
                    3
                    5
                 Done
```

The display indicates that the primes less than or equal to 6 are 2, 3, and 5.

EXAMPLE 4 Applying Divisibility Tests

In each case, decide whether the first number is divisible by the second.

(a) 2,984,094; 4   **(b)** 2,429,806,514; 9

Solution

(a) The last two digits form the number 94. Since 94 is not divisible by 4, the given number is not divisible by 4.

(b) The sum of the digits is

$$2 + 4 + 2 + 9 + 8 + 0 + 6 + 5 + 1 + 4 = 41,$$

which is not divisible by 9. The given number is, therefore, not divisible by 9. ∎

The Fundamental Theorem of Arithmetic

A *composite* number can be thought of as "composed" of smaller factors. For example, 42 is composite since $42 = 6 \cdot 7$. If the smaller factors are all primes, then we have a *prime factorization*. For example, $42 = 2 \cdot 3 \cdot 7$.

THE FUNDAMENTAL THEOREM OF ARITHMETIC

Every natural number can be expressed in one and only one way as a product of primes (if the order of the factors is disregarded). This unique product of primes is called the **prime factorization** of the natural number.

Because a prime natural number is not composed of smaller factors, its prime factorization is simply itself. For example, $17 = 17$.

EXAMPLE 5 Finding the Unique Prime Factorization of a Composite Number

Find the prime factorization of the number 1320.

Solution

We use a "factor tree." The factor tree can start with $1320 = 2 \cdot 660$, as shown below on the left. Then $660 = 2 \cdot 330$, and so on, until every branch of the tree ends with a prime. All the resulting prime factors are shown circled in the diagram.

Alternatively, the same factorization is obtained by repeated division by primes, as shown on the right. (In general, you would divide by the primes 2, 3, 5, 7, 11, and so on, each as many times as possible, until the answer is no longer composite.)

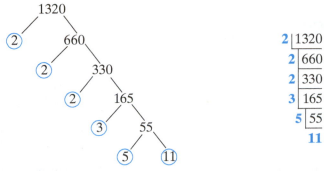

By either method, the prime factorization, in exponential form, is

$$1320 = 2^3 \cdot 3 \cdot 5 \cdot 11. \quad 2 \cdot 2 \cdot 2 = 2^3$$ ∎

FOR FURTHER THOUGHT

Prime Factor Splicing

Prime factor splicing (PFS) has received increasing attention in recent years. A good basic explanation can be found in the article "Home Primes and Their Families," in *Mathematics Teacher*, vol. 107, no. 8, April 2014. Like many topics in number theory, PFS concerns easily stated problems but quickly leads to difficult and unanswered questions. Here is the process:

- Begin with a composite natural number, say 15.
- Obtain its prime factorization: $15 = 3 \cdot 5$
- Form a new natural number by arranging all the prime factors, in ascending order: 35
- Repeat the steps until a prime number is produced, called the **home prime**. For 15, we can summarize the entire process as shown here.

$$15 = 3 \cdot 5 \rightarrow 35 = 5 \cdot 7 \rightarrow 57 = 3 \cdot 19 \rightarrow 319 = 11 \cdot 29$$

$$\rightarrow 1129 \text{ (a prime)}$$

If we call the prime 1129 the "parent," then the composite 15 is its "child." If two or more children are found to have the same parent, call them "siblings." The Sieve of Eratosthenes (see **page 179**) provides plenty of composites to get started. We consider just a few arbitrarily chosen examples here. To do very much prime factor splicing, you would want to use some computer algebra system (CAS) for the factoring. In the case above, the child is 15, the parent is 1129, and the four arrows show that it took four iterations to get from child to parent.

For Group or Individual Investigation

Complete the table, and then answer the questions that follow it.

Child	Number of Iterations	Parent
4	_____	_____
6	_____	_____
9	_____	_____
10	_____	_____
12	_____	_____
22	_____	_____
25	_____	_____
33	_____	_____
46	_____	_____
55	_____	_____

Questions

- Is it possible for a child to have more than one parent?
- Can a parent have more than one child?
- It is possible for PFS to require at least _____ iterations.
- It is possible for a family to include at least _____ siblings.
- Show that if 49 and 77 both have parents, they must be the same.

Interesting Facts

- The child 80 requires 31 iterations and has a 48-digit prime parent.
- 49 and 77 have been "spliced" to at least 110 and 109 iterations, respectively, and no one knows whether a parent exists for them.

5.1 EXERCISES

Decide whether each statement is true or false.

1. If n is a natural number and $9 \mid n$, then $3 \mid n$.

2. If n is a natural number and $5 \mid n$, then $10 \mid n$.

3. Every natural number is divisible by 1.

4. There are no even prime numbers.

5. 1 is the least prime number.

6. Every natural number is both a factor and a multiple of itself.

7. If 16 divides a natural number, then 2, 4, and 8 must also divide that natural number.

8. The prime number 53 has exactly two natural number factors.

Find all natural number factors of each number.

9. 12

10. 20

11. 28

12. 172

Use divisibility tests to decide whether the given number is divisible by each number.

(a) 2   **(b)** 3   **(c)** 4   **(d)** 5   **(e)** 6   **(f)** 8

(g) 9   **(h)** 10   **(i)** 12

13. 321

14. 540

15. 36,360

16. 123,456,789

17. (a) In constructing the Sieve of Eratosthenes for 2 through 100, we said that any composite in that range had to be a multiple of some prime less than or equal to 7 (since the next prime, 11, is greater than the square root of 100). Explain.

(b) To extend the Sieve of Eratosthenes to 200, what is the largest prime whose multiples would have to be considered?

(c) Complete this statement: In seeking prime factors of a given number, we need only consider all primes up to and including the _____ _____ of that number, since a prime factor greater than the _____ _____ can occur only if there is at least one other prime factor less than the _____ _____.

(d) Complete this statement: If no prime less than or equal to $\sqrt{n}$ divides n, then n is a _____ number.

18. (a) Continue the Sieve of Eratosthenes in **Table 1** from 101 to 200 and list the primes between 100 and 200. How many are there?

(b) From your list in part (a), verify that the numbers 197 and 199 are both prime.

19. In your list for **Exercise 18(a)**, consider the six largest primes less than 200. Which pairs of these would have products that end in the digit 7?

20. By checking your pairs of primes from **Exercise 19**, give the prime factorization of 35,657.

21. List two primes that are consecutive natural numbers. Can there be any others?

22. Can there be three primes that are consecutive natural numbers? Explain.

23. For a natural number to be divisible by both 2 and 5, what must be true about its last digit?

24. Consider the divisibility tests for 2, 4, and 8 (all powers of 2). Use inductive reasoning to predict the divisibility test for 16. Then use the test to show that 456,882,320 is divisible by 16.

25. Redraw the factor tree of **Example 5**, assuming that you first observe that $1320 = 12 \cdot 110$, then that $12 = 3 \cdot 4$ and $110 = 10 \cdot 11$. Complete the process and give the resulting prime factorization.

26. Explain how your result in **Exercise 25** illustrates the fundamental theorem of arithmetic.

Find the prime factorization of each composite number.

27. 126

28. 825

29. 1183

30. 340

Here is a divisibility test for 7.

(a) *Double the last digit of the given number, and subtract this value from the given number with the last digit omitted.*

(b) *Repeat the process of part (a) as many times as necessary until it is clear whether the number obtained is divisible by 7.*

(c) *If the final number obtained is divisible by 7, then the given number also is divisible by 7. If the final number is not divisible by 7, then neither is the given number.*

Use this divisibility test to determine whether each number is divisible by 7.

31. 226,233

32. 548,184

33. 496,312

34. 368,597

Here is a divisibility test for 11.

(a) *Starting at the left of the given number, add together every other digit.*

(b) *Add together the remaining digits.*

(c) *Subtract the smaller of the two sums from the larger. (If they are the same, the difference is 0.)*

(d) *If the final number obtained is divisible by 11, then the given number also is divisible by 11. If the final number is not divisible by 11, then neither is the given number.*

Use this divisibility test to determine whether each number is divisible by 11.

35. 6,524,846

36. 108,410,313

37. 60,128,459,358

38. 29,630,419,088

39. Consider the divisibility test for the composite number 6, and make a conjecture for the divisibility test for the composite number 15.

40. Give two factorizations of the number 75 that are not prime factorizations.

Determine all possible digit replacements for x so that the first number is divisible by the second. For example, 37,58x is divisible by 2 if

$$x = 0, 2, 4, 6, or\ 8.$$

41. 398,87x; 2   **42.** 2,45x,765; 3

43. 64,537,84x; 4   **44.** 2,143,89x; 5

45. 985,23x; 6   **46.** 23,x54,470; 10

*There is a method to determine the **number of divisors** of a composite number. To do this, write the composite number in its prime factored form, using exponents. Add 1 to each exponent and multiply these numbers. Their product gives the number of divisors of the composite number. For example,*

$$24 = 2^3 \cdot 3 = 2^3 \cdot 3^1.$$

Now add 1 to each exponent:

$$3 + 1 = 4,\ 1 + 1 = 2.$$

Multiply 4 · 2 to get 8. There are 8 divisors of 24. (Because 24 is rather small, this can be verified easily. The divisors are 1, 2, 3, 4, 6, 8, 12, and 24—a total of eight, as predicted.)

Find the number of divisors of each composite number.

47. 105   **48.** 156

49. $5^8 \cdot 29^2$ **50.** $2^4 \cdot 7^2 \cdot 13^3$

Leap years occur when the year number is divisible by 4. An exception to this occurs when the year number is divisible by 100 (that is, it ends in two zeros). In such a case, the number must be divisible by 400 in order for the year to be a leap year. Determine which years are leap years.

51. 1556   **52.** 1990

53. 2200   **54.** 2400

55. Why is the following *not* a valid divisibility test for the number 8? "A number is divisible by 8 if it is divisible by both 4 and 2." Support your answer with an appropriate example.

56. Choose any three consecutive natural numbers, multiply them together, and divide the product by 6. Repeat this several times, using different choices of three consecutive numbers. Make a conjecture concerning the result.

57. Explain why the product of three consecutive natural numbers must be divisible by 6. Include examples in your explanation.

58. Choose any 6-digit number consisting of three digits followed by the same three digits in the same order (for example, 467,467). Divide by 13. Divide by 11. Divide by 7. What do you notice? Why do you think this happens?

*One of the authors has three sons who were born, from eldest to youngest, on August 30, August 31, and October 14. For most (but not all) of each year, their ages are spaced two years from the eldest to the middle and three years from the middle to the youngest. In 2011, their ages were three consecutive prime numbers for a period of exactly 44 days. This same situation had also occurred in exactly two previous years. Use this information for **Exercises 59–62**. (Hint: Consult the Sieve of Eratosthenes on **page 179** for a listing of all primes less than 100.)*

59. Which were the "two previous years" referred to above?

60. What were the years of birth of the three sons?

61. In what year will the same situation next occur?

62. What will be the ages of the three sons at that time?

5.2   LARGE PRIME NUMBERS

OBJECTIVES

1 Understand the infinitude of primes.

2 Investigate several categories of prime numbers.

3 Learn how large primes are identified.

The Infinitude of Primes

One important basic result about prime numbers was proved by Euclid around 300 B.C., namely that there are infinitely many primes. This means that no matter how large a prime we identify, there are always others even larger. Euclid's proof remains today as one of the most elegant proofs in all of mathematics. (An *elegant* mathematical proof is one that demonstrates the desired result in a most direct, concise manner. Mathematicians strive for elegance in their proofs.) It is called a **proof by contradiction.**

A statement can be proved by contradiction as follows: Assume that the negation of the statement is true, and use that assumption to produce some sort of contradiction, or absurdity. Logically, the fact that the negation of the original statement leads to a contradiction means that the original statement must be true.

The Riemann Hypothesis is an insightful conjecture stated in the mid-1800s by Georg Friedrich Bernhard Riemann (profiled in **Chapter 9**). It concerns how the prime numbers are distributed on the number line and is undoubtedly the most important unproven claim in all of mathematics. Thousands of other "theorems" are built upon the assumption of its truth. If it were ever disproved, those results would fall apart. A proof, on the other hand, may provide sufficient understanding of the primes to demolish public key cryptography, upon which rests the security of all Internet commerce (among other things).

To better understand a particular part of the proof that there are infinitely many primes, first examine the following argument.

Suppose that $M = 2 \cdot 3 \cdot 5 \cdot 7 + 1 = 211$. Now M is the product of the first four prime numbers, plus 1. If we divide 211 by each of the primes 2, 3, 5, and 7, the remainder is always 1.

$$
\begin{array}{cccc}
105 & 70 & 42 & 30 \\
2\overline{)211} & 3\overline{)211} & 5\overline{)211} & 7\overline{)211} \\
210 & 210 & 210 & 210 \\
\mathbf{1} & \mathbf{1} & \mathbf{1} & \mathbf{1}
\end{array}
$$

All remainders are 1.

Thus 211 is not divisible by any of the primes 2, 3, 5, and 7.

Now we can present Euclid's proof that there are infinitely many primes. If *there is no largest prime number,* then there must be infinitely many primes.

EXAMPLE 1   Proving the Infinitude of Primes

Prove by contradiction that there are infinitely many primes.

Solution

Suppose there is a largest prime number, called P. Form the number M such that

$$M = p_1 \cdot p_2 \cdot p_3 \cdots P + 1,$$

where $p_1, p_2, p_3, \ldots, P$ represent all the primes less than or equal to P. Now the number M must be either prime or composite.

1. Suppose that M is prime.
   M is obviously larger than P, so if M is prime, it is larger than the assumed largest prime P. We have reached a *contradiction.*

2. Suppose that M is composite.
   If M is composite, it must have a prime factor. But none of $p_1, p_2, p_3, \ldots, P$ is a factor of M, because division by each will leave a remainder of 1. (Recall the above argument.) So if M has a prime factor, it must be greater than P. But this is a *contradiction,* because P is the assumed largest prime.

In either case 1 or case 2, we reach a contradiction. The whole argument was based on the assumption that a largest prime exists, but this leads to contradictions, so there must be no largest prime, or, equivalently, ***there are infinitely many primes.*** ■

We could never investigate all infinitely many primes directly. So, historically, people have observed properties of the smaller, familiar, ones and then tried to either "disprove" the property (usually by finding counterexamples) or prove it (usually by some deductive argument).

Here is one way to **partition** the natural numbers (divide them into disjoint subsets).

Set A: all natural numbers of the form $4k$

Set B: all natural numbers of the form $4k + 1$

Set C: all natural numbers of the form $4k + 2$

Set D: all natural numbers of the form $4k + 3$

In each case, k is some whole number, except that for set A, k cannot be 0. (Why?) Every natural number is now contained in exactly one of the sets A, B, C, or D. Any number of the form $4k + 4$ would be in subset A, because

$$4k + 4 = 4(k + 1).$$

Any number of the form $4k + 5$ would be in subset B, because

$$4k + 5 = 4(k + 1) + 1, \quad \text{and so on.}$$

Marin Mersenne (1588–1648), in his *Cogitata Physico-Mathematica* (1644), claimed that M_n was prime for $n = 2, 3,$ 5, 7, 13, 17, 19, 31, 67, 127, and 257, and composite for all other prime numbers n less than 257. Other mathematicians at the time knew that Mersenne could not have actually tested all these values, but no one else could prove or disprove them either. It was more then 300 years before all primes up to 257 were legitimately checked out, and Mersenne was finally revealed to have made five errors:

M_{61} is prime.
M_{67} is composite.
M_{89} is prime.
M_{107} is prime.
M_{257} is composite.

The subset A consists of all the multiples of 4. Each of these is divisible by 4; hence, each is not a prime. Now consider C. Because

$$4k + 2 = 2(2k + 1),$$

each member is divisible by 2; hence, each is not a prime (except for 2 itself).

Under this partitioning, all primes other than 2 must be in either B or D.

EXAMPLE 2 **Partitioning the Natural Numbers**

List the first eight members of each of the infinite sets A, B, C, and D.

Solution

$A = \{4, 8, 12, 16, 20, 24, 28, 32, \ldots\}$; $\qquad B = \{1, 5, 9, 13, 17, 21, 25, 29, \ldots\}$
$C = \{2, 6, 10, 14, 18, 22, 26, 30, \ldots\}$; $\qquad D = \{3, 7, 11, 15, 19, 23, 27, 31, \ldots\}$ ■

As mentioned earlier, sets A and C contain no primes except 2. All other primes must lie in sets B and D. (In fact, there are infinitely many primes in each.)

EXAMPLE 3 **Identifying Primes of the Forms $4k + 1$ and $4k + 3$**

Identify all primes *specifically listed* in the following infinite sets of **Example 2.**

(a) set B **(b)** set D

Solution

(a) 5, 13, 17, 29 **(b)** 3, 7, 11, 19, 23, 31 ■

Pierre de Fermat (profiled on **page 201**) proved that every prime number of the form $4k + 1$ can be expressed as the sum of two squares.

EXAMPLE 4 **Expressing Primes as Sums of Squares**

Express each prime in the solution of **Example 3(a)** as a sum of two squares.

Solution

$5 = 1^2 + 2^2$; $\qquad 13 = 2^2 + 3^2$; $\qquad 17 = 1^2 + 4^2$; $\qquad 29 = 2^2 + 5^2$ ■

The Search for Large Primes

Identifying larger and larger prime numbers and factoring large composite numbers into their prime components is of great practical importance today, because it is the basis of modern **cryptography systems.** (See the chapter opener.)

No reasonable formula has ever been found that will consistently generate prime numbers, much less "generate all primes." The most useful attempt, named to honor the French monk Marin Mersenne (1588–1648), follows.

At one time, $2^{11,213} - 1$ was the largest known **Mersenne prime.** To honor its discovery, the Urbana, Illinois, post office used the cancellation picture above.

MERSENNE NUMBERS AND MERSENNE PRIMES

For $n = 1, 2, 3, \ldots,$ the **Mersenne numbers** are those generated by the formula

$$M_n = 2^n - 1.$$

(1) If n is composite, then M_n is also composite.

(2) If n is prime, then M_n may be either prime or composite.

The prime values of M_n are called the **Mersenne primes.** Large primes being verified currently are typically Mersenne primes.

Long before Mersenne's time, there was general agreement on statement (1) in the box. (**Exercises 23–25** show how to find a factor of $2^n - 1$ whenever n is composite.) However, some early writers did not agree with statement (2), believing instead (incorrectly) that a prime n would always produce a prime M_n.

EXAMPLE 5 **Finding Mersenne Numbers**

Find each Mersenne number M_n for $n = 2, 3,$ and 5.

Solution

$$M_2 = 2^2 - 1 = 3 \qquad 2^2 = 2 \cdot 2 = 4$$

$$M_3 = 2^3 - 1 = 7 \qquad 2^3 = 2 \cdot 2 \cdot 2 = 8$$

$$M_5 = 2^5 - 1 = 31 \qquad 2^5 = 2 \cdot 2 \cdot 2 \cdot 2 \cdot 2 = 32$$

Note that all three values, 3, 7, and 31, are indeed primes. ∎

It turns out that $M_7 = 2^7 - 1 = 127$ is also a prime (see **Exercise 18(a)** of **Section 5.1**), but it was discovered in 1536 that

$$M_{11} = 2^{11} - 1 = 2047 \quad \text{is not prime (since it is } 23 \cdot 89\text{).}$$

***So prime values of n do* not *always produce prime* M_n.** Which prime values of n *do* produce prime Mersenne numbers (the so-called **Mersenne primes**)? No way was ever found to identify, in general, which prime values of n result in Mersenne primes. It is a matter of checking out each prime n value individually—not an easy task given that the Mersenne numbers rapidly become very large.

The Mersenne prime search yielded results slowly. By about 1600, M_n had been verified as prime for all prime n up to 19 (except for 11, as mentioned above). The next one was M_{31}, verified by Euler sometime between 1752 and 1772.

In 1876, French mathematician Edouard Lucas used a clever test he had developed to show that M_{127} (a 39-digit number) is prime. In the 1930s Lucas's method was further simplified by D. H. Lehmer, and the testing of Mersenne numbers for primality has been done ever since with the Lucas-Lehmer test. In 1952 an early computer verified that M_{521}, M_{607}, M_{1279}, M_{2203}, and M_{2281} are primes.

Over the last half century, most new record-breaking primes have been identified by computer algorithms devised and implemented by mathematicians and programmers. In 1996, the **Great Internet Mersenne Prime Search (GIMPS)** was launched and now involves well over 900,000 CPUs worldwide. Of the 48 Mersenne primes presently known, the GIMPS program has discovered the fourteen largest ones. The latest one found (as of summer 2014),

$$M_{57,885,161} = 2^{57,885,161} - 1,$$

has 17,425,170 digits.

During the same general period that Mersenne was thinking about prime numbers, Pierre de Fermat (about 1601–1665) conjectured that the formula

$$2^{2^n} + 1$$

would always produce a prime, for any whole number value of n. **Table 3** on the next page shows how this formula generates the first four **Fermat numbers,** which are all primes. The fifth Fermat number (from $n = 4$) is likewise prime. Fermat had verified these first five by around 1630. But the sixth Fermat number (from $n = 5$) turns out to be 4,294,967,297, which is *not* prime. (See **Exercises 15 and 16.**)

Prime Does Pay Since August 23, 2008, GIMPS has offered participants (individuals or groups) these awards:

- $3000 GIMPS research award for a new Mersenne prime with fewer than 100,000,000 (decimal) digits
- $50,000 for the first prime discovered with at least 100,000,000 (decimal) digits

This second award is one-third of $150,000 offered by the Electronic Frontier Foundation (EFF). GIMPS plans to retain another $50,000 for expenses and the final $50,000 to fund past and/or future awards. EFF seeks, by its sponsorship, to "encourage the harmonious integration of Internet innovation into the whole of society."

Distributed computing is a way of achieving great computer power by having lots of individual machines do separate parts of the computation. One example is the **Great Internet Mersenne Prime Search (GIMPS),** described in this section. Another example is **SETI@ home (Search for Extraterrestrial Intelligence),** which assigns the analysis of signal data from small patches of the "sky" to participants. (The movie *Contact* was fiction, but still the search goes on.)

A third example, **Folding@home,** based at Stanford University, investigates the folding of proteins in living organisms into complex shapes and how they interact with other biological molecules.

In July 2014, **Prime Grid,** another distributed computing project, making use of the Berkeley Open Infrastructure for Network Computing (BOINC), confirmed the largest Fermat number known to be composite,

$$F_{3329780},$$

by discovering its prime factor

$$193 \times 2^{3329782} + 1,$$

a prime with more than one million decimal digits.

Source: www.wikipedia.org

As of late 2014, no more Fermat primes had been found. All F_n were known to be composite for $5 \le n \le 32$, but only F_0 to F_{11} had been completely factored.

Table 3 The Generation of Fermat Numbers

n	2^n	2^{2^n}	$2^{2^n} + 1$
0	1	2	3
1	2	4	5
2	4	16	17
3	8	256	257

Of historical interest are a couple of polynomial formulas that produce primes. (A *polynomial* in a given variable involves adding or subtracting integer multiples of whole number powers of the variable. Discussed in **Section 7.6,** polynomials are among the most basic mathematical functions.) In 1732, Leonhard Euler offered the formula

$$n^2 - n + 41, \quad \text{Euler's formula}$$

which generates primes for n up to 40 and fails at $n = 41$. In 1879, E. B. Escott produced more primes with the formula

$$n^2 - 79n + 1601, \quad \text{Escott's formula}$$

which first fails at $n = 80$.

EXAMPLE 6 **Finding Numbers Using Euler's and Escott's Formulas**

Find the first five numbers produced by each of the polynomial formulas of Euler and Escott.

Solution

Table 4 shows the required numbers.

Table 4 A Few Polynomial-Generated Prime Numbers

n	Euler formula $n^2 - n + 41$	Escott formula $n^2 - 79n + 1601$
1	41	1523
2	43	1447
3	47	1373
4	53	1301
5	61	1231

All values found here are primes. (Use **Table 1** to verify the Euler values.) ∎

Actually, it is not hard to prove that there can be no polynomial that will consistently generate primes. More complicated mathematical formulas exist for generating primes, but none produced so far can be practically applied in a reasonable amount of time, even using the fastest computers.

Imagine working for the U.S. Department of Justice as part of the effort to control cybercrime. Your work would surely involve Internet security. A recent ransomware scheme involved encrypting victims' computer files and then demanding payment for the decryption key to regain access.)

The chapter opener described the central role of multiplying large primes and factoring their products in public key cryptography. Another key mathematical component is **modular systems.** The partitioning of the natural numbers described on **page 185** is an example. In that case every natural number n was, in a sense, equated with one of just four numbers, 0, 1, 2, 3 — the remainders when n is divided by 4. In general, the divisor, any natural number greater than 1, is called the modulus (m), and the remainder, called the **residue,** is denoted $n \pmod m$. For example,

$$849{,}657{,}221 \pmod{215}$$

denotes the remainder when 849,657,221 is divided by 215. Using a calculator may produce

$$\frac{849{,}657{,}221}{215} = 3951894.051.$$

Subtract the integer part, 3951894. (We want the remainder only, represented by the fractional part.) Now the calculator may display

$$.0511628.$$

Multiply this result by the divisor, 215, to obtain 11. (Round if necessary — A slight error may result from the calculator's inherent limited number of significant digits.) The desired remainder, or residue, is 11. This establishes that

$$849{,}657{,}221 = 3{,}951{,}894 \cdot 215 + 11.$$

Now consider an even larger number. Try to find the remainder when 11^{14} is divided by 18. The number 11^{14} is so large that the calculator displays it in exponential notation:

$$11^{14} = 3.797498336\text{E}14$$

and

$$\frac{11^{14}}{18} = 2.109721298\text{E}13.$$

The calculator cannot display enough significant digits to reveal the integer portion of the result, so we cannot proceed as before. But we can apply a rule of exponents (see **Chapter 7**) and modular arithmetic as follows.

$$11^{14} \text{ (mod 18)} = 11^{7+7} \text{ (mod 18)} \qquad 14 = 7 + 7$$

$$= (11^7 \cdot 11^7) \text{ (mod 18)} \qquad \text{Rule of exponents}$$

$$= [11^7 \text{ (mod 18)}] \cdot [11^7 \text{ (mod 18)}] \qquad \text{Modular arithmetic}$$

Now

$$11^7 = 19{,}487{,}171 \qquad \text{Use a calculator.}$$

$$\frac{19{,}487{,}171}{18} = 1082620.6\overline{1}$$

and

$$.6\overline{1} \cdot 18 = 11.$$

Therefore,

$$11^7 \text{ (mod 18)} = 11$$

$$11^{14} \text{ (mod 18)} = 11 \cdot 11 = 121$$

$$\frac{121}{18} = 6.7\overline{2}$$

$$.7\overline{2} \cdot 18 = 13.$$

The desired remainder, or residue, is 13.

These examples give a hint of the mathematics involved. But in practice the numbers would be much, much larger, and high-powered computers would be used.

5.2 EXERCISES

In Exercises 1–6 decide whether each statement is true *or* false.

1. A proof by contradiction assumes the negation of a statement and proceeds until a contradiction is encountered.

2. Marin Mersenne gave the first proof that there are infinitely many prime numbers.

3. If n is prime, then $2^n - 1$ must be prime also.

4. There are infinitely many Fermat numbers.

5. There can be no polynomial formula that consistently generates prime numbers.

6. As of late 2014, only five Fermat primes had ever been found.

7. Find the next three primes, of the form $4k + 1$, *not* listed specifically in **Example 3(a),** and express them as sums of squares.

8. Recall the first few perfect squares: 1, 4, 9, 16, 25. Try writing the numbers of the form $4k + 3$ listed in **Example 3(b)** as sums of two squares. Then complete this statement: The primes tested, of the form $4k + 3$, _____ be expressed as the sum of two squares. (Fermat claimed, but did not prove, that *no* prime of the form $4k + 3$ was the sum of two squares. Euler proved it 100 years later.)

9. Explain how the expressions $4k$, $4k + 1$, $4k + 2$, and $4k + 3$ serve to "partition" the natural numbers.

10. Explain why **Example 4** does not prove that every prime of the form $4k + 1$ can be expressed as a sum of two squares.

11. Evaluate Euler's polynomial formula for each of the following, and determine whether each value is prime or composite. If composite, give the prime factorization.

    (a) $n = 41$

    (b) $n = 42$

    (c) $n = 43$

12. Consider your answers for **Exercise 11,** and choose the correct completion: For $n > 41$, Euler's formula produces a prime

    A. never.   **B.** sometimes.   **C.** always.

13. Evaluate Escott's polynomial formula for each of the following, and determine whether each value is prime or composite. If composite, give the prime factorization.

    (a) $n = 80$

    (b) $n = 81$

    (c) $n = 82$

14. Consider your answers for **Exercise 13,** and choose the correct completion: For $n > 80$, Escott's formula produces a prime

    A. never.   **B.** sometimes.   **C.** always.

15. **(a)** Evaluate the Fermat number F_4: $2^{2^n} + 1$ for $n = 4$.

    (b) In seeking possible prime factors of the Fermat number of part (a), what is the largest potential prime factor that one would have to try? (As stated in the text, F_4 is in fact prime.)

16. **(a)** Verify the value given in the text for the "sixth" Fermat number ($2^{2^5} + 1$).

    (b) Divide this Fermat number by 641 and express it in factored form. (Euler discovered this factorization in 1732, proving that the sixth Fermat number is not prime.)

17. The 48th Mersenne prime was identified in the text. Write a short report on when, how, and by whom it was found.

18. The margin note on Marin Mersenne on **page 186** cites a 1644 claim that was not totally resolved for some 300 years. Find out when, and by whom, Mersenne's five errors were demonstrated. (*Hint:* One was mentioned in the margin note on **page 179.**)

19. Why do you suppose it normally takes up to a few years to discover each new Mersenne prime?

20. In Euclid's proof that there is no largest prime, we formed a number M by taking the product of primes and adding 1. Observe the pattern below.

$M = 2 + 1 = 3$	(3 is prime)
$M = 2 \cdot 3 + 1 = 7$	(7 is prime)
$M = 2 \cdot 3 \cdot 5 + 1 = 31$	(31 is prime)
$M = 2 \cdot 3 \cdot 5 \cdot 7 + 1 = 211$	(211 is prime)
$M = 2 \cdot 3 \cdot 5 \cdot 7 \cdot 11 + 1 = 2311$	(2311 is prime)

    It may seem as though this pattern will always yield a prime number. Now evaluate

    $$M = 2 \cdot 3 \cdot 5 \cdot 7 \cdot 11 \cdot 13 + 1.$$

21. Is the final value of M computed in **Exercise 20** prime or composite? If it is composite, give its prime factorization.

22. Explain in your own words the proof by Euclid that there is no largest prime.

The text stated that the Mersenne number M_n is composite whenever n is composite. Exercises 23–26 develop one way in which you can always find a factor of such a Mersenne number.

23. For the composite number $n = 6$, find

    $$M_n = 2^n - 1.$$

24. Notice that $p = 3$ is a prime factor of $n = 6$. Find $2^p - 1$ for $p = 3$. Is $2^p - 1$ a factor of $2^n - 1$?

25. Complete this statement: If p is a prime factor of n, then _____ is a factor of the Mersenne number $2^n - 1$.

26. Find $M_n = 2^n - 1$ for $n = 10$.

27. Use the statement of **Exercise 25** to find two distinct factors of M_{10}.

28. Do you think this procedure will always produce *prime* factors of M_n for composite n? (*Hint:* Consider $n = 22$ and its prime factor $p = 11$, and recall the statement following **Example 5.**) Explain.

29. Explain why large prime numbers are important in modern cryptography systems.

30. Describe the difference between Mersenne *numbers* and Mersenne *primes*.

5.3 SELECTED TOPICS FROM NUMBER THEORY

OBJECTIVES

1. Understand and identify perfect numbers.
2. Understand and identify deficient and abundant numbers.
3. Understand amicable (friendly) numbers.
4. State and evaluate Goldbach's conjecture.
5. Understand and identify twin primes.
6. State and evaluate Fermat's Last Theorem.

The mathematician **Albert Wilansky,** when phoning his brother-in-law, Mr. Smith, noticed an interesting property concerning Smith's phone number (493–7775). The number 4,937,775 is composite, and its prime factorization is

$$3 \cdot 5 \cdot 5 \cdot 65,837.$$

When the digits of the phone number are added, the result, 42, is equal to the sum of the digits in the prime factors:

$$3 + 5 + 5 + 6 + 5 + 8 + 3 + 7 = 42.$$

Wilansky termed such a number a **Smith number.** In 1985 it was proved that there are infinitely many Smith numbers, but there still are many unanswered questions about them.

Perfect Numbers

In **Chapter 1,** we introduced figurate numbers, a topic investigated by the Pythagoreans, a group of Greek mathematicians and musicians who held their meetings in secret. In this section we examine some of the other special numbers that fascinated the Pythagoreans and are still studied by mathematicians today.

Divisors of a natural number were covered in **Section 5.1.** The **proper divisors** of a natural number include all divisors of the number except the number itself. For example, the proper divisors of 8 are 1, 2, and 4. (8 is *not* a proper divisor of 8.)

PERFECT NUMBERS

A natural number is said to be **perfect** if it is equal to the sum of its proper divisors.

Is 8 perfect? No, because $1 + 2 + 4 = 7$, and $7 \neq 8$. The least perfect number is 6, because the proper divisors of 6 are 1, 2, and 3, and

$$1 + 2 + 3 = 6. \quad \text{6 is perfect.}$$

EXAMPLE 1 Verifying a Perfect Number

Show that 28 is a perfect number.

Solution

The proper divisors of 28 are 1, 2, 4, 7, and 14. The sum of these is 28:

$$1 + 2 + 4 + 7 + 14 = 28.$$

By the definition, 28 is perfect. ∎

The numbers 6 and 28 are the two least perfect numbers. The next two are 496 and 8128. The pattern of these first four perfect numbers led early writers to conjecture that

1. The nth perfect number contains exactly n digits.
2. The even perfect numbers end in the digits 6 and 8, alternately.

} Conjectures **NOT NECESSARILY TRUE**

(**Exercises 39–41** will help you analyze these conjectures.)

There still are many unanswered questions about perfect numbers. Euclid showed that the following is true.

If $2^n - 1$ is prime, then $2^{n-1}(2^n - 1)$ is perfect, and conversely.

Because the prime values of $2^n - 1$ are the Mersenne primes (discussed in the previous section), this means that for every new Mersenne prime discovered, another perfect number is automatically revealed. (Hence, as of summer 2014, there were 48 known perfect numbers.) Also, it is known that the following is true.

***All* even perfect numbers must take the form $2^{n-1}(2^n - 1)$.**

It is strongly suspected that no odd perfect numbers exist. (Any odd one would have at least eight different prime factors and would have at least 300 decimal digits.) Therefore, Euclid and the early Greeks most likely identified the form of all perfect numbers.

Deficient and Abundant Numbers

Earlier we saw that 8 is not perfect because it is not equal to the sum of its proper divisors (8 ≠ 7). Next we define two alternative categories for natural numbers that are *not* perfect.

DEFICIENT AND ABUNDANT NUMBERS

A natural number is **deficient** if it is greater than the sum of its proper divisors. It is **abundant** if it is less than the sum of its proper divisors.

Based on this definition, a *deficient number* is one with proper divisors that add up to less than the number itself, while an *abundant number* is one with proper divisors that add up to more than the number itself. For example, because the proper divisors of 8 (1, 2, and 4) add up to 7, which is less than 8, the number 8 is deficient.

▐ EXAMPLE 2 Identifying Deficient and Abundant Numbers

Decide whether each number is deficient or abundant.

(a) 12   **(b)** 10

Solution

(a) The proper divisors of 12 are 1, 2, 3, 4, and 6. The sum of these divisors is 16. Because 16 > 12, the number 12 is abundant.

(b) The proper divisors of 10 are 1, 2, and 5. Since $1 + 2 + 5 = 8$, and $8 < 10$, the number 10 is deficient.   ■

Amicable (Friendly) Numbers

Suppose that we add the proper divisors of 284.

$$1 + 2 + 4 + 71 + 142 = 220$$

Their sum is 220. Now, add the proper divisors of 220.

$$1 + 2 + 4 + 5 + 10 + 11 + 20 + 22 + 44 + 55 + 110 = 284$$

The sum of the proper divisors of 220 is 284, while the sum of the proper divisors of 284 is 220. Number pairs with this property are said to be *amicable*, or *friendly*.

AMICABLE OR FRIENDLY NUMBERS

The natural numbers a and b are **amicable, or friendly,** if the sum of the proper divisors of a is b, and the sum of the proper divisors of b is a.

An extension of the idea of amicable numbers results in **sociable numbers.** In a chain of sociable numbers, the sum of the proper divisors of each number is the next number in the chain, and the sum of the proper divisors of the last number in the chain is the first number. Here is a 5-link chain of sociable numbers:

12,496

14,288

15,472

14,536

14,264.

The number 14,316 starts a 28-link chain of sociable numbers.

The smallest pair of amicable numbers, 220 and 284, was known to the Pythagoreans, but it was not until 1636 that Fermat found the next pair, 17,296 and 18,416. Many more pairs were found over the next few decades, but it took a 16-year-old Italian boy named Nicolo Paganini to discover, in the year 1866, that the pair of amicable numbers 1184 and 1210 had been overlooked for centuries!

Today, powerful computers continually extend the lists of known amicable pairs. The last time we checked, nearly twelve million pairs were known. It still is unknown, however, whether there are infinitely many such pairs. No one has found an amicable pair without prime factors in common, but the possibility of such a pair has not been eliminated.

Goldbach's Conjecture

The mathematician Christian Goldbach (1690–1764) stated the following conjecture (guess), which is one of the most famous unsolved problems in mathematics. Mathematicians have tried to prove the conjecture but have not succeeded. However, the conjecture has been verified (as of late 2014) for numbers up to 4×10^{18}.

> **GOLDBACH'S CONJECTURE (NOT PROVED)**
>
> Every even number greater than 2 can be written as the sum of two prime numbers.
>
> *Examples:* $8 = 5 + 3$
> $10 = 5 + 5$ (or $10 = 7 + 3$)

A Dull Number? The Indian mathematician **Srinivasa Ramanujan** (1887–1920) developed many ideas in number theory. His friend and collaborator on occasion was G. H. Hardy, also a number theorist and professor at Cambridge University in England.

A story has been told about Ramanujan that illustrates his genius. Hardy once mentioned to Ramanujan that he had just taken a taxicab with a rather dull number: 1729. Ramanujan countered by saying that this number isn't dull at all; it is the smallest natural number that can be expressed as the sum of two cubes in two different ways:

$$1^3 + 12^3 = 1729$$
and $\quad 9^3 + 10^3 = 1729.$

Show that 85 can be written as the sum of two *squares* in two ways.

EXAMPLE 3 **Expressing Numbers as Sums of Primes**

Write each even number as the sum of two primes.

(a) 18 **(b)** 60

Solution

(a) $18 = 5 + 13$. Another way of writing it is $7 + 11$. Notice that $1 + 17$ is *not* valid because by definition 1 is not a prime number.

(b) $60 = 7 + 53$. Can you find other ways? Why is $3 + 57$ not valid? ∎

Twin Primes

Prime numbers that differ by 2 are called **twin primes.** Some twin prime pairs are

3 and 5, 5 and 7, 11 and 13, and so on.

Like Goldbach's conjecture, the following conjecture about twin primes has never been proved. Interestingly, however, on May 13, 2013, substantial progress was made toward proofs of *both* the twin prime conjecture and Goldbach's conjecture. Two mathematicians, one in New Hampshire and one in Paris, proved conditions that get us significantly closer to proofs of the two famous conjectures.

> **TWIN PRIME CONJECTURE (NOT PROVED)**
>
> There are infinitely many pairs of twin primes.

Mathematics professor Gregory Larkin, played by Jeff Bridges, woos colleague Rose Morgan (Barbra Streisand) in the 1996 film *The Mirror Has Two Faces.* Larkin's research and book focus on the **twin prime conjecture,** which he correctly states in a dinner scene. He is amazed that his nonmathematician friend actually understands what he is talking about.

You may wish to verify that there are eight such pairs less than 100, using the Sieve of Eratosthenes in **Table 1.** As of summer 2014, the largest known twin primes were

$$3{,}756{,}801{,}695{,}685 \cdot 2^{666{,}669} \pm 1. \quad \text{Each contains 200,700 digits.}$$

Recall from **Section 5.2** that Euclid's proof of the infinitude of primes used numbers of the form

$$p_1 \cdot p_2 \cdot p_3 \cdots p_n + 1,$$

that is, the product of the first n primes, plus 1. It may seem that any such number must be prime, but that is not so. (See **Exercises 20 and 21** of **Section 5.2.**) However, this form often does produce primes (as does the same form with the plus replaced by a minus).

Sophie Germain (1776–1831) studied at the École Polytechnique in Paris in a day when female students were not admitted. A **Sophie Germain prime** is a prime p for which $2p + 1$ also is prime. Lately, large Sophie Germain primes have been discovered at the rate of one or more per year. As of late 2014, the largest one known was $18{,}543{,}637{,}900{,}515 \cdot 2^{666{,}667} - 1$, which has 200,701 digits.

(*Source:* www.primes.utm.edu)

The popular animated television series *The Simpsons* provides not only humor and social commentary but also lessons in mathematics. One episode depicted the equation

$$1782^{12} + 1841^{12} = 1922^{12},$$

which, according to **Fermat's Last Theorem,** cannot be true. Your calculator may indicate that the equation is true, but this is because it cannot accurately display powers of this size. Actually, 1782^{12} must be an *even* number because *an even number to any power is even.* Also, 1841^{12} must be *odd*, because *an odd number to any power is odd.* So the sum on the left must be *odd*, because

$$even + odd = odd.$$

Similarly, 1922^{12} must be *even.* So the equation states that an odd number equals an even number, which is impossible. (See www.simpsonsmath.com)

When *all* the primes up to p_n are included, the resulting numbers, if prime, are called **primorial primes.** They are denoted

$$p\# \pm 1.$$

For example, $5\# + 1 = 2 \cdot 3 \cdot 5 + 1 = 31$ is a primorial prime. (In late 2014, the largest known primorial prime was $1{,}098{,}133\# - 1$, a number with 476,311 digits.) The primorial primes are a popular place to look for twin primes.

EXAMPLE 4   Verifying Twin Primes

Verify that the primorial formula $p\# \pm 1$ produces twin prime pairs for both **(a)** $p = 3$ and **(b)** $p = 5$.

Solution

(a) $3\# \pm 1 = 2 \cdot 3 \pm 1 = 6 \pm 1 = 5$ and 7 Twin primes

> Multiply, then add and subtract.

(b) $5\# \pm 1 = 2 \cdot 3 \cdot 5 \pm 1 = 30 \pm 1 = 29$ and 31 Twin primes   ■

Fermat's Last Theorem

In any right triangle with shorter sides (legs) a and b, and longest side (hypotenuse) c, the equation $a^2 + b^2 = c^2$ will hold true. This is the famous Pythagorean theorem. For example,

$$3^2 + 4^2 = 5^2 \qquad a = 3,\ b = 4,\ c = 5$$
$$9 + 16 = 25 \qquad \text{Apply the exponents.}$$
$$25 = 25. \qquad \text{True}$$

It is known that there are infinitely many such triples (a, b, c) that satisfy the equation $a^2 + b^2 = c^2$. Is something similar true of the equation

$$a^n + b^n = c^n$$

for natural numbers $n \geq 3$? Pierre de Fermat, who is profiled in a margin note on **page 201,** thought that not only were there not infinitely many such triples, but that there were, in fact, none. He made the following claim in the 1600s.

> **FERMAT'S LAST THEOREM (PROVED IN THE 1990s)**
>
> For *any* natural number $n \geq 3$, there are *no* triples (a, b, c) that satisfy the equation
>
> $$a^n + b^n = c^n.$$

Fermat's assertion was the object of some 350 years of attempts by mathematicians to provide a suitable proof. While it was verified for many specific cases (Fermat himself proved it for $n = 3$), a proof of the general case could not be found until the Princeton mathematician Andrew Wiles announced a proof in the spring of 1993. Although some flaws were discovered in his argument, Wiles was able, by the fall of 1994, to repair the proof.

There were probably about 100 mathematicians around the world qualified to understand the Wiles proof. *Today Fermat's Last Theorem finally is regarded by the mathematics community as officially proved.*

EXAMPLE 5   Applying a Theorem Proved by Fermat

One of the theorems legitimately proved by Fermat is as follows:

Every odd prime can be expressed as the difference of two squares in one and only one way.

Express each odd prime as the difference of two squares.

(a) 3   (b) 7

Solution

(a) $3 = 4 - 1 = 2^2 - 1^2$ (b) $7 = 16 - 9 = 4^2 - 3^2$ ∎

FOR FURTHER THOUGHT

Curious and Interesting

One of the most remarkable books on number theory is *The Penguin Dictionary of Curious and Interesting Numbers* (1986) by David Wells. This book contains fascinating numbers and their properties, including the following.

- There are only three sets of three digits that form prime numbers in all possible arrangements: {1, 1, 3}, {1, 9, 9}, {3, 3, 7}.

- Find the sum of the cubes of the digits of 136:

$$1^3 + 3^3 + 6^3 = 244.$$

Repeat the process with the digits of 244:

$$2^3 + 4^3 + 4^3 = 136.$$

We're back to where we started.

- 635,318,657 is the least number that can be expressed as the sum of two fourth powers in two ways:

$$635{,}318{,}657 = 59^4 + 158^4 = 133^4 + 134^4.$$

- The number 24,678,050 has an interesting property:

$$24{,}678{,}050 = 2^8 + 4^8 + 6^8 + 7^8 + 8^8 + 0^8$$
$$+ 5^8 + 0^8.$$

- The number 54,748 has a similar interesting property:

$$54{,}748 = 5^5 + 4^5 + 7^5 + 4^5 + 8^5.$$

- The number 3435 has this property:

$$3435 = 3^3 + 4^4 + 3^3 + 5^5.$$

For anyone whose curiosity is piqued by such facts, the book mentioned above is for you!

For Group or Individual Investigation

Have each student in the class choose a three-digit number that is a multiple of 3. Add the cubes of the digits. Repeat the process until the same number is obtained over and over. Then, have the students compare their results. What is curious and interesting about this process?

5.3   EXERCISES

In Exercises 1–10 decide whether each statement is true *or* false.

1. Given a prime number, no matter how large, there is always another prime even larger.

2. The prime numbers 2 and 3 are twin primes.

3. The first and third perfect numbers both end in the digit 6, and the second and fourth perfect numbers both end in the digits 28.

4. For every Mersenne prime, there is a corresponding perfect number.

5. All prime numbers are deficient.

6. The equation $17 + 51 = 68$ verifies Goldbach's conjecture for the number 68.

7. Even perfect numbers are more plentiful than Mersenne primes.

8. The twin prime conjecture was proved in 2013.

9. The number $2^5(2^6 - 1)$ is perfect.

10. Every natural number greater than 1 must be one of the following: prime, abundant, or deficient.

11. The proper divisors of 496 are 1, 2, 4, 8, 16, 31, 62, 124, and 248. Use this information to verify that 496 is perfect.

12. The proper divisors of 8128 are 1, 2, 4, 8, 16, 32, 64, 127, 254, 508, 1016, 2032, and 4064. Use this information to verify that 8128 is perfect.

13. As mentioned in the text, when $2^n - 1$ is prime,

$$2^{n-1}(2^n - 1)$$

is perfect. By letting $n = 2, 3, 5$, and 7, we obtain the first four perfect numbers. Show that $2^n - 1$ is prime for $n = 13$, and then find the decimal digit representation for the fifth perfect number.

14. In the summer of 2014, the largest known prime number was $2^{57,885,161} - 1$. Use the formula in **Exercise 13** to write an expression for the perfect number generated by this prime number.

15. It has been proved that the reciprocals of *all* the positive divisors of a perfect number have a sum of 2. Verify this for the perfect number 28.

16. Consider the following equations.

$$6 = 1 + 2 + 3$$
$$28 = 1 + 2 + 3 + 4 + 5 + 6 + 7$$

Show that a similar equation is valid for the third perfect number, 496.

Determine whether each number is abundant *or* deficient.

17. 32   **18.** 60   **19.** 84   **20.** 75

21. There are four abundant numbers between 1 and 25. Find them. (*Hint:* They are all even, and no prime number is abundant.)

22. Explain why a prime number must be deficient.

23. There are no odd abundant numbers less than 945. If 945 has proper divisors 1, 3, 5, 7, 9, 15, 21, 27, 35, 45, 63, 105, 135, 189, and 315, determine whether it is abundant or deficient.

24. Explain in your own words the terms *perfect number, abundant number,* and *deficient number.*

25. The proper divisors of 1184 are 1, 2, 4, 8, 16, 32, 37, 74, 148, 296, and 592. The proper divisors of 1210 are 1, 2, 5, 10, 11, 22, 55, 110, 121, 242, and 605. Verify that 16-year-old Nicolo Paganini was correct about these two numbers in 1866. See the subsection "Amicable (Friendly) Numbers."

26. An Arabian mathematician of the ninth century stated the following: "If the three numbers

$$x = 3 \cdot 2^{n-1} - 1,$$
$$y = 3 \cdot 2^n - 1,$$

and
$$z = 9 \cdot 2^{2n-1} - 1$$

are all prime and $n \geq 2$, then $2^n xy$ and $2^n z$ are amicable numbers."

(a) Use $n = 2$, and show that the result is the least pair of amicable numbers, namely 220 and 284.

(b) Use $n = 4$ to obtain another pair of amicable numbers.

Write each even number as the sum of two primes. (There may be more than one way to do this.)

27. 12   **28.** 24   **29.** 40   **30.** 54

31. Joseph Louis Lagrange (1736–1813) conjectured that every odd natural number greater than 5 can be written as a sum $a + 2b$, where a and b are both primes.

(a) Verify this for the odd natural number 11.

(b) Verify that the odd natural number 17 can be written in this form in four different ways.

32. Another unproved conjecture in number theory states that every natural number multiple of 6 can be written as the difference of two primes. Verify this for 12 and 24.

Find one pair of twin primes between the two numbers given.

33. 45, 65   **34.** 65, 85

Pierre de Fermat provided proofs of many theorems in number theory. Exercises 35–38 investigate some of these theorems.

35. If p is prime and the natural numbers a and p have no common factor except 1, then $a^{p-1} - 1$ is divisible by p.

(a) Verify this for $p = 5$ and $a = 3$.

(b) Verify this for $p = 7$ and $a = 2$.

36. Every odd prime can be expressed as the difference of two squares in one and only one way.

(a) Find this one way for the prime number 5.

(b) Find this one way for the prime number 11.

37. There is only one solution in natural numbers for $a^2 + 2 = b^3$, and it is $a = 5$, $b = 3$. Verify this solution.

38. There are only two solutions in integers for $a^2 + 4 = b^3$. One solution is $a = 2$, $b = 2$. Find the other solution.

The first four perfect numbers were identified in the text: 6, 28, 496, and 8128. The next two are 33,550,336 and 8,589,869,056. Use this information about perfect numbers to work Exercises 39–41.

39. Verify that each of these six perfect numbers ends in either 6 or 28. (In fact, this is true of all even perfect numbers.)

40. Is conjecture (1) in the text (that the nth perfect number contains exactly n digits) true or false? Explain.

41. Is conjecture (2) in the text (that the even perfect numbers end in the digits 6 and 8, alternately) true or false? Explain.

*According to the Web site www.shyamsundergupta.com/ amicable.htm, a natural number is **happy** if the process of repeatedly summing the squares of its decimal digits finally ends in 1. For example, the least natural number (greater than 1) that is happy is 7, as shown here.*

$$7^2 = 49, \quad 4^2 + 9^2 = 97, \quad 9^2 + 7^2 = 130,$$
$$1^2 + 3^2 + 0^2 = 10, \quad 1^2 + 0^2 = 1.$$

*An amicable pair is a **happy amicable pair** if and only if both members of the pair are happy numbers. (The first 5000 amicable pairs include only 111 that are happy amicable pairs.) For each amicable pair, determine whether neither, one, or both of the members are happy, and whether the pair is a happy amicable pair.*

42. 220 and 284

43. 1184 and 1210

44. 10,572,550 and 10,854,650

45. 35,361,326 and 40,117,714

46. If the early Greeks knew the form of all even perfect numbers, namely $2^{n-1}(2^n - 1)$, then why did they not discover all the ones that are known today?

47. Explain why the primorial formula $p\# \pm 1$ does not result in a pair of twin primes for the prime value $p = 2$.

48. (a) What two numbers does the primorial formula produce for $p = 7$?

(b) Which, if either, of these numbers is prime?

49. Choose the correct completion: The primorial formula produces twin primes

A. never.   **B.** sometimes.   **C.** always.

*See the margin note (on **page 195**) defining a Sophie Germain prime, and complete this table.*

	p	$2p+1$	Is p a Sophie Germain prime?
50.	2	___	___
51.	3	___	___
52.	5	___	___
53.	7	___	___

Factorial primes *are of the form* $n! \pm 1$ *for natural numbers* n. ($n!$ *denotes "n factorial," the product of all natural numbers up to n, not just the primes as in the primorial primes. For example,* $4! = 1 \cdot 2 \cdot 3 \cdot 4 = 24$.) *As of late 2014, the largest verified factorial prime was* $150,209! + 1$, *which has 712,355 digits. Find the missing entries in the following table.*

n	$n!$	$n! - 1$	$n! + 1$	Is $n! - 1$ prime?	Is $n! + 1$ prime?
2	2	1	3	no	yes
54. 3	___	___	___	___	___
55. 4	___	___	___	___	___
56. 5	___	___	___	___	___

57. Explain why the factorial prime formula does not give twin primes for $n = 2$.

Based on the preceding table, complete each statement with one of the following:

A. *never,*   **B.** *sometimes, or*   **C.** *always.*

When applied to particular values of n, the factorial prime formula $n! \pm 1$ produces

58. no primes _____   **59.** exactly one prime _____

60. twin primes _____

*Because it does not equal the sum of its proper divisors, an abundant number is not perfect, but it is called **pseudoperfect** if it is equal to the sum of a subset of its proper divisors. For example, 12, an abundant number, has proper divisors* 1, 2, 3, 4, *and* 6. $12 \neq 1 + 2 + 3 + 4 + 6$, *but* $12 = 2 + 4 + 6$. *(Also,* $12 = 1 + 2 + 3 + 6$.)
   Show that each number is pseudoperfect.

61. 18   (Show it with two different sums.)

62. 24   (Show it with five different sums.)

*Abundant numbers are so commonly pseudoperfect that when we find one that isn't, we call it **weird**. There are no weird numbers less than 70. Do the following exercises to investigate 70.*

63. Show that 70 is abundant.

64. Show that 70 is *not* pseudoperfect and must therefore be the smallest weird number. (Among the first 10,000 counting numbers, there are only seven weird ones: 70, 863, 4030, 5830, 7192, 7912, and 9272.)

The following number is known as Belphegor's prime.

One nonillion, sixty-six quadrillion, six hundred trillion, one

Belphegor, who is referred to in many literary works, is one of the seven princes of hell, known for tempting men toward particular evils. Exercises 65–68 refer to Belphegor's prime.

65. Write out the decimal form of the number. (It is a "palindrome"—that is, it reads the same backward and forward. And in keeping with its name, it actually is a prime.)

66. What are the "middle three" digits? (The biblical book of Revelation calls this "the number of the beast.")

67. How many zeroes are on each side of the middle three digits?

68. With what digit does Belphegor's prime begin and end?

5.4   GREATEST COMMON FACTOR AND LEAST COMMON MULTIPLE

OBJECTIVES

1 Find the greatest common factor by several methods.

2 Find the least common multiple by several methods.

Greatest Common Factor

The *greatest common factor* is defined as follows.

> **GREATEST COMMON FACTOR**
>
> The **greatest common factor (GCF)** of a group of natural numbers is the largest natural number that is a factor of all the numbers in the group.
>
> *Examples:*   18 is the GCF of 36 and 54, because 18 is the largest natural number that divides both 36 and 54.
>
>   1 is the GCF of 7 and 16.

The greatest common factor is often called the *greatest common divisor*. Greatest common factors can be found by using prime factorizations. To determine the GCF of 36 and 54, first write the prime factorization of each number.

$$36 = 2^2 \cdot 3^2$$
$$54 = 2^1 \cdot 3^3$$

The GCF is the product of the primes common to the factorizations, with each prime raised to the power indicated by the *least* exponent that it has in any factorization. Here, the prime 2 has 1 as the least exponent (in $54 = 2^1 \cdot 3^3$), while the prime 3 has 2 as the least exponent (in $36 = 2^2 \cdot 3^2$).

$$\text{GCF} = 2^1 \cdot 3^2 = 2 \cdot 9 = 18$$

We summarize as follows.

> **FINDING THE GREATEST COMMON FACTOR (PRIME FACTORS METHOD)**
>
> **Step 1**   Write the prime factorization of each number.
>
> **Step 2**   Choose all primes common to *all* factorizations, with each prime raised to the *least* exponent that it has in any factorization.
>
> **Step 3**   Form the product of all the numbers in Step 2. This product is the greatest common factor.

EXAMPLE 1   **Finding the Greatest Common Factor by the Prime Factors Method**

Find the greatest common factor of 360 and 2700.

Solution

Write the prime factorization of each number.

$$360 = 2^3 \cdot 3^2 \cdot 5 \qquad 2700 = 2^2 \cdot 3^3 \cdot 5^2$$

Find the primes common to both factorizations, with each prime having as its exponent the *least* exponent from either product.

Use the least exponents.

$$\text{GCF} = 2^2 \cdot 3^2 \cdot 5 = 180$$

The greatest common factor of 360 and 2700 is 180.   ∎

gcd(360,2700)
                    180

The calculator shows that the greatest common divisor (factor) of 360 and 2700 is 180. Compare with **Example 1.**

EXAMPLE 2 Finding the Greatest Common Factor by the Prime Factors Method

Find the greatest common factor of 720, 1000, and 1800.

Solution

Write the prime factorization for each number.

$$720 = 2^4 \cdot 3^2 \cdot 5$$
$$1000 = 2^3 \cdot 5^3$$
$$1800 = 2^3 \cdot 3^2 \cdot 5^2$$

Use the smallest exponent on each prime common to the factorizations.

$$\text{GCF} = 2^3 \cdot 5 = 40$$

(The prime 3 is not used in the greatest common factor because it does not appear in the prime factorization of 1000.) ∎

EXAMPLE 3 Finding the Greatest Common Factor by the Prime Factors Method

Find the greatest common factor of 80 and 63.

Solution

$$80 = 2^4 \cdot 5$$
$$63 = 3^2 \cdot 7$$

There are no primes in common here, so the GCF is 1. The number 1 is the largest number that will divide into both 80 and 63. ∎

Two numbers, such as 80 and 63, with a greatest common factor of 1 are called **relatively prime numbers**—that is, they are prime *relative* to one another. (They have no common factors other than 1.)

Another method of finding the greatest common factor involves dividing the numbers by common prime factors.

FINDING THE GREATEST COMMON FACTOR (DIVIDING BY PRIME FACTORS METHOD)

Step 1   Write the numbers in a row.

Step 2   Divide each of the numbers by a common prime factor. Try 2, then try 3, and so on.

Step 3   Divide the quotients by a common prime factor. Continue until no prime will divide into all the quotients.

Step 4   The product of the primes in Steps 2 and 3 is the greatest common factor.

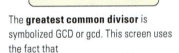

gcd(12,gcd(18,60))

6

The **greatest common divisor** is symbolized GCD or gcd. This screen uses the fact that

gcd(a, b, c) = gcd(a, gcd(b, c)).

Compare with **Example 4.**

EXAMPLE 4   Finding the Greatest Common Factor by Dividing by Prime Factors

Find the greatest common factor of 12, 18, and 60.

Solution

Write the numbers in a row and divide by 2.

$$
\begin{array}{r|rrr}
2 & 12 & 18 & 60 \\
\hline
 & 6 & 9 & 30
\end{array}
$$

The numbers 6, 9, and 30 are not all divisible by 2, but they are divisible by 3.

$$
\begin{array}{r|rrr}
2 & 12 & 18 & 60 \\
3 & 6 & 9 & 30 \\
\hline
 & 2 & 3 & 10
\end{array}
$$

No prime divides into 2, 3, and 10, so the greatest common factor of the numbers 12, 18, and 60 is given by the product of the primes on the left, 2 and 3.

$$
\begin{array}{r|rrr}
2 & 12 & 18 & 60 \\
3 & 6 & 9 & 30 \\
\hline
 & 2 & 3 & 10
\end{array}
$$

$$2 \cdot 3 = 6$$

The GCF of 12, 18, and 60 is 6. ∎

Another method of finding the greatest common factor of two numbers (but not more than two) is called the **Euclidean algorithm.***

EXAMPLE 5   Finding the Greatest Common Factor Using the Euclidean Algorithm

Use the Euclidean algorithm to find the greatest common factor of 90 and 168.

Solution

Step 1 Begin by dividing the larger, 168, by the smaller, 90. Disregard the quotient, but note the remainder.

$$
\begin{array}{r}
1 \\
90\overline{)168} \\
\underline{90} \\
78
\end{array}
$$

Step 2 Divide the smaller of the two numbers by the remainder obtained in Step 1. Once again, note the remainder.

$$
\begin{array}{r}
1 \\
78\overline{)90} \\
\underline{78} \\
12
\end{array}
$$

Step 3 Continue dividing the successive remainders as many times as necessary to obtain a remainder of 0.

$$
\begin{array}{r}
6 \\
12\overline{)78} \\
\underline{72} \\
6
\end{array}
$$

6 Greatest common factor

Step 4 The *last positive remainder* in this process is the greatest common factor of 90 and 168. It can be seen that their GCF is 6.

$$
\begin{array}{r}
2 \\
6\overline{)12} \\
\underline{12} \\
0
\end{array}
$$

∎

Pierre de Fermat (about 1601–1665), a government official who did not interest himself in mathematics until he was past 30, devoted leisure time to its study. He was a worthy scholar, best known for his work in number theory. His other major contributions involved certain applications in geometry and his original work in probability.

Much of Fermat's best work survived only on loose sheets or jotted, without proof, in the margins of works that he read.

---

*For a proof that this process does indeed give the greatest common factor, see *Elementary Introduction to Number Theory, Second Edition*, by Calvin T. Long, pp. 34–35.

The Euclidean algorithm is particularly useful if the two numbers are difficult to factor into primes. We summarize the algorithm here.

> **FINDING THE GREATEST COMMON FACTOR (EUCLIDEAN ALGORITHM)**
>
> To find the greatest common factor of two unequal numbers, divide the larger by the smaller. Note the remainder, and divide the previous divisor by this remainder. Continue the process until a remainder of 0 is obtained. The greatest common factor is the last positive remainder obtained.

Least Common Multiple

Closely related to the idea of the greatest common factor is the concept of the *least common multiple.*

> **LEAST COMMON MULTIPLE**
>
> The **least common multiple (LCM)** of a group of natural numbers is the smallest natural number that is a multiple of all the numbers in the group.
>
> *Example:* 30 is the LCM of 15 and 10 because 30 is the smallest number that appears in both sets of multiples.
>
> Multiples of 15: $\{15, \mathbf{30}, 45, 60, 75, 90, 105, \ldots\}$
>
> Multiples of 10: $\{10, 20, \mathbf{30}, 40, 50, 60, 70, \ldots\}$

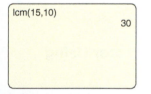

The least common multiple of 15 and 10 is 30.

The set of natural numbers that are multiples of *both* 15 and 10 form the set of *common multiples:*

$$\{30, 60, 90, 120, \ldots\}.$$

While there are infinitely many common multiples, the *least* common multiple is observed to be 30.

A method similar to the first one given for the greatest common factor may be used to find the least common multiple of a group of numbers.

> **FINDING THE LEAST COMMON MULTIPLE (PRIME FACTORS METHOD)**
>
> *Step 1* Write the prime factorization of each number.
>
> *Step 2* Choose all primes belonging to *any* factorization, with each prime raised to the power indicated by the *greatest* exponent that it has in any factorization.
>
> *Step 3* Form the product of all the numbers in Step 2. This product is the least common multiple.

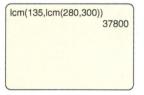

The least common multiple of 135, 280, and 300 is 37,800. Compare with **Example 6.**

EXAMPLE 6 **Finding the Least Common Multiple by the Prime Factors Method**

Find the least common multiple of 135, 280, and 300.

Solution

Write the prime factorizations:

$$135 = 3^3 \cdot 5, \quad 280 = 2^3 \cdot 5 \cdot 7, \quad \text{and} \quad 300 = 2^2 \cdot 3 \cdot 5^2.$$

Form the product of all the primes that appear in *any* of the factorizations. Use the *greatest* exponent from any factorization.

> Use the greatest exponents.

$$LCM = 2^3 \cdot 3^3 \cdot 5^2 \cdot 7 = 37,800$$

The least natural number divisible by 135, 280, and 300 is 37,800.   ∎

The least common multiple of a group of numbers can also be found by dividing by prime factors. The process is slightly different from that for finding the GCF.

FINDING THE LEAST COMMON MULTIPLE (DIVIDING BY PRIME FACTORS METHOD)

Step 1   Write the numbers in a row.

Step 2   Divide each of the numbers by a common prime factor. Try 2, then try 3, and so on.

Step 3   Divide the quotients by a common prime factor. When no prime will divide all quotients, but a prime will divide some of them, divide where possible and bring any nondivisible quotients down. Continue until no prime will divide any two quotients.

Step 4   The product of all prime divisors from Steps 2 and 3 as well as all remaining quotients is the least common multiple.

EXAMPLE 7   **Finding the Least Common Multiple by Dividing by Prime Factors**

Find the least common multiple of 12, 18, and 60.

Solution

Proceed just as in **Example 4** to obtain the following.

$$
\begin{array}{r|rrr}
2 & 12 & 18 & 60 \\
\hline
3 & 6 & 9 & 30 \\
\hline
 & 2 & 3 & 10
\end{array}
$$

Now, even though no prime will divide 2, 3, and 10, the prime 2 will divide 2 and 10. Divide the 2 and the 10 and bring down the 3.

$$
\begin{array}{r|rrr}
2 & 12 & 18 & 60 \\
\hline
3 & 6 & 9 & 30 \\
\hline
2 & 2 & 3 & 10 \\
\hline
 & 1 & 3 & 5
\end{array}
\qquad 2 \cdot 3 \cdot 2 \cdot 1 \cdot 3 \cdot 5 = 180
$$

The LCM of 12, 18, and 60 is 180.   ∎

The least common multiple of two numbers *m* and *n* can be obtained by dividing their product by their greatest common factor.

FINDING THE LEAST COMMON MULTIPLE (FORMULA)

The least common multiple of *m* and *n* can be computed as follows.

$$LCM = \frac{m \cdot n}{\text{GCF of } m \text{ and } n}$$

(This method works only for two numbers, not for more than two.)

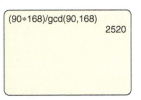

This supports the result in **Example 8.**

EXAMPLE 8 Finding the Least Common Multiple by Formula

Use the formula to find the least common multiple of 90 and 168.

Solution

In **Example 5** we used the Euclidean algorithm to find that the greatest common factor of 90 and 168 is 6. Therefore, the formula gives us

$$\text{Least common multiple of 90 and 168} = \frac{90 \cdot 168}{6} = 2520. \qquad \blacksquare$$

Problem-Solving Strategy

Problems that deal with questions such as "How many objects will there be in each group if each group contains the same number of objects?" and "When will two events occur at the same time?" can sometimes be solved using the ideas of greatest common factor and least common multiple.

EXAMPLE 9 Finding Common Starting Times of Movie Cycles

The King Theatre and the Star Theatre run movies continuously, and each starts its first feature at 1:00 P.M. If the movie shown at the King lasts 80 minutes and the movie shown at the Star lasts 2 hours, when will the two movies again start at the same time?

Solution

First, convert 2 hours to 120 minutes. The question can be restated as follows: "What is the smallest number of minutes it will take for the two movies to start at the same time again?" This is equivalent to asking, "What is the least common multiple of 80 and 120?"

Using any of the methods described in this section, we find that the least common multiple of 80 and 120 is 240. Therefore, it will take 240 minutes, or $\frac{240}{60} = 4$ hours, for the movies to start at the same time again. By adding 4 hours to 1:00 P.M., we find that they will start together again at 5:00 P.M. $\qquad \blacksquare$

EXAMPLE 10 Finding the Greatest Common Size of Stacks of Cards

Joshua has 450 football cards and 840 baseball cards. He wants to place them in stacks on a table so that each stack has the same number of cards, and no stack has different types of cards within it. What is the greatest number of cards that he can have in each stack?

Solution

Here, we are looking for the greatest number that will divide evenly into 450 and 840. This is, of course, the greatest common factor of 450 and 840. Using any of the methods described in this section, we find that

$$\text{greatest common factor of 450 and 840} = 30.$$

Therefore, the greatest number of cards he can have in each stack is 30. He will have 15 stacks of 30 football cards and 28 stacks of 30 baseball cards. $\qquad \blacksquare$

WHEN Will I Ever USE This ?

Addition and subtraction of fractions normally involve identifying the least common denominator, which is the LCM of all denominators. Suppose, for example, that a nurse needs the total displacement (volume) of the following quantities combined.

$$\frac{1}{5} \text{ teacup}, \quad \frac{2}{3} \text{ cup}, \quad 5 \text{ teaspoons}, \quad \text{and} \quad 3 \text{ tablespoons}$$

(In practice, the fractions would normally first be converted to decimal equivalents in metric units, but we illustrate the math here with the common fractions and non-metric units.)

The conversions necessary to obtain common units (in this case, ounces) are shown in the table.

1 teaspoon	=	$\frac{1}{6}$ ounce
1 tablespoon	=	$\frac{1}{2}$ ounce
1 teacup	=	6 ounces
1 cup	=	8 ounces

Thus

$$\frac{1}{5} \text{ teacup} = \frac{1}{5}(6 \text{ ounces}) = \frac{6}{5} \text{ ounces,}$$

$$\frac{2}{3} \text{ cup} = \frac{2}{3}(8 \text{ ounces}) = \frac{16}{3} \text{ ounces,}$$

$$5 \text{ teaspoons} = 5\left(\frac{1}{6} \text{ ounce}\right) = \frac{5}{6} \text{ ounces,}$$

$$3 \text{ tablespoons} = 3\left(\frac{1}{2} \text{ ounce}\right) = \frac{3}{2} \text{ ounces,}$$

and we need the sum $\frac{6}{5} + \frac{16}{3} + \frac{5}{6} + \frac{3}{2}$. One method of finding the least common denominator is dividing by prime factors.

$$
\begin{array}{r|cccc}
2 & 2 & 3 & 5 & 6 \\
3 & 1 & 3 & 5 & 3 \\
5 & 1 & 1 & 5 & 1 \\
\hline
& 1 & 1 & 1 & 1
\end{array}
\quad \text{LCM} = 2 \cdot 3 \cdot 5 \cdot 1 \cdot 1 \cdot 1 \cdot 1 = 30
$$

Using 30 as the least common denominator, we obtain

$$
\frac{6}{5} + \frac{16}{3} + \frac{5}{6} + \frac{3}{2} = \frac{6}{5} \cdot \frac{6}{6} + \frac{16}{3} \cdot \frac{10}{10} + \frac{5}{6} \cdot \frac{5}{5} + \frac{3}{2} \cdot \frac{15}{15}
$$

$$
= \frac{36}{30} + \frac{160}{30} + \frac{25}{30} + \frac{45}{30}
$$

$$
= \frac{36 + 160 + 25 + 45}{30}
$$

$$
= \frac{266}{30} = 8\frac{26}{30}.
$$

The combined total is $8\frac{26}{30}$ ounces, or, in decimal form, about 8.867 ounces.

5.4 EXERCISES

Decide whether each statement is true or false.

1. No two even natural numbers can be relatively prime.

2. Two different prime numbers must be relatively prime.

3. If p is a prime number, then the greatest common factor of p and p^2 is p^2.

4. If p is a prime number, then the least common multiple of p and p^2 is p^3.

5. There is no prime number p such that the greatest common factor of p and 2 is 2.

6. The set of all common multiples of two given natural numbers is infinite.

7. Any two natural numbers have at least one common factor.

8. The least common multiple of two different primes is their product.

9. No two composite numbers can be relatively prime.

10. The product of any two natural numbers is equal to the product of their least common multiple and their greatest common factor.

Use the prime factors method to find the greatest common factor of each group of numbers.

11. 84 and 140

12. 315 and 90

13. 275 and 132

14. 264 and 504

15. 68, 102, and 425

16. 765, 780, and 990

Use the method of dividing by prime factors to find the greatest common factor of each group of numbers.

17. 150 and 260

18. 237 and 395

19. 600 and 90

20. 330 and 255

21. 84, 90, and 210

22. 585, 1680, and 990

Use the Euclidean algorithm to find the greatest common factor of each group of numbers.

23. 18 and 60

24. 77 and 84

25. 36 and 90

26. 72 and 90

27. 945 and 450

28. 200 and 350

Use the prime factors method to find the least common multiple of each group of numbers.

29. 48 and 60

30. 21 and 35

31. 81 and 45

32. 84 and 98

33. 20, 30, and 50

34. 15, 21, and 45

Use the method of dividing by prime factors to find the least common multiple of each group of numbers.

35. 27 and 36

36. 21 and 56

37. 63 and 99

38. 16, 120, and 216

39. 48, 54, and 60

40. 154, 165, and 2310

*Use the formula given in the text on **page 203** and the results of **Exercises 23–28** to find the least common multiple of each group of numbers.*

41. 18 and 60

42. 77 and 84

43. 36 and 90

44. 72 and 90

45. 945 and 450

46. 200 and 350

47. Explain in your own words how to find the greatest common factor of a group of numbers.

48. Explain in your own words how to find the least common multiple of a group of numbers.

49. If p, q, and r are different primes, and a, b, and c are natural numbers such that $a < b < c$,

    (a) what is the greatest common factor of $p^a q^c r^b$ and $p^b q^a r^c$?

    (b) what is the least common multiple of $p^b q^a$, $q^b r^c$, and $p^a r^b$?

50. Find **(a)** the greatest common factor and **(b)** the least common multiple of $2^{25} \cdot 5^{17} \cdot 7^{21}$ and $2^{28} \cdot 5^{22} \cdot 7^{13}$. Leave your answers in prime factored form.

It is possible to extend the Euclidean algorithm in order to find the greatest common factor of more than two numbers. For example, if we wish to find the greatest common factor of 150, 210, and 240, we can first use the algorithm to find the greatest common factor of two of these (say, for example, 150 and 210). Then we find the greatest common factor of that result and the third number, 240. The final result is the greatest common factor of the original group of numbers.

Use the Euclidean algorithm as just described to find the greatest common factor of each group of numbers.

51. 90, 105, and 315

52. 48, 315, and 450

53. 144, 180, and 192

54. 180, 210, and 630

55. Suppose that the least common multiple of p and q is pq. What can we say about p and q?

56. Suppose that the least common multiple of p and q is q. What can we say about p and q?

57. Recall some of your early experiences in mathematics (for example, in the elementary grade classroom). What topic involving fractions required the use of the least common multiple? Give an example.

58. Recall some of your experiences in elementary algebra. What topics required the use of the greatest common factor? Give an example.

*Refer to **Examples 9 and 10** to solve each problem.*

59. *Inspecting Calculators* Jameel and Fahima work on an assembly line, inspecting electronic calculators. Jameel inspects the electronics of every sixteenth calculator, while Fahima inspects the workmanship of every thirty-sixth calculator. If they both start working at the same time, which calculator will be the first that they both inspect?

60. *Night Off for Security Guards* Tomas and Jenny work as security guards at a factory. Tomas has every sixth night off, and Jenny has every tenth night off. If both are off on July 1, what is the next night that they will both be off together?

61. *Stacking Coins* Suyín has 240 pennies and 288 nickels. She wants to place them all in stacks so that each stack has the same number of coins, and each stack contains only one denomination of coin. What is the greatest number of coins that she can place in each stack?

62. *Bicycle Racing* Aki and Felipe are in a bicycle race, following a circular track. If they start at the same place and travel in the same direction, and Aki completes a revolution every 40 seconds, which Felipe takes 45 seconds to complete each revolution, how long will it take them before they reach the starting point again simultaneously?

63. *Selling Books* Azad sold some books at $24 each and used the money to buy some concert tickets at $50 each. He had no money left over after buying the tickets. What is the least amount of money he could have earned from selling the books? What is the least number of books he could have sold?

64. *Sawing Lumber* Terri has some pieces of two-by-four lumber. Some are 60 inches long, and some are 72 inches long. All of them must be sawn into shorter pieces. If all sawn pieces must be the same length, what is the longest such piece so that no lumber is left over?

5.5 THE FIBONACCI SEQUENCE AND THE GOLDEN RATIO

OBJECTIVES

1 Work with the Fibonacci sequence.

2 Understand the golden ratio.

3 See relationships between the Fibonacci sequence and the golden ratio.

The solution of **Fibonacci's rabbit problem** is examined in **Chapter 1**, pages 20–21.

The **Fibonacci Association** is a research organization dedicated to investigation into the **Fibonacci sequence** and related topics. Check your library to see if it has the journal *Fibonacci Quarterly*. The first two journals of 1963 contain a basic introduction to the Fibonacci sequence.

The Fibonacci Sequence

One of the most famous problems in elementary mathematics comes from the book *Liber Abaci,* written in 1202 by Leonardo of Pisa, a.k.a. Fibonacci. The problem is as follows:

> A man put a pair of rabbits in a cage. During the first month the rabbits produced no offspring, but each month thereafter produced one new pair of rabbits. If each new pair thus produced reproduces in the same manner, how many pairs of rabbits will there be at the end of one year?

The solution of this problem leads to a sequence of numbers known as the **Fibonacci sequence.** Here are the first fifteen terms of the Fibonacci sequence:

$$1, 1, 2, 3, 5, 8, 13, 21, 34, 55, 89, 144, 233, 377, 610.$$

After the first two terms (both 1) in the sequence, each term is obtained by adding the two previous terms. For example, the third term is obtained by adding $1 + 1$ to get 2, the fourth term is obtained by adding $1 + 2$ to get 3, and so on. This can be described by a mathematical formula known as a **recursion formula.**

If F_n represents the Fibonacci number in the nth position in the sequence, then

$$F_1 = 1$$
$$F_2 = 1$$
$$F_n = F_{n-2} + F_{n-1}, \quad \text{for } n \geq 3.$$

Using the recursion formula $F_n = F_{n-2} + F_{n-1}$, we obtain

$$F_3 = F_1 + F_2 = 1 + 1 = 2, \quad F_4 = F_2 + F_3 = 1 + 2 = 3, \quad \text{and so on.}$$

The Fibonacci sequence exhibits many interesting patterns, and by inductive reasoning we can make many conjectures about these patterns. However, simply observing a finite number of examples does not provide a proof of a statement. Proofs of the properties of the Fibonacci sequence often involve mathematical induction (covered in college algebra texts). Here we simply observe some of the patterns and do not attempt to provide proofs.

EXAMPLE 1 Observing a Pattern of the Fibonacci Numbers

Find the sum of the squares of the first n Fibonacci numbers for $n = 1, 2, 3, 4, 5$, and examine the pattern. Generalize this relationship.

Solution

$$1^2 = 1 = 1 \cdot 1 = F_1 \cdot F_2$$
$$1^2 + 1^2 = 2 = 1 \cdot 2 = F_2 \cdot F_3$$
$$1^2 + 1^2 + 2^2 = 6 = 2 \cdot 3 = F_3 \cdot F_4$$
$$1^2 + 1^2 + 2^2 + 3^2 = 15 = 3 \cdot 5 = F_4 \cdot F_5$$
$$1^2 + 1^2 + 2^2 + 3^2 + 5^2 = 40 = 5 \cdot 8 = F_5 \cdot F_6$$

The sum of the squares of the first n Fibonacci numbers seems to always be the product of F_n and F_{n+1}. This has been proved to be true, in general, using mathematical induction.

The following program for the TI-83 Plus utilizes the *Binet form* of the *n*th Fibonacci number (see **Exercises 33–36**) to determine its value.

```
PROGRAM: FIB
: ClrHome
: Disp "WHICH TERM"
: Disp "OF THE"
: Disp "SEQUENCE DO"
: Disp "YOU WANT?"
: Input N
: (1 + √(5))/2 → A
: (1 − √(5))/2 → B
: (A^N − B^N)/√(5) → F
: Disp F
```

```
WHICH TERM
OF THE
SEQUENCE DO
YOU WANT?
?20
                6765
                Done
```

This screen indicates that the twentieth Fibonacci number is 6765.

Fibonacci Fun To observe one of the many interesting properties of the Fibonacci sequence, do the following.

1. Choose any term after the first and square it.
2. Multiply the terms before and after the term chosen in Step 1.
3. Subtract the smaller value from the larger.
4. What is your result?

Try this procedure beginning with several different Fibonacci terms. **The result is always the same.**

EXAMPLE 2 Observing the Fibonacci Sequence in a Long Division Problem

Observe the steps of the long-division algorithm used to find the first few decimal places of the reciprocal of 89, the eleventh Fibonacci number. Locate occurrences of the terms of the Fibonacci sequence in the algorithm.

Solution

$$
\begin{array}{r}
.011235\ldots \\
89)\overline{1.000000\ldots} \\
89 \\
\hline
110 \\
89 \\
\hline
210 \\
178 \\
\hline
320 \\
267 \\
\hline
530 \\
445 \\
\hline
850\ldots
\end{array}
$$

Notice that after the 0 in the tenths place, the next five digits are the first five terms of the Fibonacci sequence. In addition, as indicated in color in the process, the digits 1, 1, 2, 3, 5, 8 appear in the division steps. Now, look at the digits next to the ones in color, beginning with the second "1"; they, too, are 1, 1, 2, 3, 5, ...

If the division process is continued past the final step shown above, the pattern seems to stop, because to ten decimal places, $\frac{1}{89} \approx 0.0112359551$. (The decimal representation actually begins to repeat later in the process, since $\frac{1}{89}$ is a rational number.) However, the sum below indicates how the Fibonacci numbers are actually "hidden" in this decimal.

$$
\begin{array}{r}
0.01\mathbf{1} \\
0.001\mathbf{1} \\
0.0002 \\
0.00003 \\
0.000005 \\
0.0000008 \\
0.00000013 \\
0.000000021 \\
0.0000000034 \\
0.00000000055 \\
0.000000000089 \\
\hline
\frac{1}{89} = 0.0112359550.\ldots
\end{array}
$$

Fibonacci patterns have been found in numerous places in nature. For example, male honeybees (drones) hatch from eggs that have not been fertilized, so a male bee has only one parent, a female. On the other hand, female honeybees hatch from fertilized eggs, so a female has two parents, one male and one female. **Figure 1** shows several generations of ancestors for a male honeybee.

Notice that in the first generation, starting at the bottom, there is 1 bee, in the second there is 1 bee, in the third there are 2 bees, and so on. These are the terms of the Fibonacci sequence. Furthermore, beginning with the second generation, the numbers of female bees form the sequence, and beginning with the third generation, the numbers of male bees also form the sequence.

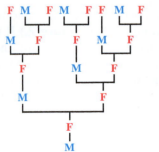

Figure 1

A fraction such as

$$1 + \cfrac{1}{1 + \cfrac{1}{1 + \cfrac{1}{1 + \cdots}}}$$

is called a **continued fraction.** This continued fraction can be evaluated as follows.

Let $x = 1 + \cfrac{1}{1 + \cfrac{1}{1 + \cdots}}$

Then $x = 1 + \dfrac{1}{x}$

$$x^2 = x + 1$$
$$x^2 - x - 1 = 0.$$

By the quadratic formula from algebra,

$$x = \frac{1 \pm \sqrt{1 - 4(1)(-1)}}{2(1)}$$

$$x = \frac{1 \pm \sqrt{5}}{2}.$$

Notice that the positive solution is the **golden ratio.**

Successive terms in the Fibonacci sequence also appear in some plants. For example, the photo on the left below shows the double spiraling of a daisy head, with 21 clockwise spirals and 34 counterclockwise spirals. These numbers are successive terms in the sequence.

Most pineapples (see the photo on the right below) exhibit the Fibonacci sequence in the following way: Count the spirals formed by the "scales" of the cone, first counting from lower left to upper right. Then count the spirals from lower right to upper left. You should find that in one direction you get 8 spirals, and in the other you get 13 spirals, once again successive terms of the Fibonacci sequence. Many pinecones exhibit 5 and 8 spirals, and the cone of the giant sequoia has 3 and 5 spirals.

The Golden Ratio

If we consider the quotients of successive Fibonacci numbers, a pattern emerges.

$\dfrac{1}{1} = 1$	$\dfrac{5}{3} = 1.666\ldots$	$\dfrac{21}{13} \approx 1.615384615$	$\dfrac{89}{55} = 1.618181818\ldots$
$\dfrac{2}{1} = 2$	$\dfrac{8}{5} = 1.6$	$\dfrac{34}{21} \approx 1.619047619$	
$\dfrac{3}{2} = 1.5$	$\dfrac{13}{8} = 1.625$	$\dfrac{55}{34} \approx 1.617647059$	

These quotients seem to be approaching some "limiting value" close to 1.618. In fact, as we go farther into the sequence, these quotients approach the number

$$\frac{1 + \sqrt{5}}{2}, \quad \text{Golden ratio}$$

known as the **golden ratio** and often symbolized by $\phi,$ the Greek letter phi.

The golden ratio appears over and over in art, architecture, music, and nature. Its origins go back to the days of the ancient Greeks, who thought that a golden rectangle exhibited the most aesthetically pleasing proportions.

A **golden rectangle** is one that can be divided into a square and another (smaller) rectangle the same shape as the original rectangle. (See **Figure 2.**) If we let the smaller rectangle have length L and width W, as shown in the figure, then we see that the original rectangle has length $L + W$ and width L. Both rectangles (being "golden") have their lengths and widths in the golden ratio, ϕ, so we have

$$\frac{L}{W} = \frac{L + W}{L}$$

$$\frac{L}{W} = \frac{L}{L} + \frac{W}{L} \qquad \text{Write the right side as the sum of two fractions.}$$

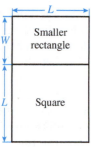

Figure 2

(Figure labels: L, W, Smaller rectangle, L, Square)

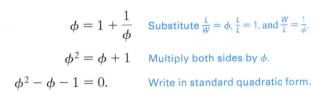

$$\phi = 1 + \frac{1}{\phi} \qquad \text{Substitute } \tfrac{L}{W} = \phi, \tfrac{L}{L} = 1, \text{ and } \tfrac{W}{L} = \tfrac{1}{\phi}.$$

$$\phi^2 = \phi + 1 \qquad \text{Multiply both sides by } \phi.$$

$$\phi^2 - \phi - 1 = 0. \qquad \text{Write in standard quadratic form.}$$

Using the quadratic formula from algebra, the positive solution of this equation is found to be $\frac{1 + \sqrt{5}}{2} \approx 1.618033989$, the golden ratio.

The Parthenon (see the photo), built on the Acropolis in ancient Athens during the fifth century B.C., is an example of architecture exhibiting many distinct golden rectangles.

To see an interesting connection between the terms of the Fibonacci sequence, the golden ratio, and a phenomenon of nature, we can start with a rectangle measuring 89 by 55 units. (See **Figure 3.**) This is a very close approximation to a golden rectangle. Within this rectangle a square is then constructed, 55 units on a side. The remaining rectangle is also approximately a golden rectangle, measuring 55 units by 34 units. Each time this process is repeated, a square and an approximate golden rectangle are formed.

As indicated in **Figure 3,** vertices of the squares may be joined by a smooth curve known as a *spiral*. This spiral resembles the outline of a cross section of the shell of the chambered nautilus, as shown in the photo.

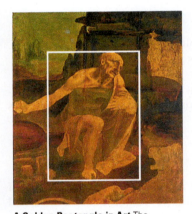

A Golden Rectangle in Art The rectangle outlining the figure in *St. Jerome* by Leonardo da Vinci is an example of a golden rectangle.

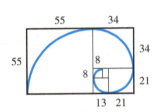

Figure 3

FOR FURTHER THOUGHT

Mathematical Animation

The 1959 animated film *Donald in Mathmagic Land* has endured for over 50 years as a classic. It provides a 25-minute trip with Donald Duck, led by the Spirit of Mathematics, through the world of mathematics. Several minutes of the film are devoted to the golden ratio (or, as it is termed there, the golden section).

Disney provides animation to explain the golden ratio in a way that the printed word simply cannot do. The golden ratio is seen in architecture, nature, and the human body.

For Group or Individual Investigation

1. Verify the following Fibonacci pattern in the conifer family. Obtain a pineapple, and count spirals formed by the "scales" of the cone, first counting from lower left to upper right. Then count the spirals from lower right to upper left. What do you find?

2. Two popular sizes of index cards are 3″ by 5″ and 5″ by 8″. Why do you think that these are industry-standard sizes?

3. Divide your height by the height to your navel. Find a class average. What value does this come close to?

5.5   EXERCISES

Answer each question concerning the Fibonacci sequence or the golden ratio.

1. The fifteenth Fibonacci number is 610 and the seventeenth Fibonacci number is 1597. What is the value of the sixteenth Fibonacci number?

2. Recall that F_n represents the Fibonacci number in the nth position in the sequence. Find all values of n such that $F_n = n$.

3. Write the "parity" (o for odd, e for even) of each of the first twelve terms of the Fibonacci sequence.

4. From the pattern shown in the parity sequence, determine the parity of the one-hundredth term.

5. What is the exact value of the golden ratio?

6. What is the approximate value of the golden ratio to the nearest thousandth?

In each of Exercises 7–14, a pattern is established involving terms of the Fibonacci sequence. Use inductive reasoning to make a conjecture concerning the next equation in the pattern, and verify it. You may wish to refer to the first few terms of the sequence given in the text.

7. $1 = 2 - 1$
   $1 + 1 = 3 - 1$
   $1 + 1 + 2 = 5 - 1$
   $1 + 1 + 2 + 3 = 8 - 1$

8. $1 = 2 - 1$
   $1 + 3 = 5 - 1$
   $1 + 3 + 8 = 13 - 1$
   $1 + 3 + 8 + 21 = 34 - 1$

9. $1 = 1$
   $1 + 2 = 3$
   $1 + 2 + 5 = 8$
   $1 + 2 + 5 + 13 = 21$

10. $1^2 + 1^2 = 2$
    $1^2 + 2^2 = 5$
    $2^2 + 3^2 = 13$
    $3^2 + 5^2 = 34$

11. $2^2 - 1^2 = 3$
    $3^2 - 1^2 = 8$
    $5^2 - 2^2 = 21$
    $8^2 - 3^2 = 55$

12. $2^3 + 1^3 - 1^3 = 8$
    $3^3 + 2^3 - 1^3 = 34$
    $5^3 + 3^3 - 2^3 = 144$
    $8^3 + 5^3 - 3^3 = 610$

13. $1 = 1^2$
    $1 - 2 = -1^2$
    $1 - 2 + 5 = 2^2$
    $1 - 2 + 5 - 13 = -3^2$

14. $1 - 1 = -1 + 1$
    $1 - 1 + 2 = 1 + 1$
    $1 - 1 + 2 - 3 = -2 + 1$
    $1 - 1 + 2 - 3 + 5 = 3 + 1$

15. Every natural number can be expressed as a sum of Fibonacci numbers, where no number is used more than once. For example,

    $$25 = 21 + 3 + 1.$$

    Express each of the following in this way.
    (a) 39   (b) 59   (c) 99

16. It has been shown that if m divides n, then F_m is a factor of F_n. Show that this is true for the following values of m and n.
    (a) $m = 3, n = 6$ (b) $m = 4, n = 12$
    (c) $m = 5, n = 15$

17. It has been shown that if the greatest common factor of m and n is r, then the greatest common factor of F_m and F_n is F_r. Show that this is true for the following values of m and n.
    (a) $m = 10, n = 4$ (b) $m = 12, n = 6$
    (c) $m = 14, n = 6$

18. For any prime number p except 2 or 5, either F_{p+1} or F_{p-1} is divisible by p. Show that this is true for the following values of p.
    (a) $p = 3$ (b) $p = 7$ (c) $p = 11$

19. Earlier we saw that if a term of the Fibonacci sequence is squared and then the product of the terms on each side of the term is found, there will always be a difference of 1. Follow the steps below, choosing the sixth Fibonacci number, 8.

    (a) Square 8. Multiply the terms of the sequence two positions away from 8 (i.e., 3 and 21). Subtract the smaller result from the larger, and record your answer.

    (b) Square 8. Multiply the terms of the sequence three positions away from 8. Once again, subtract the smaller result from the larger, and record your answer.

    (c) Repeat the process, moving four terms away from 8.

    (d) Make a conjecture about what will happen when you repeat the process, moving five terms away. Verify your answer.

20. *A Number Trick*   Here is a number trick that you can perform. Ask someone to pick any two numbers at random and to write them down. Ask the person to determine a third number by adding the first and second, a fourth number by adding the second and third, and so on, until ten numbers are determined. Then ask the person to add these ten numbers. You will be able to give the sum before the person even completes the list, because the sum will always be 11 times the seventh number in the list. Verify that this is true, by using x and y as the first two numbers arbitrarily chosen. (*Hint:* Remember the distributive property from algebra.)

Another Fibonacci-type sequence that has been studied by mathematicians is the **Lucas sequence,** *named after a French mathematician of the nineteenth century. The first nine terms of the Lucas sequence are*

$$1, 3, 4, 7, 11, 18, 29, 47, 76.$$

21. What is the tenth term of the Lucas sequence?

22. Choose any term of the Lucas sequence and square it. Then multiply the terms on either side of the one you chose. Subtract the smaller result from the larger. Repeat this for a different term of the sequence. Do you get the same result? Make a conjecture about this pattern.

23. The first term of the Lucas sequence is 1. Add the first and third terms. Record your answer. Now add the first, third, and fifth terms and record your answer. Continue this pattern, each time adding another term that is in an *odd* position in the sequence. What do you notice about all of your sums?

24. The second term of the Lucas sequence is 3. Add the second and fourth terms. Record your answer. Now add the second, fourth, and sixth terms and record your answer. Continue this pattern, each time adding another term that is in an *even* position of the sequence. What do you notice about all of your sums?

25. Many interesting patterns exist among the terms of the Fibonacci sequence and the Lucas sequence. Make a conjecture about the next equation that would appear in each of the lists, and then verify it.

(a) $1 \cdot 1 = 1$
$1 \cdot 3 = 3$
$2 \cdot 4 = 8$
$3 \cdot 7 = 21$
$5 \cdot 11 = 55$

(b) $1 + 2 = 3$
$1 + 3 = 4$
$2 + 5 = 7$
$3 + 8 = 11$
$5 + 13 = 18$

(c) $1 + 1 = 2 \cdot 1$
$1 + 3 = 2 \cdot 2$
$2 + 4 = 2 \cdot 3$
$3 + 7 = 2 \cdot 5$
$5 + 11 = 2 \cdot 8$

(d) $1 + 4 = 5 \cdot 1$
$3 + 7 = 5 \cdot 2$
$4 + 11 = 5 \cdot 3$
$7 + 18 = 5 \cdot 5$
$11 + 29 = 5 \cdot 8$

26. In the text we illustrate that the quotients of successive terms of the Fibonacci sequence approach the golden ratio. Make a similar observation for the terms of the Lucas sequence; that is, find the decimal approximations for the quotients

$$\frac{3}{1}, \frac{4}{3}, \frac{7}{4}, \frac{11}{7}, \frac{18}{11}, \frac{29}{18},$$

and so on, using a calculator. Then make a conjecture about what seems to be happening.

Recall the **Pythagorean theorem** *from geometry: If a right triangle has legs of lengths a and b and hypotenuse of length c, then*

$$a^2 + b^2 = c^2.$$

Suppose that we choose any four successive terms of the Fibonacci sequence. Multiply the first and fourth. Double the product of the second and third. Add the squares of the second and third. The three results obtained form a **Pythagorean triple** *(three numbers that satisfy the equation $a^2 + b^2 = c^2$). Find the Pythagorean triple obtained this way, using the four given successive terms of the Fibonacci sequence.*

27. 1, 1, 2, 3   **28.** 2, 3, 5, 8   **29.** 5, 8, 13, 21

30. Look at the values of the hypotenuse (c) in the answers to **Exercises 27–29.** What do you notice about each of them?

31. The following array of numbers is called **Pascal's triangle.**

```
              1
            1   1
          1   2   1
        1   3   3   1
      1   4   6   4   1
    1   5  10  10   5   1
  1   6  15  20  15   6   1
```

This array is important in the study of counting techniques and probability (see later chapters) and appears in algebra in the binomial theorem. If the triangular array is written in a different form, as follows, and the sums along the diagonals as indicated by the dashed lines are found, there is an interesting occurrence. What do you find when the numbers are added?

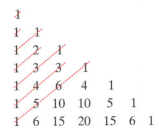

32. Write a paragraph explaining some of the occurrences of the Fibonacci sequence and the golden ratio in your everyday surroundings.

Exercises 33–36 require a scientific calculator.

33. The positive solution of $x^2 - x - 1 = 0$ is

$$\frac{1 + \sqrt{5}}{2},$$

as indicated in the text. The negative solution is

$$\frac{1 - \sqrt{5}}{2}.$$

Find the decimal approximations for both. What similarity do you notice between the two decimals?

34. In some cases, writers define the golden ratio to be the *reciprocal of* $\frac{1 + \sqrt{5}}{2}$. Find a decimal approximation for the reciprocal of $\frac{1 + \sqrt{5}}{2}$. What similarity do you notice between the decimals for $\frac{1 + \sqrt{5}}{2}$ and its reciprocal?

A remarkable relationship exists between the two solutions of the equation $x^2 - x - 1 = 0$, which are

$$\phi = \frac{1 + \sqrt{5}}{2} \quad and \quad \overline{\phi} = \frac{1 - \sqrt{5}}{2},$$

and the Fibonacci numbers. To find the nth Fibonacci number without using the recursion formula, use a calculator to evaluate

$$\frac{\phi^n - \overline{\phi}^n}{\sqrt{5}}.$$

Thus, to find the thirteenth Fibonacci number, evaluate

$$\frac{\left(\frac{1 + \sqrt{5}}{2}\right)^{13} - \left(\frac{1 - \sqrt{5}}{2}\right)^{13}}{\sqrt{5}}.$$

This form is known as the **Binet form** *of the nth Fibonacci number. Use the Binet form and a calculator to find the nth Fibonacci number for each of the following values of n.*

35. $n = 16$ **36.** $n = 24$

CHAPTER 5   SUMMARY

KEY TERMS

5.1

number theory
prime number
composite number
divisibility
divisibility tests
factor
divisor
multiple
factorization
prime factorization

Sieve of Eratosthenes
fundamental theorem
    of arithmetic
cryptography
modular systems

5.2

proof by contradiction
Mersenne number
Mersenne prime
partition

5.3

perfect number
deficient number
abundant number
amicable (friendly)
    numbers
Goldbach's conjecture
twin primes
primordial primes
Fermat's Last Theorem

5.4

greatest common factor (GCF)
relatively prime numbers
Euclidean algorithm
least common multiple (LCM)

5.5

Fibonacci sequence
Fibonacci numbers
golden ratio
golden rectangle

NEW SYMBOLS

$a \mid b$ *a* divides *b*

ϕ the golden ratio

TEST YOUR WORD POWER

See how well you have learned the vocabulary in this chapter.

1. In mathematics, **number theory** in general deals with
  A. the properties of the rational numbers.
  B. operations on fractions.
  C. the properties of the natural numbers.
  D. only those numbers that are prime.

2. A **prime number** is
  A. divisible by 2, whereas a composite number is not.
  B. divisible by itself and 1 only.
  C. one of the finite set of numbers discovered by Marin Mersenne.
  D. any number expressible in the form $2^p - 1$, where p is a prime.

3. A **partition** of a set is
   A. a rule that separates half of its elements from the other half.
   B. a subset that contains some but not all of the elements of the set.
   C. a rule by which all elements of the set can be identified.
   D. a division of all elements of the set into disjoint subsets.

4. A natural number that is less than the sum of its **proper divisors** is
   A. an **abundant** number.
   B. a **perfect** number.
   C. a **deficient** number.
   D. a **Fermat number.**

5. An **amicable** pair consists of two numbers that are
   A. either both prime or both composite.
   B. either both odd or both even.
   C. each equal to the sum of the proper divisors of the other.
   D. one Fermat number and one Mersenne number.

6. **Goldbach's conjecture**
   A. was formulated over 300 years before it was proved.
   B. has to do with two *n*th powers summing to a third *n*th power.

C. has to do with expressing natural numbers as products of primes.
D. has to do with expressing natural numbers as sums of primes.

7. The so-called **twin primes**
   A. always occur in pairs that differ by 2.
   B. always occur in pairs that differ by 1.
   C. cannot exceed 10^{100}.
   D. have finally all been identified.

8. **Fermat's Last Theorem**
   A. was proved by Euler about 100 years after Fermat stated it.
   B. has never been proved, but most mathematicians believe it is true.
   C. was proved by Wiles in the 1990s.
   D. says that the equation $a^n + b^n = c^n$ cannot be true for $n = 2$.

ANSWERS
1. C **2.** B **3.** D **4.** A
5. C **6.** D **7.** A **8.** C

QUICK REVIEW

Concepts	Examples

5.1 Prime and Composite Numbers

The natural number a is divisible by the natural number b, or, equivalently, "b divides a" (denoted $b \mid a$), if there exists a natural number k such that $a = bk$.

$2 \mid 14$, since $14 = 2 \cdot 7$ ($k = 7$). 2 is a factor of 14, and 14 is a multiple of 2.

In the equation above, we say that b is a **factor** (or **divisor**) of a and that a is a multiple of b.

$2 \nmid 13$, since there is no suitable k. Here $13 = 2 \cdot 6.5$ (and $k = 6.5$ is *not* a natural number).

A **factorization** of a natural number expresses it as a product of some, or all, of its divisors.

$2 \cdot 12, \; 3 \cdot 8, \; 4 \cdot 6, \; 2 \cdot 2 \cdot 6, \; 2 \cdot 3 \cdot 4, \; 2 \cdot 2 \cdot 2 \cdot 3$
Factorizations of 24

A **prime number** is greater than 1 and has only itself and 1 as factors.

17 is prime. Its only factors are 1 and 17.

A **composite number** is greater than 1 and is *not* prime—that is, it has at least one factor other than itself and 1.

10 is composite. In addition to 1 and 10, it has the factors 2 and 5.

Divisibility tests can help determine the factors of a given natural number.

54,086,226 is divisible by 3 since the sum of its digits
$$5 + 4 + 0 + 8 + 6 + 2 + 2 + 6 = 33$$
is divisible by 3.

258,784 is *not* divisible by 3 since the sum of its digits
$$2 + 5 + 8 + 7 + 8 + 4 = 34$$
is *not* divisible by 3.

Concepts	Examples
A factorization with prime factors only is a number's **prime factorization**.	The prime factorization of 24 is $$2 \cdot 2 \cdot 2 \cdot 3.$$
The **fundamental theorem of arithmetic** states that the prime factorization of a number is unique (if the order of the factors is disregarded).	24 has no other prime factorization except rearrangements of the factors shown above, such as $2 \cdot 3 \cdot 2 \cdot 2$ and $3 \cdot 2 \cdot 2 \cdot 2$.
A natural number's prime factorization can be determined by using a **factor tree** or by using **repeated division by primes**.	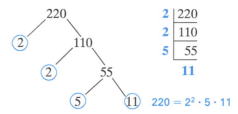 $220 = 2^2 \cdot 5 \cdot 11$

5.2 Large Prime Numbers

Concepts	Examples
Modern **cryptography systems** are based on large prime numbers.	Cryptography systems, in practice, use primes so huge that the most powerful computers can multiply them but *cannot* factor their product.
Large primes are found today mainly among the **Mersenne numbers**.	The largest Mersenne prime discovered, as of summer 2014, is $2^{57,885,161} - 1$. It has 17,425,170 digits.

5.3 Selected Topics from Number Theory

Concepts	Examples
The **proper divisors** of a natural number are all of its divisors except the number itself.	42 has proper divisors 1, 2, 3, 6, 7, 14, and 21.
A natural number is **perfect** if the sum of its proper divisors equals the number itself.	28 has proper divisors **1, 2, 4, 7,** and **14**. $$28 = \mathbf{1 + 2 + 4 + 7 + 14}$$
All even perfect numbers are of the form $\mathbf{2^{n-1}(2^n - 1)}$, where $2^n - 1$ is prime.	The number $2^3 - 1 = 7$ is prime, so $$2^{3-1}(2^3 - 1) = 4(7) = 28 \quad \text{is perfect.}$$
A natural number is **abundant** if the sum of its proper divisors is greater than the number itself.	$1 + 2 + 3 + 6 + 7 + 14 + 21 = 54 > \mathbf{42} \leftarrow$ Abundant
A natural number is **deficient** if the sum of its proper divisors is less than the number itself.	$1 + 2 + 4 + 8 = 15 < \mathbf{16} \leftarrow$ Deficient
Two natural numbers are **amicable**, or **friendly**, if each equals the sum of the other's proper divisors.	The smallest amicable pair is 220 and 284. Proper divisors of **220**: 1, 2, 4, 5, 10, 11, 20, 22, 44, 55, 110 Proper divisors of **284**: 1, 2, 4, 71, 142 $$1 + 2 + 4 + 5 + 10 + 11 + 20 + 22 + 44 + 55 + 110 = \mathbf{284}$$ $$1 + 2 + 4 + 71 + 142 = \mathbf{220}$$
Goldbach's conjecture (which is not proved) states that every even number greater than 2 can be written as the sum of two prime numbers.	$12 = 5 + 7 \qquad 46 = 3 + 43 \qquad$ Goldbach's conjecture
Twin primes are two primes that differ by 2.	17 and 19 29 and 31 Twin primes
Fermat's Last Theorem (proved in the 1990s) states that there are no triples of natural numbers (a, b, c) that satisfy the equation $a^n + b^n = c^n$ for any natural number $n \geq 3$.	$2^3 + 6^3 = 8 + 216 = 224$, but 224 is not a perfect cube.

Concepts	**Examples**

5.4 Greatest Common Factor and Least Common Multiple

The **greatest common factor (GCF)** of a group of natural numbers is the largest natural number that is a factor of all the numbers in the group.

Prime factors method:

$$450 = \mathbf{2} \cdot 3^2 \cdot 5^2$$

$$660 = 2^2 \cdot \mathbf{3} \cdot \mathbf{5} \cdot 11$$

$$\text{GCF} = \mathbf{2} \cdot \mathbf{3} \cdot \mathbf{5} = 30$$

Dividing by prime factors method:

```
2 | 450   660
3 | 225   330
5 |  75   110
      15    22
```

GCF = 2 · 3 · 5 = 30

Euclidean algorithm:

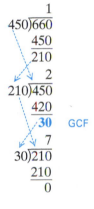

The **least common multiple (LCM)** of a group of natural numbers is the smallest natural number that is a multiple of all the numbers in the group.

Prime factors method:

$$450 = 2 \cdot \mathbf{3^2} \cdot \mathbf{5^2}$$

$$660 = \mathbf{2^2} \cdot 3 \cdot 5 \cdot \mathbf{11}$$

$$\text{LCM} = \mathbf{2^2} \cdot \mathbf{3^2} \cdot \mathbf{5^2} \cdot \mathbf{11} = 9900$$

Dividing by prime factors method:

```
2 | 450   660
2 | 225   330
3 | 225   165
3 |  75    55
5 |  25    55
      5    11
```

LCM = 2 · 2 · 3 · 3 · 5 · 5 · 11 = 9900

Formula:

$$\text{LCM} = \frac{450 \cdot 660}{\text{GCF of 450 and 660}} = \frac{297{,}000}{30} = 9900$$

5.5 The Fibonacci Sequence and the Golden Ratio

The **Fibonacci sequence** begins with two 1s, and each succeeding term is the sum of the preceding two terms.

1, 1, 2, 3, 5, 8, 13, 21, 34, 55, 89, 144, 233, 377, . . .

Concepts	Examples
The **golden ratio,** $$\frac{1 + \sqrt{5}}{2},$$ arises in many ways. One is as the limit of the ratio of adjacent terms in the Fibonacci sequence.	The ratios (to three decimal places), $\frac{1}{1} = 1.000$, $\frac{2}{1} = 2.000$, $\frac{3}{2} = 1.500$, $\frac{5}{3} \approx 1.667$, $\frac{8}{5} = 1.600, \dots$ are approaching $\frac{1 + \sqrt{5}}{2} \approx 1.618$.

CHAPTER 5 TEST

In Exercises 1–6, decide whether each statement is true *or* false.

1. For all natural numbers n, 1 is a factor of n, and n is a multiple of n.

2. If a natural number is not perfect, then it must be abundant.

3. If a natural number is divisible by 9, then it must also be divisible by 3.

4. If p and q are different primes, 1 is their greatest common factor, and pq is their least common multiple.

5. No two prime numbers differ by 1.

6. There are infinitely many prime numbers.

7. Use divisibility tests to determine whether the number

$$540{,}487{,}608$$

is divisible by each of the following.
   (a) 2   **(b)** 3   **(c)** 4
   (d) 5   **(e)** 6   **(f)** 8
   (g) 9   **(h)** 10   **(i)** 12

8. Decide whether each number is prime, composite, or neither.
   (a) 91   **(b)** 101   **(c)** 1

9. Give the prime factorization of 2970.

10. In your own words state the fundamental theorem of arithmetic.

11. Decide whether each number is perfect, deficient, or abundant.
   (a) 18   **(b)** 32   **(c)** 28

12. Which of the following statements is false?
   A. There are no known odd perfect numbers.
   B. Every even perfect number must end in 8 or 26.
   C. Goldbach's conjecture for the number 12 is illustrated by the equation $12 = 7 + 5$.

13. Give a pair of twin primes between 60 and 80.

14. Find the greatest common factor of 99, 135, and 216.

15. Find the least common multiple of 91 and 154.

16. *Day Off for Fast-food Workers* Both Katherine and Josh work at a fast-food outlet. Katherine has every sixth day off, and Josh has every fourth day off. If they are both off on Wednesday of this week, what will be the day of the week that they are next off together?

17. The twenty-first Fibonacci number is 10,946 and the twenty-second Fibonacci number is 17,711. What is the twenty-third Fibonacci number?

18. Make a conjecture about the next equation in the following list, and verify it.

$$8 - (1 + 1 + 2 + 3) = 1$$
$$13 - (1 + 2 + 3 + 5) = 2$$
$$21 - (2 + 3 + 5 + 8) = 3$$
$$34 - (3 + 5 + 8 + 13) = 5$$

19. Choose the correct completion of this statement: If p is a prime number, then $2^p - 1$ is _____ prime.
   A. never   **B.** sometimes   **C.** always

20. Determine the eighth term of a Fibonacci-type sequence with first term 1 and second term 4.

21. Choose any term after the first in the sequence described in **Exercise 20.** Square it. Multiply the two terms on either side of it. Subtract the smaller result from the larger. Now repeat the process with a different term. Make a conjecture about what this process will yield for any term of the sequence.

22. Which one of the following is the *exact* value of the golden ratio?

   A. $\dfrac{1 + \sqrt{5}}{2}$ **B.** $\dfrac{1 - \sqrt{5}}{2}$ **C.** 1.6   **D.** 1.618

The Real Numbers and Their Representations

6

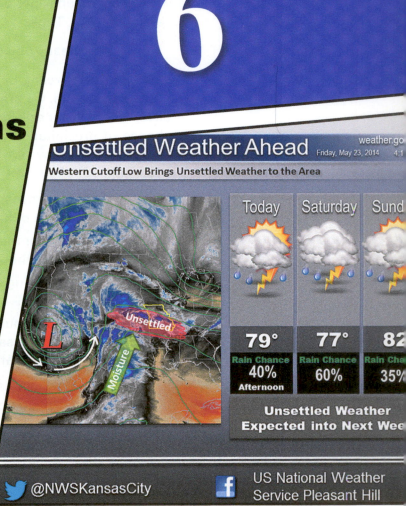

In the March 12, 2014, issue of the *New Orleans Times Picayune*, columnist Jarvis DeBerry wrote the following in an article titled "Old Math or Common Core Math, Which Troubles You More?"

After a February board meeting for my alma mater's student newspaper, I told some students that I began Washington University in the engineering school. One asked if my previous focus on math and science ever proves useful in journalism. I joked: "Only when a colleague needs help figuring out the percent something has increased." Another board member who, like me, graduated from WashU and worked in journalism, responded with something close to awe: "You can do percentages? I mean, seriously?"

In the development of the real numbers (the topic of this chapter), the unit number 1 is the place at which everything starts. (0 showed up much later in our number system.) The concept of percent has become a standard by which portions of one whole thing are described.

Percent means "per one hundred," and the symbol % is used to represent percent. Thus,

1% means one one-hundredth, or $\dfrac{1}{100}$, or 0.01.

One hundred percent, or 100%, means 1 whole thing.

As consumers, we are faced with the use of percent on a daily basis: advertisements for sale items, interest on loans and savings, poll numbers, chance of precipitation in daily weathercasts, and so on. Occupations that otherwise may not require formal mathematics often require the basic concept of percent.

- Retail managers marking up an item to sell might be required to find a given percent of the wholesale cost.
- The same retailer may need to use a percent to determine how much to subtract from the usual selling price to mark an item that is on sale.
- Teachers assigning grades may need to find a percent of a total number of points.
- A manager in a publishing company may need to determine the percent change of sales from one edition of a text to the next, or the percent of sales that a particular title captures in its market.

Some people, such as DeBerry, have a natural talent—or a learned ability—for determining percents, while others don't have a clue. Those who are proficient use many different methods to find simple percents. For example, if you are asked to find

20% of 50,

you might mentally calculate using one of the following methods.

1. You think, "Well, 20% means $\frac{1}{5}$, and to find $\frac{1}{5}$ of something I divide by 5, so 50 divided by 5 is 10. The answer is 10."

2. You think, "20% is twice 10%, and to find 10% of something I move the decimal point one place to the left. So 10% of 50 is 5.0, or just 5, and I simply double 5 to get 10. The answer is 10."

For those who don't think this way, there is the tried-and-true method of multiplying 50 by 0.20.

One whole dollar ($1.00) is the basic monetary unit in the United States. One one-hundredth of a dollar is a penny, or one cent. Notice the correlation between this concept and that of percent. One cent is represented $0.01, or just $.01. *Then how do we represent 99 cents?* The answer is **$.99.** But so many times we see signs such as the one here from Wendy's, indicating a price of **.99¢.** Because ¢ is a symbol for "cents," this posted price actually means that the spicy nuggets sell for less than one penny. Think about it. *Do you understand why?*

If you don't have the knack for computing percents using shortcuts, there are sound mathematical procedures for working with this concept. This chapter introduces the real number system, the operations of arithmetic with real numbers, and selected applications of fractions, decimals, and percents in occupations and trades.

6.1 REAL NUMBERS, ORDER, AND ABSOLUTE VALUE

OBJECTIVES

1 Represent a number on a number line.

2 Identify a number as positive, negative, or zero.

3 Identify a number as belonging to one or more sets of numbers.

4 Given two numbers a and b, determine whether $a = b$, $a < b$, or $a > b$.

5 Given a number a, determine its additive inverse and absolute value.

6 Interpret signed numbers in tables of economic and occupations data.

Sets of Real Numbers

The mathematician Leopold Kronecker (1823–1891) once made the statement, "God made the integers, all the rest is the work of man." The *natural numbers* are those numbers with which we count discrete objects. By including 0 in the set, we obtain the set of *whole numbers*.

NATURAL NUMBERS

$\{1, 2, 3, 4, \ldots\}$ is the set of **natural numbers.**

WHOLE NUMBERS

$\{0, 1, 2, 3, \ldots\}$ is the set of **whole numbers.**

These numbers, along with many others, can be represented on **number lines** like the one pictured in **Figure 1.** We draw a number line by locating any point on the line and labeling it 0. This is the **origin.** Choose any point to the right of 0 and label it 1. The distance between 0 and 1 gives a unit of measure used to locate other points, as shown in **Figure 1.** The numbers labeled and those continuing in the same way to the right correspond to the set of whole numbers.

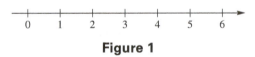

Figure 1

All the whole numbers starting with 1 are located to the right of 0 on the number line. But numbers may also be placed to the left of 0. These numbers, written $-1, -2, -3$, and so on, are shown in **Figure 2.** (The negative sign is used to show that the numbers are located to the *left* of 0.)

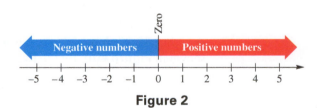

Figure 2

The numbers to the *left* of 0 are **negative numbers.** The numbers to the *right* of 0 are **positive numbers.** Positive numbers and negative numbers are called **signed numbers.** The number 0 itself is neither positive nor negative.

There are many practical applications of negative numbers.

- Temperatures may fall below zero. The lowest temperature ever recorded in meteorological records was $-128.6°F$ at Vostok, Antarctica, on July 22, 1983.

- Altitudes below sea level can be represented by negative numbers. The shore surrounding the Dead Sea is 1312 feet below sea level, written -1312 feet.

The Origins of Zero The Mayan Indians of Mexico and Central America had one of the earliest numeration systems that included a symbol for zero. The very early Babylonians had a positional system, but they placed only a space between "digits" to indicate a missing power. When the Greeks absorbed Babylonian astronomy, they used the letter omicron, o, of their alphabet or ō to represent "no power," or "zero." The Greek numeration system was gradually replaced by the Roman numeration system.

The original Hindu word for zero was *sunya*, meaning "void." The Arabs adopted this word as *sifr*, or "vacant." The word *sifr* passed into Latin as *zephirum*, which over the years became *zevero*, *zepiro*, and finally, *zero*.

The Origins of Negative Numbers
Negative numbers can be traced back to the Chinese between 200 B.C. and A.D. 220. Mathematicians at first found negative numbers ugly and unpleasant, even though they kept cropping up in the solutions of problems. For example, an Indian text of about A.D. 1150 gives the solution of an equation as −5 and then makes fun of anything so useless.

Leonardo of Pisa (Fibonacci), while working on a financial problem, was forced to conclude that the solution must be a negative number (that is, a financial loss). In A.D. 1545, the rules governing operations with negative numbers were published by **Girolamo Cardano** in his *Ars Magna* (Great Art).

The natural numbers, their negatives, and zero make up the set of *integers*.

INTEGERS

$\{\ldots, -3, -2, -1, 0, 1, 2, 3, \ldots\}$ is the set of **integers.**

Not all numbers are integers. For example, $\frac{1}{2}$ is a number *halfway* between the integers 0 and 1. Several numbers that are not integers are *graphed* in **Figure 3.** The **graph** of a number is a point on the number line representing that number. Think of the graph of a set of numbers as a "picture" of the set. All the numbers indicated in **Figure 3** can be written as quotients of integers and are examples of *rational numbers*. An integer, such as 2, is also a rational number. This is true because $2 = \frac{2}{1}$.

Figure 3

RATIONAL NUMBERS

$\{x \mid x$ is a quotient of two integers, with denominator not equal to $0\}$ is the set of **rational numbers.**

(Read the part in the braces as "the set of all numbers x such that x is a quotient of two integers, with denominator not equal to 0.")

The set symbolism used in the definition of rational numbers,

$$\{x \mid x \text{ has a certain property}\},$$

is called **set-builder notation.** This notation is convenient to use when it is not possible, or practical, to list all the elements of the set.

There are other numbers on the number line that are not rational. The first such number ever to be discovered was $\sqrt{2}$. The Pythagorean theorem, named for the Greek mathematician Pythagoras and covered in more detail in **Chapter 9,** says that in a right triangle, the length of the side opposite the right angle (the hypotenuse) is equal to the square root of the sum of the squares of the two perpendicular sides (the legs). In **Figure 4(a),** this is symbolized

$$c = \sqrt{a^2 + b^2}.$$

A right triangle having both legs a and b of length 1 is placed on a number line as shown in **Figure 4(b).** By the Pythagorean theorem, the length of the hypotenuse c is

$$c = \sqrt{1^2 + 1^2}$$

$$c = \sqrt{1 + 1}$$

$$c = \sqrt{2}.$$

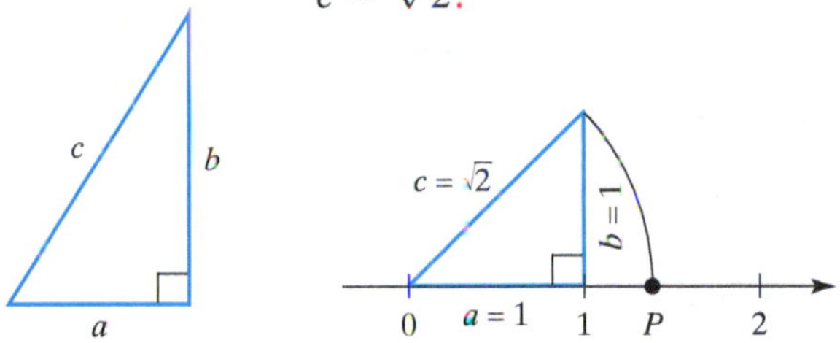

(a)

(b)

Figure 4

Figure 4(b) also shows an arc of a circle, centered at 0 with radius $\sqrt{2}$, intersecting the number line at point P. The coordinate of P is $\sqrt{2}$. The Greeks discovered that this number cannot be expressed as a quotient of integers—it is *irrational*.

IRRATIONAL NUMBERS

$\{x \mid x$ is a number on the number line that is not rational$\}$ is the set of **irrational numbers.**

Irrational numbers include $\sqrt{3}$, $\sqrt{7}$, $-\sqrt{10}$, and π, which is the ratio of the distance around a circle (its *circumference*) to the distance across it (its *diameter*). All numbers that can be represented by points on the number line (the union of the sets of rational and irrational numbers) are called *real numbers*.

REAL NUMBERS

$\{x \mid x$ is a number that can be represented by a point on the number line$\}$ is the set of **real numbers.**

Real numbers can be written as decimal numbers. Any rational number can be written as a decimal that will either come to an end (terminate) or repeat in a fixed "block" of digits. For example,

$$\frac{2}{5} = 0.4 \quad \text{and} \quad \frac{27}{100} = 0.27 \quad \text{are terminating decimals.}$$

$$\frac{1}{3} = 0.3333\ldots \quad \text{and} \quad \frac{3}{11} = 0.27272727\ldots \quad \text{are repeating decimals.}$$

The decimal representation of an irrational number neither terminates nor repeats. Decimal representations of rational and irrational numbers will be discussed further later in this chapter.

Figure 5 illustrates two ways to represent the relationships among the various sets of real numbers.

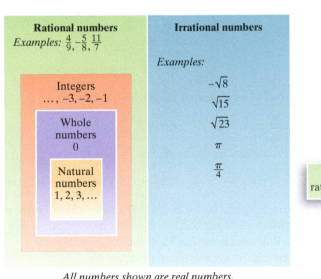

All numbers shown are real numbers.

(a)

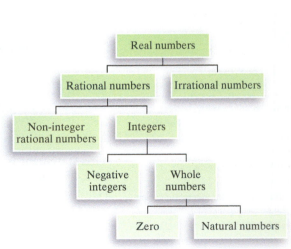

(b)

Figure 5

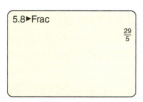

The TI-83/84 Plus calculator will convert a decimal to a fraction. See **Example 1(d).**

EXAMPLE 1 Identifying Elements of a Set of Numbers

List the numbers in the set that belong to each set of numbers.

$$\left\{ -5, -\frac{2}{3}, 0, \sqrt{2}, \frac{13}{4}, 5, 5.8 \right\}$$

(a) natural numbers   **(b)** whole numbers   **(c)** integers

(d) rational numbers   **(e)** irrational numbers   **(f)** real numbers

Solution

(a) The only natural number in the set is 5.

(b) The whole numbers consist of the natural numbers and 0, so the elements of the set that are whole numbers are 0 and 5.

(c) The integers in the set are $-5, 0,$ and 5.

(d) The rational numbers are $-5, -\frac{2}{3}, 0, \frac{13}{4}, 5,$ and 5.8, because each of these numbers *can* be written as the quotient of two integers. For example, $5.8 = \frac{58}{10} = \frac{29}{5}$.

(e) The only irrational number in the set is $\sqrt{2}$.

(f) All the numbers in the set are real numbers. ∎

Order in the Real Numbers

Suppose that a and b represent two real numbers. If their graphs on the number line are the same point, then a **is equal to** b. If the graph of a lies to the left of b, then a **is less than** b, and if the graph of a lies to the right of b, then a **is greater than** b. The **law of trichotomy** says that for two numbers a and b, one and only one of the following is true.

$$a = b, \quad a < b, \quad \text{or} \quad a > b$$

When read from left to right, the inequality $a < b$ is read "a is less than b."

$$7 < 8 \qquad \text{7 is less than 8.}$$

The inequality $a > b$ means "a is greater than b."

$$8 > 2 \qquad \text{8 is greater than 2.}$$

Notice that the symbol always points to the lesser number.

$$\text{Lesser number} \longrightarrow \quad 8 < 15$$

The symbol $\leq$, read from left to right, means "is less than or equal to."

$$5 \leq 9 \qquad \text{5 is less than or equal to 9.}$$

This statement is true because $5 < 9$ is true. **If either the $<$ part or the $=$ part is true, then the inequality $\leq$ is true.** Also, $8 \leq 8$ is true because $8 = 8$ is true. But it is not true that $13 \leq 9$ because neither $13 < 9$ nor $13 = 9$ is true.

The symbol $\geq$ means "is greater than or equal to."

$$9 \geq 5 \qquad \text{9 is greater than or equal to 5.}$$

This statement is true because $9 > 5$ is true.

The calculator returns a 1 for each of these statements of inequality, signifying that each is true.

The symbol for equality, =, was first introduced by the Englishman **Robert Recorde** in his 1557 algebra text *The Whetstone of Witte*. He used two parallel line segments, because he claimed, no two things can be more equal.

The symbols for order relationships, $<$ and $>$, were first used by **Thomas Harriot** (1560–1621), another Englishman. These symbols were not immediately adopted by other mathematicians.

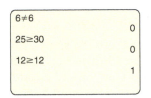

EXAMPLE 2 **Comparing Real Numbers**

Determine whether each statement is *true* or *false*.

(a) $6 \neq 6$ **(b)** $5 < 19$ **(c)** $15 \leq 20$ **(d)** $25 \geq 30$ **(e)** $12 \geq 12$

Solution

(a) The statement $6 \neq 6$ is false because 6 *is equal to* 6.

(b) Because 5 is indeed less than 19, this statement is true.

(c) The statement $15 \leq 20$ is true because $15 < 20$.

(d) Both $25 > 30$ and $25 = 30$ are false, so $25 \geq 30$ is false.

(e) Because $12 = 12$, the statement $12 \geq 12$ is true. ■

Additive Inverses and Absolute Value

For any nonzero real number x, there is exactly one number on the number line the same distance from 0 as x but on the opposite side of 0. In **Figure 6,** the numbers 3 and -3 are the same distance from 0 but are on opposite sides of 0. Thus, 3 and -3 are called **additive inverses, negatives,** or **opposites** of each other.

Figure 6

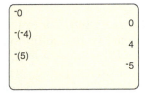

The additive inverse of the number 0 is 0 itself. This makes 0 the only real number that is its own additive inverse. Other additive inverses occur in distinct pairs. For example, 4 and -4 are additive inverses, as are 6 and -6. Several pairs of additive inverses are shown in **Figure 7.**

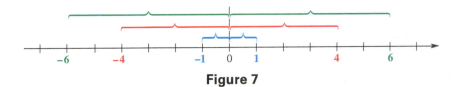

Figure 7

The additive inverse of a number can be indicated by writing the symbol $-$ in front of the number. With this symbol, the additive inverse of 7 is written -7. The additive inverse of -4 is written $-(-4)$ and can be read "the opposite of -4" or "the negative of -4." **Figure 7** suggests that 4 is an additive inverse of -4. A number can have only one additive inverse, so the symbols 4 and $-(-4)$ must represent the same number, which means that

$$-(-4) = 4.$$

DOUBLE NEGATIVE RULE

For any real number x, the following is true.

$$-(-x) = x$$

Table 1

Number	Additive Inverse
−4	−(−4) or 4
0	0
19	−19
$-\dfrac{2}{3}$	$\dfrac{2}{3}$

Table 1 shows several numbers and their additive inverses. An important property of additive inverses will be studied later in this chapter:

$$a + (-a) = (-a) + a = 0, \quad \text{for all real numbers } a.$$

As mentioned above, additive inverses are numbers that are the same distance from 0 (but in opposite directions) on the number line. See **Figure 7** on the previous page. This idea can also be expressed by saying that a number and its additive inverse have the same *absolute value*. The **absolute value** of a real number can be defined as the undirected distance between 0 and the number on the number line.

The symbol for the absolute value of the number x is $|x|$, read **"the absolute value of x."** For example, the distance between 0 and 2 on the number line is 2 units, so

$$|2| = 2.$$

Because the distance between 0 and −2 on the number line is also 2 units,

$$|-2| = 2.$$

Absolute value is a measure of undirected distance, so *the absolute value of a number is never negative.* Because 0 is a distance of 0 units from 0, $|0| = 0.$

FORMAL DEFINITION OF ABSOLUTE VALUE

For any real number x, the absolute value of x is defined as follows.

$$|x| = \begin{cases} x & \text{if } x \geq 0 \\ -x & \text{if } x < 0 \end{cases}$$

If x is a positive number or 0, then its absolute value is x itself. For example, because 8 is a positive number, $|8| = 8$. By the second part of the definition, *if x is a negative number, then its absolute value is the additive inverse of x.* For example, if $x = -9$, then $|-9| = -(-9) = 9$, since the additive inverse of −9 is 9.

The formal definition of absolute value can be confusing if it is not read carefully. The "$-x$" in the second part of the definition *does not* represent a negative number. Since x is negative in the second part, $-x$ represents the opposite of a negative number, which is a positive number.

EXAMPLE 3   Using Absolute Value

Simplify by finding the absolute value.

(a) $|5|$　　　(b) $|-5|$　　　(c) $-|5|$

(d) $-|-14|$　　(e) $|8 - 2|$　　(f) $-|8 - 2|$

Solution

(a) $|5| = 5$　　　　　　(b) $|-5| = -(-5) = 5$

(c) $-|5| = -(5) = -5$　　(d) $-|-14| = -(14) = -14$

(e) $|8 - 2| = |6| = 6$　　(f) $-|8 - 2| = -|6| = -6$ ■

```
|(-5)|
                5
-|-14|
              -14
-|8-2|
               -6
```

This screen supports the results of Example 3(b), (d), and (f).

Example 3(e) shows that absolute value bars also serve as grouping symbols. *Perform any operations that appear inside absolute value symbols before finding the absolute value.*

Applications of Real Numbers

EXAMPLE 4 Interpreting Signed Numbers in a Table

The consumer price index (CPI) measures the average change in prices of goods and services purchased by urban consumers in the United States. **Table 2** shows the percent change in the CPI for selected categories of goods and services from 2009 to 2010 and from 2010 to 2011. Use the table to answer each question.

Table 2

Category	Change from 2009 to 2010	Change from 2010 to 2011
Appliances	−4.5	−1.2
Education	4.4	4.2
Gasoline	18.4	26.4
Housing	−0.4	1.3
Medical care	3.4	3.0

Source: U.S. Bureau of Labor Statistics.

(a) Which category in which years represents the greatest percent decrease?

(b) Which category in which years represents the least change?

Solution

(a) We must find the negative number with the greatest absolute value. The number that satisfies this condition is −4.5, so the greatest percent decrease was shown by appliances from 2009 to 2010.

(b) We must find the number (either positive, negative, or zero) with the least absolute value. From 2009 to 2010, housing showed the least change, a decrease of 0.4 percent. ∎

In the next example, we see how the need for certain occupations during the upcoming years is changing, some for better and some for worse.

EXAMPLE 5 Comparing Occupational Rates of Change

Table 3 shows the total rates of change projected to occur in demand for some of the fastest-growing and some of the most rapidly-declining occupations from 2012 to 2022.

Table 3

Occupation (2012–2022)	Total Rate of Change (in percent)
Interpreters and translators	46.1
Personal care aides	48.8
Physicians' assistants	38.4
Word processors and typists	−25.1
Postal service clerks	−31.8
Locomotive firers	−42.0

Source: Bureau of Labor Statistics.

Which occupation in **Table 3** on the preceding page is expected to see

(a) the greatest change?   **(b)** the least change?

Solution

(a) We want the greatest change, without regard to whether the change is an increase or a decrease. Look for the number with the greatest absolute value. That number is for personal care aides:

$$|48.8| = 48.8.$$

(b) Using similar reasoning, the least change is for word processors and typists:

$$|-25.1| = 25.1.$$ ∎

6.1   EXERCISES

Give a number that satisfies the given condition.

1. An integer between 4.5 and 5.5

2. A rational number between 2.8 and 2.9

3. A whole number that is not positive and is less than 1

4. A whole number greater than 4.5

5. An irrational number that is between $\sqrt{13}$ and $\sqrt{15}$

6. A real number that is neither negative nor positive

Decide whether each statement is true *or* false.

7. Every natural number is positive.

8. Every whole number is positive.

9. Every integer is a rational number.

10. Every rational number is a real number.

List all numbers from each set that are **(a)** *natural numbers;* **(b)** *whole numbers;* **(c)** *integers;* **(d)** *rational numbers;* **(e)** *irrational numbers;* **(f)** *real numbers.*

11. $\left\{ -9, -\sqrt{7}, -1\frac{1}{4}, -\frac{3}{5}, 0, \sqrt{5}, 3, 5.9, 7 \right\}$

12. $\left\{ -5.3, -5, -\sqrt{3}, -1, -\frac{1}{9}, 0, 1.2, 1.8, 3, \sqrt{11} \right\}$

Use an integer or decimal to express each number in boldface print representing a change or measurement in each of the following.

13. *U.S. Population*   Between July 1, 2011, and July 1, 2012, the population of the United States increased by approximately **2,259,000.** (*Source:* U.S. Census Bureau.)

14. *Number of Movie Screens*   Between 2011 and 2012, the number of movie screens in the United States increased by **209.** (*Source:* Motion Picture Association of America.)

15. *World Series Attendance*   From 2011 to 2012, attendance at the first game of the World Series went from 46,406 to 42,982, a decrease of **3424.** (*Source:* Major League Baseball.)

16. *Number of U. S. Banks*   In 1935, there were 15,295 banks in the United States. By 2012, the number was 7083, representing a decrease of **8212.** (*Source:* Federal Deposit Insurance Corporation.)

17. *Dow Jones Average*   On Friday, August 23, 2013, the Dow Jones Industrial Average (DJIA) closed at 15,010.51. On the previous day, it had closed at 14,963.74. Thus, on Friday it closed up **46.77** points. (*Source: The Washington Post.*)

18. *NASDAQ*   On Tuesday, August 27, 2013, the NASDAQ closed at 3578.52. On the previous day, it had closed at 3657.57. Thus, on Tuesday it closed down **79.05** points. (*Source: The Washington Post.*)

19. *Height of Mt. Arenal in Costa Rica*   The height of Mt. Arenal, an active volcano in Costa Rica, is **5436** feet above sea level. (*Source: The New York Times Almanac.*)

20. *Elevation of New Orleans*   The city of New Orleans, devastated by Hurricane Katrina in 2005, lies **8** feet below sea level. (*Source:* U.S. Geological Survey, *Elevations and Distances in the United States.*)

21. *Melting Point of Fluorine*   The melting point of fluorine gas is **220°** below 0° Celsius.

22. *Boiling Point of Chlorine*   The boiling point of chlorine is approximately **30°** below 0° Fahrenheit.

Depths and Heights of Seas and Mountains *The chart gives selected depths and heights of bodies of water and mountains.*

Bodies of Water	Average Depth in Feet (as a negative number)	Mountains	Altitude in Feet (as a positive number)
Pacific Ocean	−14,040	McKinley	20,237
South China Sea	−4802	Point Success	14,164
Gulf of California	−2375	Matlalcueyetl	14,636
Caribbean Sea	−8448	Ranier	14,416
Indian Ocean	−12,800	Steele	16,642

Source: The World Almanac and Book of Facts.

23. List the bodies of water in order, starting with the deepest and ending with the shallowest.

24. List the mountains in order, starting with the lowest and ending with the highest.

25. *True or false:* The absolute value of the depth of the Pacific Ocean is greater than the absolute value of the depth of the Indian Ocean.

26. *True or false:* The absolute value of the depth of the Gulf of California is greater than the absolute value of the depth of the Caribbean Sea.

Graph each group of numbers on a number line.

27. $-2, -6, -4, 3, 4$

28. $-5, -3, -2, 0, 4$

29. $\frac{1}{4}, 2\frac{1}{2}, -3\frac{4}{5}, -4, -1\frac{5}{8}$

30. $5\frac{1}{4}, 4\frac{5}{9}, -2\frac{1}{3}, 0, -3\frac{2}{5}$

31. Match each expression in Column I with its value in Column II. Some choices in Column II may not be used.

I	II
(a) $\lvert -7 \rvert$	**A.** 7
(b) $-(-7)$	**B.** -7
(c) $-\lvert -7 \rvert$	**C.** neither A nor B
(d) $-\lvert -(-7) \rvert$	**D.** both A and B

32. Fill in the blanks with the correct values: The opposite of -2 is ____, while the absolute value of -2 is ____. The additive inverse of -2 is ____, while the additive inverse of the absolute value of -2 is ____.

Find **(a)** *the additive inverse (or opposite) of each number and* **(b)** *the absolute value of each number.*

33. -2

34. -8

35. 6

36. 11

37. $7 - 4$

38. $8 - 3$

39. $7 - 7$

40. $3 - 3$

Select the lesser of the two given numbers.

41. $-12, -4$

42. $-9, -14$

43. $-8, -1$

44. $-15, -16$

45. $3, \lvert -4 \rvert$

46. $5, \lvert -2 \rvert$

47. $\lvert -3 \rvert, \lvert -4 \rvert$

48. $\lvert -8 \rvert, \lvert -9 \rvert$

49. $-\lvert -6 \rvert, -\lvert -4 \rvert$

50. $-\lvert -2 \rvert, -\lvert -3 \rvert$

51. $\lvert 5 - 3 \rvert, \lvert 6 - 2 \rvert$

52. $\lvert 7 - 2 \rvert, \lvert 8 - 1 \rvert$

Decide whether each statement is true *or* false.

53. $6 > -(-2)$

54. $-8 > -(-2)$

55. $-4 \le -(-5)$

56. $-6 \le -(-3)$

57. $\lvert -6 \rvert < \lvert -9 \rvert$

58. $\lvert -12 \rvert < \lvert -20 \rvert$

59. $-\lvert 8 \rvert > \lvert -9 \rvert$

60. $-\lvert 12 \rvert > \lvert -15 \rvert$

61. $-\lvert -5 \rvert \ge -\lvert -9 \rvert$

62. $-\lvert -12 \rvert \le -\lvert -15 \rvert$

63. $\lvert 6 - 5 \rvert \ge \lvert 6 - 2 \rvert$

64. $\lvert 13 - 8 \rvert \le \lvert 7 - 4 \rvert$

65. ***Population Change*** The table shows the percent change in population from 2000 to 2012 for selected metropolitan areas.

Metropolitan Area	Percent Change
Las Vegas	45.4
San Francisco	8.0
Chicago	4.7
New Orleans	−8.3
Phoenix	33.1
Detroit	−3.6

Source: U. S. Census Bureau.

(a) Which metropolitan area had the greatest percent change in population? What was this change? Was it an increase or a decrease?

(b) Which metropolitan area had the least change in population? What was this change? Was it an increase or a decrease?

66. *Trade Balance*   The table gives the net trade balance, in millions of dollars, for selected U.S. trade partners for October 2013.

Country	Trade Balance (in millions of dollars)
India	−1967
China	−28,862
The Netherlands	2325
France	−1630
Turkey	396

Source: U.S. Census Bureau.

A negative balance means that imports to the United States exceeded exports from the United States. A positive balance means that exports exceeded imports.

(a) Which country had the greatest discrepancy between exports and imports? Explain.

(b) Which country had the least discrepancy between exports and imports? Explain.

(c) Which two countries were closest in their trade balance figures?

67. *CPI Data*   Refer to **Table 2** in **Example 4.** Of the data for 2009 to 2010, which of the categories Appliances or Housing shows the greater change (without regard to sign)?

68. *Change in Occupations*   Refer to **Table 3** in **Example 5.** Did the demand for physicians' assistants or for postal service clerks show the lesser change (without regard to sign)?

6.2   OPERATIONS, PROPERTIES, AND APPLICATIONS OF REAL NUMBERS

OBJECTIVES

1 Perform the operations of addition, subtraction, multiplication, and division of signed numbers.

2 Apply the rules for order of operations.

3 Identify and apply properties of addition and multiplication of real numbers.

4 Determine change in investment and meteorological data using subtraction and absolute value.

Operations on Real Numbers

The result of adding two numbers is called their **sum.** The numbers being added are called **addends** (or **terms**).

ADDING REAL NUMBERS

Like Signs   Add two numbers with the *same* sign by adding their absolute values. The sign of the sum (either + or −) is the same as the sign of the two addends.

Unlike Signs   Add two numbers with *different* signs by subtracting the lesser absolute value from the greater to find the absolute value of the sum. The sum is positive if the positive number has the greater absolute value. The sum is negative if the negative number has the greater absolute value.

For example, to add −12 and −8, first find their absolute values.

$$|-12| = 12 \quad \text{and} \quad |-8| = 8$$

−12 and −8 have the *same* sign, so add their absolute values: $12 + 8 = 20$. Give the sum the sign of the two numbers. Because both numbers are negative, the sum is negative and

$$-12 + (-8) = -20.$$

Find $-17 + 11$ by subtracting the absolute values, because these numbers have different signs.

$$|-17| = 17 \quad \text{and} \quad |11| = 11$$
$$17 - 11 = 6$$

Give the result the sign of the number with the larger absolute value.

$$-17 + 11 = -6$$

Negative because $|-17| > |11|$

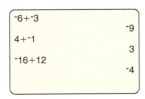

The calculator supports the results of
Example 1(a), (c), and (e).

EXAMPLE 1   Adding Signed Numbers

Find each sum.

(a) $-6 + (-3)$ **(b)** $-12 + (-4)$ **(c)** $4 + (-1)$

(d) $-9 + 16$ **(e)** $-16 + 12$ **(f)** $-4 + 4$

Solution

(a) $-6 + (-3) = -(6 + 3) = -9$

(b) $-12 + (-4) = -(12 + 4) = -16$

(c) $4 + (-1) = 3$

(d) $-9 + 16 = 7$

(e) $-16 + 12 = -4$

(f) $-4 + 4 = 0$ The sum of additive inverses is 0.   ■

    The result of subtracting two numbers is called their **difference.** In $a - b$, a is called the **minuend,** and b is called the **subtrahend.** Compare the two statements below.

$$7 - 5 = 2 \quad \text{and} \quad 7 + (-5) = 2$$

In a similar way,

$$9 - 3 = 9 + (-3).$$

That is, to subtract 3 from 9, add the additive inverse of 3 to 9. These examples suggest the following rule for subtraction.

DEFINITION OF SUBTRACTION

For all real numbers a and b,

$$a - b = a + (-b).$$

(Change the sign of the subtrahend and add.)

EXAMPLE 2   Subtracting Signed Numbers

Find each difference.

(a) $6 - 8$ **(b)** $-12 - 4$ **(c)** $-10 - (-7)$ **(d)** $15 - (-3)$

Solution

Change to addition.

Change sign of the subtrahend and add.

(a) $6 - 8 = 6 + (-8) = -2$

Change to addition.

Sign is changed.

(b) $-12 - 4 = -12 + (-4) = -16$

(c) $-10 - (-7) = -10 + [-(-7)]$ This step can be omitted.

$$= -10 + 7$$

$$= -3$$

(d) $15 - (-3) = 15 + 3 = 18$ ■

Practical Arithmetic From the time of Egyptian and Babylonian merchants, practical aspects of arithmetic complemented mystical (or "Pythagorean") tendencies. This was certainly true in the time of **Adam Riese** (1489–1559), a "reckon master" influential when commerce was growing in Northern Europe. He championed new methods of reckoning using Hindu-Arabic numerals and quill pens. (The Roman methods then in common use moved counters on a ruled board.) Riese thus fulfilled Fibonacci's efforts 300 years earlier to supplant Roman numerals and methods.

The result of multiplying two numbers is called their **product.** The two numbers being multiplied are called **factors.**

Any rules for multiplication with negative real numbers should be consistent with the usual rules for multiplication of positive real numbers and zero. Observe the pattern of products below.

$$4 \cdot 5 = 20$$
$$4 \cdot 4 = 16$$
$$4 \cdot 3 = 12$$
$$4 \cdot 2 = 8$$
$$4 \cdot 1 = 4$$
$$4 \cdot 0 = 0$$
$$4 \cdot (-1) = \text{?}$$

Products decrease by 4 in each line.

What number must be assigned as the product $4 \cdot (-1)$ so that the pattern is maintained? The numbers just to the left of the equality symbols decrease by 1 each time, and the products to the right decrease by 4 each time. To maintain the pattern, the number to the right in the bottom equation must be 4 less than 0, which is -4, so

$$4 \cdot (-1) = -4.$$

The pattern continues with

$$4 \cdot (-2) = -8$$
$$4 \cdot (-3) = -12$$
$$4 \cdot (-4) = -16,$$

and so on. In the same way,

$$-4 \cdot 2 = -8$$
$$-4 \cdot 3 = -12$$
$$-4 \cdot 4 = -16,$$

and so on. A similar observation can be made about the product of two negative real numbers. Look at the pattern that follows.

$$-5 \cdot 4 = -20$$
$$-5 \cdot 3 = -15$$
$$-5 \cdot 2 = -10$$
$$-5 \cdot 1 = -5$$
$$-5 \cdot 0 = 0$$
$$-5 \cdot (-1) = \text{?}$$

Products increase by 5 in each line.

The numbers just to the left of the equality symbols decrease by 1 each time. The products on the right increase by 5 each time. To maintain the pattern, the product $-5 \cdot (-1)$ must be 5 more than 0, so it seems reasonable for the following to be true.

$$-5 \cdot (-1) = 5$$

Continuing this pattern gives the following.

$$-5 \cdot (-2) = 10$$
$$-5 \cdot (-3) = 15$$
$$-5 \cdot (-4) = 20$$
$$\vdots$$

$$(+) \cdot (+) = +$$
$$(-) \cdot (-) = +$$
$$(+) \cdot (-) = -$$
$$(-) \cdot (+) = -$$

MULTIPLYING REAL NUMBERS

Like Signs Multiply two numbers with the *same* sign by multiplying their absolute values to find the absolute value of the product. The product is positive.

Unlike Signs Multiply two numbers with *different* signs by multiplying their absolute values to find the absolute value of the product. The product is negative.

EXAMPLE 3 Multiplying Signed Numbers

Find each product.

(a) $-9 \cdot 7$ **(b)** $14 \cdot (-5)$ **(c)** $-8 \cdot (-4)$

Solution

(a) $-9 \cdot 7 = -63$ **(b)** $14 \cdot (-5) = -70$ **(c)** $-8 \cdot (-4) = 32$ ∎

```
-9*7
            -63
14*-5
            -70
-8*-4
             32
```

The symbol (*) represents multiplication on this screen. The display supports the results of **Example 3**.

The result of dividing two numbers is called their **quotient.** In the quotient $a \div b$ (or $\frac{a}{b}$), where $b \neq 0$, a is called the **dividend** (or **numerator**), and b is called the **divisor** (or **denominator**). For real numbers a, b, and c,

$$\text{if} \quad \frac{a}{b} = c, \quad \text{then} \quad a = b \cdot c.$$

To illustrate this, consider the quotient $\frac{10}{-2}$. The value of this quotient is obtained by asking, "What number multiplied by -2 gives 10?" From our discussion of multiplication, the answer to this question must be "-5." Therefore,

$$\frac{10}{-2} = -5, \quad \text{because} \quad 10 = -2 \cdot (-5).$$

Similarly,

$$\frac{-10}{2} = -5 \quad \text{and} \quad \frac{-10}{-2} = 5.$$

These facts, along with the fact that the quotient of two positive numbers is positive, lead to the following rules for division.

$$(+)/(+) = +$$
$$(-)/(-) = +$$
$$(+)/(-) = -$$
$$(-)/(+) = -$$

DIVIDING REAL NUMBERS

Like Signs Divide two numbers with the *same* sign by dividing their absolute values to find the absolute value of the quotient. The quotient is positive.

Unlike Signs Divide two numbers with *different* signs by dividing their absolute values to find the absolute value of the quotient. The quotient is negative.

EXAMPLE 4 Dividing Signed Numbers

Find each quotient.

(a) $\dfrac{15}{-5}$ **(b)** $\dfrac{-100}{-25}$ **(c)** $\dfrac{-60}{3}$

```
15/-5
             -3
-100/-25
              4
-60/3
            -20
```

The division operation is represented by a slash (/). This screen supports the results of **Example 4**.

Solution

(a) $\dfrac{15}{-5} = -3$ **(b)** $\dfrac{-100}{-25} = 4$ **(c)** $\dfrac{-60}{3} = -20$ ∎

This is true because $-5 \cdot (-3) = 15$.

If 0 is divided by a nonzero number, the quotient is 0.

$$\frac{0}{a} = 0, \qquad \text{for } a \neq 0$$

This is true because $a \cdot 0 = 0$. However, we cannot divide by 0. There is a good reason for this. Whenever a division is performed, we want to obtain one and only one quotient. Now consider this division problem.

$$\frac{7}{0} = ?$$

We must ask ourselves,

"What number multiplied by 0 gives 7?"

There is no such number because the product of 0 and any number is zero. Now consider this quotient.

$$\frac{0}{0} = ?$$

There are infinitely many answers to the question, "What number multiplied by 0 gives 0?" Division by 0 does not yield a *unique* quotient, and thus it is not permitted.

```
ERR:DIVIDE BY 0
1: Quit
2: Goto
```

Dividing by zero leads to this message on the TI-83/84 Plus.

DIVISION INVOLVING ZERO

$\dfrac{a}{0}$ **is undefined for all a.** $\qquad \dfrac{0}{a} = 0$ **for all nonzero a.**

Order of Operations

Given an expression such as

$$5 + 2 \cdot 3,$$

```
5+2*3
```

What result does the calculator give? The order of operations determines the answer. (See **Example 5(a)**.)

should 5 and 2 be added first, or should 2 and 3 be multiplied first? When an expression involves more than one operation, we use the following rules for **order of operations.**

ORDER OF OPERATIONS

If parentheses or square brackets are present:

Step 1 Work separately above and below any **fraction bar.**

Step 2 Use the rules below within each set of **parentheses or square brackets.** Start with the innermost set and work outward.

If no parentheses or brackets are present:

Step 1 Apply any **exponents.**

Step 2 Do any **multiplications or divisions** in the order in which they occur, working from left to right.

Step 3 Do any **additions or subtractions** in the order in which they occur, working from left to right.

The sentence **"Please excuse my dear Aunt Sally"** is often used to help us remember the rules for order of operations. The letters **P, E, M, D, A, S** are the first letters of the words of the sentence, and they stand for *parentheses, exponents, multiply, divide, add, subtract* (Remember also that M and D have equal priority, as do A and S. Operations with equal priority are performed in order from left to right.)

When evaluating an exponential expression that involves a negative sign, be aware that $(-a)^n$ and $-a^n$ do not necessarily represent the same quantity. For example, if $a = 2$ and $n = 6$, then

$$(-2)^6 = (-2)(-2)(-2)(-2)(-2)(-2) = 64 \qquad \text{The base is } -2.$$

while

$$-2^6 = -(2 \cdot 2 \cdot 2 \cdot 2 \cdot 2 \cdot 2) = -64. \qquad \text{The base is } 2.$$

EXAMPLE 5 Using the Order of Operations

Use the order of operations to simplify each expression.

(a) $5 + 2 \cdot 3$ **(b)** $4 \cdot 3^2 + 7 - (2 + 8)$

(c) $\dfrac{2(8 - 12) - 11(4)}{5(-2) - 3}$ **(d)** -6^4

(e) $(-6)^4$ **(f)** $(-8)(-3) - [4 - (3 - 6)]$

Solution   *Be careful! Multiply first.*

(a) $\quad 5 + \mathbf{2 \cdot 3} = 5 + \mathbf{6}$ Multiply.

$\qquad\qquad\qquad = 11$ Add.

(b) $\quad 4 \cdot 3^2 + 7 - \mathbf{(2 + 8)} = 4 \cdot 3^2 + 7 - \mathbf{10}$ Work within parentheses first.

3^2 means $3 \cdot 3$, not $3 \cdot 2$.

$\qquad\qquad\qquad\qquad\quad = 4 \cdot \mathbf{9} + 7 - 10$ Apply the exponent.

$\qquad\qquad\qquad\qquad\quad = 36 + 7 - 10$ Multiply.

$\qquad\qquad\qquad\qquad\quad = 43 - 10$ Add.

$\qquad\qquad\qquad\qquad\quad = 33$ Subtract.

(c) $\quad \dfrac{2(8 - 12) - 11(4)}{5(-2) - 3} = \dfrac{2(-4) - 11(4)}{5(-2) - 3}$ Work separately above and below fraction bar.

$\qquad\qquad\qquad\qquad\quad = \dfrac{-8 - 44}{-10 - 3}$ Multiply.

$\qquad\qquad\qquad\qquad\quad = \dfrac{-52}{-13}$ Subtract.

$\qquad\qquad\qquad\qquad\quad = 4$ Divide.

(d) $\quad -6^4 = -(6 \cdot 6 \cdot 6 \cdot 6) = -1296$

The base is 6, not -6.

(e) $\quad (-6)^4 = (-6)(-6)(-6)(-6) = 1296$

The base is -6 here.

(f) $\quad -8(-3) - [4 - (\mathbf{3 - 6})] = -8(-3) - [4 - (\mathbf{-3})]$ Work within parentheses.

Start here.

$\qquad\qquad\qquad\qquad\qquad = -8(-3) - [4 + 3]$ Definition of subtraction

$\qquad\qquad\qquad\qquad\qquad = -8(-3) - 7$ Work within brackets.

$\qquad\qquad\qquad\qquad\qquad = 24 - 7$ Multiply.

$\qquad\qquad\qquad\qquad\qquad = 17$ Subtract. ■

DETERMINING CHANGE FROM *a* TO *b*

Suppose that *a* and *b* are real numbers. The amount (without regard to sign) of change *from a to b* is given by

$$|a - b|, \quad \text{or, equivalently,} \quad |b - a|.$$

The change is interpreted to be a **positive change** if $a < b$, while it is a **negative change** if $a > b$. If $a = b$, there is no change.

EXAMPLE 7 Interpreting Change from a Table

Suppose that you are an employee of a fund management company and you come across the results shown in **Figure 8** in a trade publication. Amounts are rounded to the nearest percent.

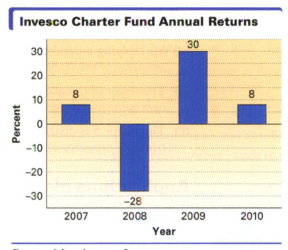

Source: Morningstar, Invesco.

Figure 8

Illustrate by subtraction and absolute value how to determine the change in percent of annual returns from **(a)** 2007 to 2008 and **(b)** 2008 to 2009.

Solution

(a) From 2007 to 2008, the annual return went from 8 percent to −28 percent. This is a decrease, so our answer will be *negative*. The amount of the decrease is

$$|8 - (-28)| = |8 + 28| = |36| = 36.$$

(Note that we could have subtracted in the order $-28 - 8 = -36$, which has the same absolute value as 36.) The change from 2007 to 2008 is

−36 percent.

The negative sign indicates a decrease. ——— 36 is the absolute value of the difference of the two amounts.

(b) To find the change from 2008 to 2009, which by inspection is an increase, find the absolute value of the difference of the two amounts in either order, and interpret it as a *positive* amount.

$$|-28 - 30| = |-58| = 58 \quad \text{or} \quad |30 - (-28)| = |58| = 58$$

The change from 2008 to 2009 is +58 percent.

The + sign is understood if not specifically indicated.

EXAMPLE 8 **Interpreting Temperature Change from a Map**

The daily highs for selected cities during a particularly cold two-day period in Minnesota are shown in **Figure 9.** Determine the change in temperature from Monday to Tuesday for each city listed.

Figure 9

Solution

In each case we determine the sign of the change by observation. If the temperature on Tuesday *is greater than* that on Monday, the change is *positive.* If Tuesday's temperature *is less than* that on Monday, the change is *negative.* Equal temperatures indicate *no change.*

In Ely, the high temperature was less on Tuesday than on Monday because $-26 < -22$, so the change is negative. The amount is

$$|-26 - (-22)| = |-26 + 22| = |-4| = 4.$$

Thus the change is -4 degrees.

In Bemidji, the high temperature went from -20 to -18, which is an increase. Subtracting these temperatures in either order and taking the absolute value gives

$$|-20 - (-18)| = |-20 + 18| = |-2| = 2.$$

Thus, the change is $+2$ degrees. The other changes can be found similarly.

Duluth: $+1$ degree	St. Cloud: no change
Minneapolis: $+5$ degrees	Mankato: $+4$ degrees
Rochester: -1 degree	

6.2   EXERCISES

Complete each statement and give an example.

1. The sum of two positive numbers is a _____ number.

2. The sum of two negative numbers is a _____ number.

3. The sum of a positive number and a negative number is negative if the negative number has the _____ absolute value.

4. The sum of a positive number and a negative number is positive if the positive number has the _____ absolute value.

5. The difference between two positive numbers is negative if _____.

6. The difference between two negative numbers is negative if _____.

7. The sum of a positive number and a negative number is 0 if the numbers are _____ .

8. The product of two numbers with the same sign is _____ .

9. The product of two numbers with different signs is _____ .

10. The quotient formed by any nonzero number divided by 0 is _____, and the quotient formed by 0 divided by any nonzero number is _____ .

Perform the indicated operations, using the order of operations as necessary.

11. $-12 + (-8)$

12. $-5 + (-2)$

13. $12 + (-16)$

14. $-6 + 17$

15. $-12 - (-1)$

16. $-3 - (-8)$

17. $-5 + 11 + 3$

18. $-9 + 16 + 5$

19. $12 - (-3) - (-5)$

20. $15 - (-6) - (-8)$

21. $(-12)(-2)$

22. $(-3)(-5)$

23. $9(-12)(-4)(-1)(3)$

24. $-5(-17)(2)(-2)(4)$

25. $\dfrac{-18}{-3}$

26. $\dfrac{-100}{-50}$

27. $\dfrac{36}{-6}$

28. $\dfrac{52}{-13}$

29. $\dfrac{0}{12}$

30. $\dfrac{0}{-7}$

31. $-6 + [5 - (3 + 2)]$

32. $-8[4 + (7 - 8)]$

33. $-4 - 3(-2) + 5^2$

34. $-6 - 5(-8) + 3^2$

35. $\dfrac{2(-5) + (-3)(-2^2)}{-3^2 + 9}$

36. $\dfrac{3(-4) + (-5)(-2)}{2^3 - 2 + (-6)}$

37. $\dfrac{(-8 + 6) \cdot (-5)}{-5 - 5}$

38. $\dfrac{(-10 + 4) \cdot (-3)}{-7 - 2}$

39. $-8(-2) - [(4^2) + (7 - 3)]$

40. $-7(-3) - [2^3 - (3 - 4)]$

41. $\dfrac{(-6 + 3) \cdot (-4)}{-5 - 1} - \dfrac{(-9 + 6) \cdot (-3)}{-4 + 3}$

42. $\dfrac{2(-5 + 3)}{-2^2} - \dfrac{(-3^2 + 2)(3)}{3 - (-4)}$

43. $-\dfrac{1}{4}[3(-5) + 7(-5) + 1(-2)]$

44. $\dfrac{5 - 3\left(\dfrac{-5 - 9}{-7}\right) - 6}{-9 - 11 + 3 \cdot 7}$

45. Which of the following expressions are undefined?

A. $\dfrac{8}{0}$ **B.** $\dfrac{9}{6 - 6}$ **C.** $\dfrac{4 - 4}{5 - 5}$ **D.** $\dfrac{0}{-1}$

46. If you have no money in your pocket and you divide it equally among your three siblings, how much does each get? Use this situation to explain division of zero by a positive integer.

Identify the property illustrated by each statement.

47. $6 + 9 = 9 + 6$

48. $8 \cdot 4 = 4 \cdot 8$

49. $9 + (-9) = 0$

50. $12 + 0 = 12$

51. $9 \cdot 1 = 9$

52. $\left(\dfrac{1}{-3}\right) \cdot (-3) = 1$

53. $0 + 283 = 283$

54. $(3 \cdot 5) \cdot 4 = 4 \cdot (3 \cdot 5)$

55. $6 \cdot (4 \cdot 2) = (6 \cdot 4) \cdot 2$

56. $0 = -8 + 8$

57. $19 \cdot 12$ is a real number.

58. $19 + 12$ is a real number.

59. $7 + (2 + 5) = (7 + 2) + 5$

60. $2 \cdot (4 + 3) = 2 \cdot 4 + 2 \cdot 3$

61. $9 \cdot 6 + 9 \cdot 8 = 9 \cdot (6 + 8)$

Provide short answers in Exercises 62–66.

62. One of the authors received an email message from an old friend, Frank Capek. Frank said that his grandson had to evaluate

$$9 + 15 \div 3.$$

Frank and his wife, Barbara, said that the answer is 8, but the grandson said that the correct answer is 14. The grandson's reasoning is "There is a rule called the Order of Process so that you proceed from right to left rather than from left to right."

(a) What is the correct answer?

(b) Is his grandson's reasoning correct? Explain.

63. The following conversation actually took place between one of the authors of this text and his son, Jack, when Jack was four years old.

DADDY: "Jack, what is $3 + 0$?"
JACK: "3"
DADDY: "Jack, what is $4 + 0$?"
JACK: "4 . . . and Daddy, *string* plus zero equals *string!*"

What property of addition of real numbers did Jack recognize?

64. Many everyday activities are commutative. The order in which they occur does not affect the outcome. For example, "putting on your shirt" and "putting on your pants" are commutative operations. Decide whether the given activities are commutative.

(a) putting on your shoes; putting on your socks

(b) getting dressed; taking a shower

(c) combing your hair; brushing your teeth

65. Many everyday occurrences can be thought of as operations that have opposites or inverses. For example, the inverse operation for "going to sleep" is "waking up." For each of the given activities, specify its inverse activity.

(a) cleaning up your room

(b) earning money

(c) increasing the volume on your MP3 player

66. The distributive property holds for multiplication with respect to addition. Does the distributive property hold for addition with respect to multiplication? That is, is

$$a + (b \cdot c) = (a + b) \cdot (a + c)$$

true for all values of a, b, and c? (*Hint:* Let $a = 2$, $b = 3$, and $c = 4$.)

Each expression in Exercises 67–74 is equal to either 81 or −81. Decide which of these is the correct value in each case.

67. -3^4 **68.** $-(3^4)$ **69.** $(-3)^4$

70. $-(-3^4)$ **71.** $-(-3)^4$ **72.** $[-(-3)]^4$

73. $-[-(-3)]^4$ **74.** $-[-(-3^4)]$

75. *S&P 500 Index Fund* The graph shows annual returns (to the nearest percent) for Class A shares of the Invesco S&P 500 Index Fund from 2009 to 2013. Use subtraction and absolute value to determine the change in returns from one year to the next. Give your answer as both a signed number (in percent) and a word description.

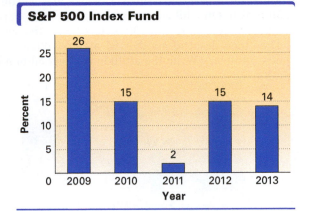

(a) 2009 to 2010   (b) 2010 to 2011

(c) 2011 to 2012   (d) 2012 to 2013

76. *Company Profits and Losses* The graph shows profits and losses for a private company for the years 2009 through 2013. Use subtraction and absolute value to determine the change in profit or loss from one year to the next. Give your answer as both a signed number (in thousands of dollars) and a word description.

(a) 2009 to 2010   (b) 2010 to 2011

(c) 2011 to 2012   (d) 2012 to 2013

Heights of Mountains and Depths of Trenches The table shows the heights of some selected mountains and the depths of some selected ocean trenches.

Mountain	Height (in feet)	Trench	Depth (in feet, as a negative number)
Foraker	17,400	Philippine	−32,995
Wilson	14,246	Cayman	−24,721
Pikes Peak	14,110	Java	−23,376

Source: World Almanac and Book of Facts.

Use the information given to answer Exercises 77–82.

77. What is the difference between the height of Mt. Foraker and the depth of the Philippine Trench?

78. What is the difference between the height of Pikes Peak and the depth of the Java Trench?

79. How much deeper is the Cayman Trench than the Java Trench?

80. How much deeper is the Philippine Trench than the Cayman Trench?

81. How much higher is Mt. Wilson than Pikes Peak?

82. If Mt. Wilson and Pikes Peak were stacked one on top of the other, how much higher would they be than Mt. Foraker?

Golf Scores *The table gives scores above or below par (that is, above or below the score "standard") for selected golfers during the 2013 PGA Tour Championship. Write a signed number that represents the total score for the four rounds (above or below par) for each golfer.*

	Golfer	Round 1	Round 2	Round 3	Round 4
83.	Steve Stricker	−4	+1*	−2	−5
84.	Phil Mickelson	+1	−3	0	−2
85.	Charl Schwartzel	−2	+9	+7	−4
86.	Kevin Streelman	−1	+2	+4	−3

*Golf scoring commonly includes a + sign with a score over par.
Source: PGA.

Temperature Change *During a cold two-day period in North Dakota, the daily highs for selected cities were as shown in the figure. Determine the change in temperature from Thursday to Friday for each city listed.* **All temperatures in Exercises 87–96 are in degrees Fahrenheit.**

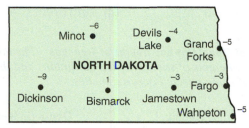

Thursday

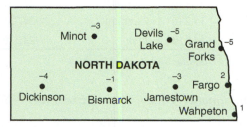

Friday

87. Dickinson

88. Minot

89. Bismarck

90. Devils Lake

91. Jamestown

92. Grand Forks

93. Fargo

94. Wahpeton

95. *24-Hour Temperature Change* Television weather forecasters often show maps indicating how the current temperature compares to that of 24 hours earlier. Refer to the map for Thursday used in **Exercises 87–94,** and imagine that the temperatures listed indicated the change in temperature compared to 24 hours earlier. Use the table to determine the temperature in each city 24 hours earlier. (When no sign is shown on the map, + is understood.)

City	Current Temperature	Temperature 24 Hours Earlier
Dickinson	34	
Minot	28	
Bismarck	30	
Devils Lake	32	
Jamestown	35	
Grand Forks	31	
Fargo	34	
Wahpeton	36	

96. *24-Hour Temperature Change* Repeat **Exercise 95** for the Friday map used in **Exercises 87–94.**

97. *Breaching of Humpback Whales* Humpback whale researchers Mark and Debbie noticed that one of their favorite whales, "Pineapple," breached 15 feet above the surface of the ocean while her mate cruised 12 feet below the surface. What is the difference between these two levels?

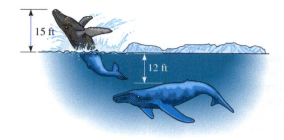

98. *Altitude of Hikers* The surface, or rim, of a canyon is at altitude 0. On a hike down into the canyon, a party of hikers stop for a rest at 130 meters below the surface. They then descend another 54 meters. What is their new altitude? (Write the altitude as a signed number.)

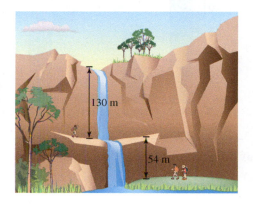

99. Birth Date of a Greek Mathematician A certain Greek mathematician was born in 428 B.C. Her father was born 41 years earlier. In what year was her father born?

100. Birth Date of a Roman Philosopher A certain Roman philosopher was born in 325 B.C. Her mother was born 35 years earlier. In what year was her mother born?

101. Student Loans In 2005, undergraduate college students had an average of $4906 in student loans. This average increased by $788 by 2010 and then dropped $154 by 2012. What was the average amount of such loans in 2012? (*Source:* The College Board.)

102. Entertainment Expenditures Among entertainment expenditures, the average annual spending per U.S. household on fees and admissions was $526 in 2001. This amount increased $80 by 2006 and then decreased $12 by 2011. What was the average household expenditure for fees and admissions in 2011? (*Source:* U.S. Bureau of Labor Statistics.)

103. Record Low Temperatures The lowest temperature ever recorded in Illinois was −36°F. The lowest temperature ever recorded in Utah was 33°F lower than Illinois's record low. What is the record low temperature in Utah? (*Source:* National Climatic Data Center.)

104. Record Low Temperatures The lowest temperature ever recorded in South Carolina was −19°F. The lowest temperature ever recorded in Wisconsin was 36°F lower than South Carolina's record low. What is the record low temperature in Wisconsin? (*Source:* National Climatic Data Center.)

105. Temperature Extremes The lowest temperature ever recorded in Arkansas was −29°F. The highest temperature ever recorded there was 149°F more than the lowest. What is this highest temperature? (*Source:* National Climatic Data Center.)

106. Drastic Temperature Change On January 23, 1943, the temperature rose 49°F in two minutes in Spearfish, South Dakota. If the starting temperature was −4°F, what was the temperature two minutes later? (*Source:* National Climatic Data Center.)

6.3 RATIONAL NUMBERS AND DECIMAL REPRESENTATION

OBJECTIVES

1 Define and identify rational numbers.

2 Write a rational number in lowest terms.

3 Add, subtract, multiply, and divide rational numbers in fraction form.

4 Solve a carpentry problem using operations with fractions.

5 Apply the density property and find arithmetic mean.

6 Convert a rational number in fraction form to a decimal number.

7 Convert a terminating or repeating decimal to a rational number in fraction form.

Definition and the Fundamental Property

Recall from **Section 6.1** that quotients of integers are called **rational numbers.** Think of the rational numbers as being made up of all the fractions (quotients of integers with denominator not equal to zero) and all the integers. Any integer can be written as a quotient of two integers.

For example, the integer 9 can be written as the quotient

$$\frac{9}{1}, \quad \text{or} \quad \frac{18}{2}, \quad \text{or} \quad \frac{27}{3}, \quad \text{and so on.}$$

Also, −5 can be expressed as a quotient of integers as

$$\frac{-5}{1}, \quad \text{or} \quad \frac{-10}{2}, \quad \text{and so on.}$$

(*How can the integer 0 be written as a quotient of integers?*)

RATIONAL NUMBERS

Rational numbers = $\{x \mid x \text{ is a quotient of two integers, with denominator not } 0\}$

A rational number is in **lowest terms** if the greatest common factor of the numerator (top number) and the denominator (bottom number) is 1. Rational numbers are written in lowest terms using the following property.

FUNDAMENTAL PROPERTY OF RATIONAL NUMBERS

If a, b, and k are integers with $b \neq 0$ and $k \neq 0$, then the following is true.

$$\frac{a \cdot k}{b \cdot k} = \frac{a}{b}$$

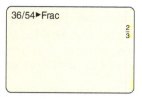

The calculator gives 36/54 in lowest terms, as illustrated in **Example 1.**

EXAMPLE 1 Writing a Fraction in Lowest Terms

Write $\frac{36}{54}$ in lowest terms.

Solution

The greatest common factor of 36 and 54 is 18.

$$\frac{36}{54} = \frac{2 \cdot 18}{3 \cdot 18} = \frac{2}{3} \quad \text{Use the fundamental property with } k = 18.$$

In **Example 1,** we see that $\frac{36}{54} = \frac{2}{3}$. If we multiply the numerator of the fraction on the left by the denominator of the fraction on the right, we obtain $36 \cdot 3 = 108$. If we multiply the denominator of the fraction on the left by the numerator of the fraction on the right, we obtain $54 \cdot 2 = 108$. The result is the same in both cases.

One way of determining whether two fractions are equal is to perform this test. If the product of the **"extremes"** (36 and 3 in this case) equals the product of the **"means"** (54 and 2), the fractions are equal. This is the **cross-product test.**

CROSS-PRODUCT TEST FOR EQUALITY OF RATIONAL NUMBERS

For rational numbers $\frac{a}{b}$ and $\frac{c}{d}$, where $b \neq 0$, $d \neq 0$, the following is true.

$$\frac{a}{b} = \frac{c}{d} \quad \text{if and only if} \quad a \cdot d = b \cdot c$$

Operations with Rational Numbers

The operation of addition of rational numbers can be illustrated by the sketches in **Figure 10.** The rectangle at the top left is divided into three equal portions, with one of the portions in color. The rectangle at the top right is divided into five equal parts, with two of them in color.

The total of the areas in color is represented by the sum

$$\frac{1}{3} + \frac{2}{5}.$$

To evaluate this sum, the areas in color must be redrawn in terms of a common unit. Since the least common multiple of 3 and 5 is 15, redraw both rectangles with 15 parts. See **Figure 11.** In the figure, 11 of the small rectangles are in color, so

$$\frac{1}{3} + \frac{2}{5} = \frac{5}{15} + \frac{6}{15} = \frac{11}{15}.$$

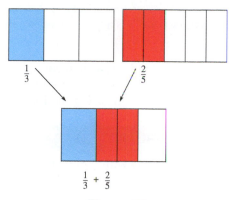

Figure 10

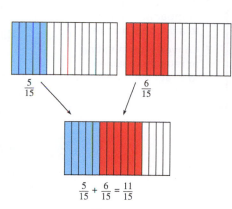

Figure 11

ADDING AND SUBTRACTING RATIONAL NUMBERS

If $\frac{a}{b}$ and $\frac{c}{d}$ are rational numbers, then the following are true.

$$\frac{a}{b} + \frac{c}{d} = \frac{ad + bc}{bd} \quad \text{and} \quad \frac{a}{b} - \frac{c}{d} = \frac{ad - bc}{bd}$$

This formal definition is seldom used in practice. We usually first rewrite the fractions with the least common multiple of their denominators, called the **least common denominator (LCD).**

EXAMPLE 2 Adding and Subtracting Rational Numbers

Perform each operation.

(a) $\dfrac{2}{15} + \dfrac{1}{10}$ (b) $\dfrac{173}{180} - \dfrac{69}{1200}$

Solution

```
2/15+1/10▶Frac
                7
               30
173/180-69/1200▶Frac
             3253
             3600
```

The results of **Example 2** are illustrated in this screen.

(a) Since $30 \div 15 = 2$, $\dfrac{2}{15} = \dfrac{2 \cdot 2}{15 \cdot 2} = \dfrac{4}{30}$

and since $30 \div 10 = 3$, $\dfrac{1}{10} = \dfrac{1 \cdot 3}{10 \cdot 3} = \dfrac{3}{30}.$

The LCD is 30.

Thus, $\dfrac{2}{15} + \dfrac{1}{10} = \dfrac{4}{30} + \dfrac{3}{30} = \dfrac{7}{30}.$

(b) The least common multiple of 180 and 1200 is 3600.

$$\frac{173}{180} - \frac{69}{1200} = \frac{3460}{3600} - \frac{207}{3600} = \frac{3460 - 207}{3600} = \frac{3253}{3600}$$

Fractions that are greater than 1 may be expressed in **mixed number** form.

The mixed number form of $\dfrac{3}{2}$ is $1\dfrac{1}{2}.$

It is understood that the whole number part and the fraction part of a mixed number are added (even though no addition symbol is shown).

EXAMPLE 3 Conversions Involving Mixed Numbers

Perform each conversion.

(a) $2\dfrac{5}{8}$ to a fraction   (b) $\dfrac{86}{5}$ to a mixed number

Solution

(a) $2\dfrac{5}{8} = 2 + \dfrac{5}{8}$ The addition is understood.

$= \dfrac{16}{8} + \dfrac{5}{8}$ Write 2 as a fraction with denominator 8.

$= \dfrac{21}{8}$ Add the fractions.

A shortcut for writing $2\frac{5}{8}$ as a fraction leads to the same result. We multiply the denominator (8) by the whole number part (2) and add the numerator (5) to find the numerator of the mixed number.

$$(8 \times 2) + 5 = \textbf{21}$$

Then use the original denominator of the fraction part (8) as the denominator.

$$2\frac{5}{8} = \frac{\textbf{21}}{\textbf{8}}$$

(b) To convert $\frac{86}{5}$ to a mixed number, we use long division. The quotient is the whole number part of the mixed number. The remainder is the numerator of the fraction part, while the divisor is the denominator.

$$
\begin{array}{r}
17 \quad \leftarrow \text{ Whole number part} \\
\text{Denominator of the fraction} \rightarrow \quad 5\overline{)86} \\
\underline{5} \quad\quad\quad \\
36 \quad\quad \\
\underline{35} \quad\quad \\
1 \quad \leftarrow \text{ Numerator of the fraction}
\end{array}
$$

Thus $\frac{86}{5} = 17\frac{1}{5}$. ∎

MULTIPLYING RATIONAL NUMBERS

If $\frac{a}{b}$ and $\frac{c}{d}$ are rational numbers, then the following is true.

$$\frac{a}{b} \cdot \frac{c}{d} = \frac{ac}{bd}$$

EXAMPLE 4 **Multiplying Rational Numbers**

Find each product. If it is greater than 1, write as a mixed number.

(a) $\dfrac{3}{4} \cdot \dfrac{7}{10}$ **(b)** $\dfrac{5}{18} \cdot \dfrac{3}{10}$ **(c)** $2\dfrac{1}{3} \cdot 1\dfrac{1}{2}$

Solution

(a) $\dfrac{3}{4} \cdot \dfrac{7}{10} = \dfrac{3 \cdot 7}{4 \cdot 10} = \dfrac{21}{40}$

(b) $\dfrac{5}{18} \cdot \dfrac{3}{10} = \dfrac{5 \cdot 3}{18 \cdot 10} = \dfrac{15}{180} = \dfrac{1 \cdot \textbf{15}}{12 \cdot \textbf{15}} = \dfrac{1}{12}$

In practice, a multiplication problem such as this is often solved by using slash marks to indicate that common factors have been divided out of the numerator and denominator.

$$\frac{\overset{1}{\cancel{5}}}{\underset{6}{\cancel{18}}} \cdot \frac{\overset{1}{\cancel{3}}}{\underset{2}{\cancel{10}}} = \frac{1}{6} \cdot \frac{1}{2} \qquad \begin{array}{l} \text{3 is divided out of 3 and 18.} \\ \text{5 is divided out of 5 and 10.} \end{array}$$

$$= \frac{1}{12}$$

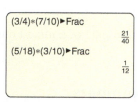

(3/4)*(7/10)▶Frac
                    $\frac{21}{40}$
(5/18)*(3/10)▶Frac
                    $\frac{1}{12}$

To illustrate the results of **Example 4(a)** and **(b)**, we use parentheses around the fraction factors.

(c) $2\dfrac{1}{3} \cdot 1\dfrac{1}{2} = \dfrac{7}{3} \cdot \dfrac{3}{2}$ Convert mixed numbers to fractions.

$= \dfrac{7}{\cancel{3}} \cdot \dfrac{\overset{1}{\cancel{3}}}{2}$ Divide out the common factor 3.

$= \dfrac{7}{2}$ Multiply the fractions.

$= 3\dfrac{1}{2}$ Write as a mixed number.   ∎

A fraction bar indicates the operation of division. The multiplicative inverse of the nonzero number b is $\dfrac{1}{b}$. We define division using the multiplicative inverse.

DEFINITION OF DIVISION

If a and b are real numbers, where $b \neq 0$, then the following is true.

$$\frac{a}{b} = a \cdot \frac{1}{b}$$

Early U.S. cents and **half cents** used fractions to denote their denominations. The half cent used $\frac{1}{200}$ and the cent used $\frac{1}{100}$. (See **Exercise 28** for a photo of an interesting error coin.)

The coins shown here were part of the collection of Louis E. Eliasberg, Sr. **Louis Eliasberg** was the only person ever to assemble a complete collection of United States coins. The Eliasberg gold coins were auctioned in 1982, while the copper, nickel, and silver coins were auctioned in two sales in 1996 and 1997. The half cent pictured sold for $506,000 and the cent sold for $27,500. The cent shown in **Exercise 28** went for a mere $2970.

You probably have heard the rule, "To divide fractions, invert the divisor and multiply." To illustrate this rule, suppose that you have $\dfrac{7}{8}$ of a gallon of milk and you wish to find how many quarts you have. A quart is $\dfrac{1}{4}$ of a gallon, so you must ask yourself, *"How many $\dfrac{1}{4}$s are there in $\dfrac{7}{8}$?"* This would be interpreted as

$$\frac{7}{8} \div \frac{1}{4}, \quad \text{or} \quad \frac{\dfrac{7}{8}}{\dfrac{1}{4}}.$$

The fundamental property of rational numbers can be extended to rational number values of a, b, and k.

$$\frac{a}{b} = \frac{a \cdot k}{b \cdot k} = \frac{\dfrac{7}{8} \cdot 4}{\dfrac{1}{4} \cdot 4} = \frac{\dfrac{7}{8} \cdot 4}{1} = \frac{7}{8} \cdot \frac{4}{1}$$ Let $a = \frac{7}{8}$, $b = \frac{1}{4}$, and $k = 4 \left(\text{the reciprocal of } b = \frac{1}{4} \right)$.

To divide $\dfrac{7}{8}$ by $\dfrac{1}{4}$ is equivalent to multiplying $\dfrac{7}{8}$ by the reciprocal, $\dfrac{4}{1}$.

$$\frac{7}{8} \cdot \frac{4}{1} = \frac{28}{8} = \frac{7}{2}$$ Multiply and reduce to lowest terms.

Thus there are $\dfrac{7}{2}$, or $3\dfrac{1}{2}$, quarts in $\dfrac{7}{8}$ gallon.

DIVIDING RATIONAL NUMBERS

If $\dfrac{a}{b}$ and $\dfrac{c}{d}$ are rational numbers, where $\dfrac{c}{d} \neq 0$, then the following is true.

$$\frac{a}{b} \div \frac{c}{d} = \frac{a}{b} \cdot \frac{d}{c} = \frac{ad}{bc}$$

EXAMPLE 5 Dividing Rational Numbers

Find each quotient.

(a) $\dfrac{3}{5} \div \dfrac{7}{15}$ **(b)** $\dfrac{-4}{7} \div \dfrac{3}{14}$ **(c)** $\dfrac{2}{9} \div 4$ **(d)** $-9 \div \dfrac{3}{5}$

Solution

(a) $\dfrac{3}{5} \div \dfrac{7}{15} = \dfrac{3}{5} \cdot \dfrac{15}{7} = \dfrac{45}{35} = \dfrac{9 \cdot 5}{7 \cdot 5} = \dfrac{9}{7}$

(b) $\dfrac{-4}{7} \div \dfrac{3}{14} = \dfrac{-4}{7} \cdot \dfrac{14}{3} = \dfrac{-56}{21} = \dfrac{-8 \cdot 7}{3 \cdot 7} = \dfrac{-8}{3} = -\dfrac{8}{3}$

> $\dfrac{-a}{b}, \dfrac{a}{-b},$ and $-\dfrac{a}{b}$ are all equal.

(c) $\dfrac{2}{9} \div 4 = \dfrac{2}{9} \div \dfrac{4}{1} = \dfrac{2}{9} \cdot \dfrac{1}{4} = \dfrac{\overset{1}{2}}{9} \cdot \dfrac{1}{\underset{2}{4}} = \dfrac{1}{18}$

(d) $-9 \div \dfrac{3}{5} = \dfrac{\overset{-3}{-9}}{1} \cdot \dfrac{5}{\underset{1}{3}} = -15$

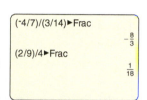

```
(-4/7)/(3/14)▶Frac
                    -8/3
(2/9)/4▶Frac
                    1/18
```

This screen supports the results in
Example 4(b) and (c).

EXAMPLE 6 Applying Operations with Fractions to Carpentry

A carpenter has a board 72 inches long that he must cut into 10 pieces of equal length. This will require 9 cuts. See **Figure 12.**

(a) If each cut causes a waste of $\dfrac{3}{16}$ inch, how many inches of actual board will remain after the cuts?

(b) What will be the length of each of the resulting pieces?

Figure 12

Solution

(a) We start with 72 inches and must subtract $\dfrac{3}{16}$ nine times.

Rather than writing out each of these, we represent the amount subtracted by $9\left(\dfrac{3}{16}\right)$.

$$72 - 9\left(\dfrac{3}{16}\right) = 72 - \left(\dfrac{9}{1} \cdot \dfrac{3}{16}\right) \qquad 9 = \dfrac{9}{1}$$

$$= 72 - \dfrac{27}{16} \qquad \text{Multiply the fractions.}$$

$$= \dfrac{1152}{16} - \dfrac{27}{16} \qquad \text{Use a common denominator.}$$

$$= \dfrac{1125}{16} \qquad \text{Subtract fractions.}$$

$$= 70\dfrac{5}{16} \qquad \text{Write as a mixed number.}$$

Thus $70\dfrac{5}{16}$ inches of actual board will remain.

(b) The $70\frac{5}{16}$, or $\frac{1125}{16}$, inches of board from part (a) will be divided into 10 pieces of equal length.

$$\frac{1125}{16} \div 10 = \frac{1125}{16} \cdot \frac{1}{10} \qquad \text{Definition of division}$$

$$= \frac{\overset{225}{\cancel{1125}}}{16} \cdot \frac{1}{\underset{2}{\cancel{10}}} \qquad \text{Divide out the common factor 5.}$$

$$= \frac{225}{32} \qquad \text{Multiply.}$$

$$= 7\frac{1}{32} \qquad \text{Convert to a mixed number.}$$

Each piece will measure $7\frac{1}{32}$ inches. ∎

Density and the Arithmetic Mean

There is no integer between two consecutive integers, such as 3 and 4. However, a rational number can always be found between any two distinct rational numbers. For this reason, the set of rational numbers is said to be *dense*.

Benjamin Banneker (1731–1806) spent the first half of his life tending a farm in Maryland. He gained a reputation locally for his mechanical skills and his ability in mathematical problem solving. In 1772 he acquired astronomy books from a neighbor and devoted himself to learning astronomy, observing the skies, and making calculations. In 1789 Banneker joined the team that surveyed what is now the District of Columbia.

Banneker published almanacs yearly from 1792 to 1802. He sent a copy of his first almanac to **Thomas Jefferson,** along with an impassioned letter against slavery. Jefferson subsequently championed the cause of this early African-American mathematician.

DENSITY PROPERTY OF THE RATIONAL NUMBERS

If r and t are distinct rational numbers, with $r < t$, then there exists a rational number s such that

$$r < s < t.$$

Repeated applications of the density property lead to the following conclusion.

There are infinitely many rational numbers between two distinct rational numbers.

One example of a rational number that is between two distinct rational numbers is the *arithmetic mean* of the two numbers. To find the **arithmetic mean,** or **average,** of n numbers, we add the numbers and then divide the sum by n. For two numbers, the number that lies halfway between them is their average.

Table 4

Academic Year	Amount of Tuition and Fees
2007–08	$21,427
2008–09	22,036
2009–10	21,908
2010–11	22,771
2011–12	23,479

Source: National Center for Education Statistics.

EXAMPLE 7 Finding an Arithmetic Mean (Average)

Table 4 shows the average amount of tuition and fees at private 4-year institutions for five academic years. What is the average cost for this period?

Solution

To find this average, divide the sum of the amounts by the number of amounts, **5**.

$$\frac{21,427 + 22,036 + 21,908 + 22,771 + 23,479}{5} = \frac{111,621}{5}$$

$$= 22,324\frac{1}{5}$$

The average amount for the period is $22,324 (rounded to the nearest dollar). ∎

EXAMPLE 8 **Finding a Measurement in a Socket Wrench Set**

A carpenter owns a socket wrench set that has measurements in the English system. As is the custom in carpentry, measurements are given in half-, quarter-, eighth-, and sixteenth-inches. He finds

that his $\frac{3}{8}$-inch socket is just a bit too small for his job, while his $\frac{1}{2}$-inch socket is just a bit too large. He suspects that he will need to use the socket with the measure that is halfway between these (their arithmetic mean, or average). What measure socket must he use?

Solution

The carpenter needs to find the average of $\frac{3}{8}$ and $\frac{1}{2}$. To do this, find their sum and divide by 2.

$$\frac{3}{8} + \frac{1}{2} = \frac{3}{8} + \frac{4}{8} \qquad \text{Find a common denominator.}$$

$$= \frac{7}{8} \qquad \text{Add the numerators, and keep the same denominator.}$$

Now divide $\frac{7}{8}$ by 2.

$$\frac{7}{8} \div 2 = \frac{7}{8} \cdot \frac{1}{2} \qquad \text{Definition of division}$$

$$= \frac{7}{16} \qquad \text{Multiply the numerators and multiply the denominators.}$$

He must use the $\frac{7}{16}$-inch socket. ∎

Simon Stevin (1548–1620) worked as a bookkeeper in Belgium and became an engineer in the Netherlands army. He is usually given credit for the development of **decimals.**

Decimal Form of Rational Numbers

Rational numbers can be expressed as decimals. Decimal numerals have place values that are powers of 10. The place values are as shown here.

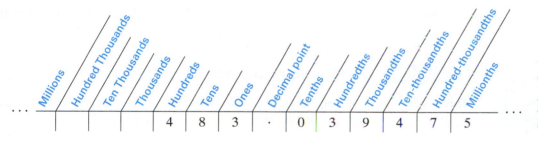

The decimal numeral 483.039475 is read "four hundred eighty-three and thirty-nine thousand, four hundred seventy-five millionths."

A rational number in the form $\frac{a}{b}$ can be expressed as a decimal most easily by entering it into a calculator. For example, to write $\frac{3}{8}$ as a decimal, enter 3, then enter the operation of division, then enter 8. Press the equals key to find the following equivalence.

$$\frac{3}{8} = 0.375$$

$$
\begin{array}{r}
0.375 \\
8\overline{)3.000} \\
\underline{24} \\
60 \\
\underline{56} \\
40 \\
\underline{40} \\
0
\end{array}
\qquad
\begin{array}{r}
0.3636\ldots \\
11\overline{)4.00000\ldots} \\
\underline{33} \\
70 \\
\underline{66} \\
40 \\
\underline{33} \\
70 \\
\underline{66} \\
40 \\
\therefore
\end{array}
$$

While 2/3 has a repeating decimal representation ($2/3 = 0.\overline{6}$), the calculator rounds off in the final decimal place displayed.

```
5/11
         .4545454545
1/3
         .3333333333
5/6
         .8333333333
```

Although only ten decimal digits are shown, all three fractions have decimals that repeat endlessly.

This same result may be obtained by long division, as shown in the margin. By this result, the rational number $\frac{3}{8}$ is the same as the decimal 0.375. A decimal such as 0.375, which stops, is called a **terminating decimal.**

$$\frac{1}{4} = 0.25, \quad \frac{7}{10} = 0.7, \quad \text{and} \quad \frac{89}{1000} = 0.089 \quad \text{Examples of terminating decimals}$$

Not all rational numbers can be represented by terminating decimals. For example, convert $\frac{4}{11}$ into a decimal by dividing 11 into 4 using a calculator. The display shows

$$0.3636363636, \quad \text{or perhaps} \quad 0.363636364.$$

However, the long division process shown in the margin indicates that we will actually get 0.3636 . . . , with the digits 36 repeating over and over indefinitely. To indicate this, we write a bar (called a *vinculum*) over the "block" of digits that repeats.

$$\frac{4}{11} = 0.\overline{36} \quad 0.\overline{36} \text{ means } 0.3636\ldots.$$

A decimal such as $0.\overline{36}$, which continues indefinitely, is called a **repeating decimal.**

$$\frac{5}{11} = 0.\overline{45}, \quad \frac{1}{3} = 0.\overline{3}, \quad \text{and} \quad \frac{5}{6} = 0.8\overline{3} \quad \text{Examples of repeating decimals}$$

While we distinguish between *terminating* and *repeating* decimals in this text, some mathematicians prefer to consider all rational numbers as repeating decimals. This can be justified by thinking this way: If the division process leads to a remainder of 0, then zeros repeat without end in the decimal form. For example, we can consider the decimal form of $\frac{3}{4}$ as follows.

$$\frac{3}{4} = 0.75\overline{0}$$

By considering the possible remainders that may be obtained when converting a quotient of integers to a decimal, we can draw an important conclusion about the decimal form of rational numbers. If the remainder is never zero, the division will produce a repeating decimal. This happens because each step of the division process must produce a remainder that is less than the divisor. Since the number of different possible remainders is less than the divisor, the remainders must eventually begin to repeat. This makes the digits of the quotient repeat, producing a repeating decimal.

DECIMAL REPRESENTATION OF RATIONAL NUMBERS

Any rational number can be expressed as either a terminating decimal or a repeating decimal.

To determine whether the decimal form of a quotient of integers will terminate or repeat, we use the following rule. Justification of this rule is based on the fact that the prime factors of 10 are 2 and 5, and the decimal system uses ten as its base.

CRITERIA FOR TERMINATING AND REPEATING DECIMALS

A rational number $\frac{a}{b}$ in *lowest terms* results in a **terminating decimal** if the only prime factor of the denominator is 2 or 5 (or both).

A rational number $\frac{a}{b}$ in *lowest terms* results in a **repeating decimal** if a prime other than 2 or 5 appears in the prime factorization of the denominator.

Year	Mike Lowell	Jacoby Ellsbury
2007	$\frac{191}{589} = .324$	$\frac{41}{116} = .353$
2008	$\frac{115}{419} = .274$	$\frac{155}{554} = .280$
Two-year total	$\frac{306}{1008} = .304$	$\frac{196}{670} = .293$

In both individual years, Ellsbury had a higher average, but for the two-year period, Lowell had the higher average. This is an example of **Simpson's paradox** from statistics. (The authors thank Carol Merrigan for researching this information.)

EXAMPLE 9 **Determining Whether a Decimal Terminates or Repeats**

Determine whether the decimal form terminates or repeats.

(a) $\frac{7}{8}$ **(b)** $\frac{13}{150}$ **(c)** $\frac{6}{75}$

Solution

(a) The rational number $\frac{7}{8}$ is in lowest terms. Its denominator is 8, and since 8 factors as 2^3, the decimal form will terminate. No primes other than 2 or 5 divide the denominator.

(b) The rational number $\frac{13}{150}$ is in lowest terms with denominator

$$150 = 2 \cdot 3 \cdot 5^2.$$

Because **3** appears as a prime factor of the denominator, the decimal form will repeat.

(c) First write the rational number $\frac{6}{75}$ in lowest terms.

$$\frac{6}{75} = \frac{2}{25} \quad \text{Denominator is 25.}$$

Because $25 = 5^2$, the decimal form will terminate. ■

We have seen that a rational number will be represented by either a terminating or a repeating decimal. Must a terminating decimal or a repeating decimal represent a rational number? The answer is *yes*. For example, the terminating decimal 0.6 represents a rational number.

$$0.6 = \frac{6}{10} = \frac{3}{5}$$

EXAMPLE 10 **Writing Decimals as Quotients of Integers**

Write each decimal as a quotient of integers.

(a) 0.437 **(b)** 8.2 **(c)** $0.\overline{85}$

Solution

(a) $0.437 = \frac{437}{1000}$ Read as "four hundred thirty-seven thousandths" and then write as a fraction.

(b) $8.2 = 8 + \frac{2}{10} = \frac{82}{10} = \frac{41}{5}$ Read as a decimal, write as a sum, and then add.

(c) To convert a repeating decimal to a quotient of integers, we use some simple algebra.

Step 1 Let $x = 0.\overline{85}$, so $x = 0.858585\ldots$.

Step 2 Multiply both sides of the equation $x = 0.858585\ldots$ by 100. (Use 100 since there are **two** digits in the part that repeats, and $100 = 10^2$.)

$$x = 0.858585\ldots$$
$$100x = 100(0.858585\ldots)$$
$$100x = 85.858585\ldots$$

.437▶Frac
$\frac{437}{1000}$
8.2▶Frac
$\frac{41}{5}$

The results of **Example 10(a) and (b)** are supported in this screen.

1 = 0.99999999...

Terminating or Repeating? One of the most baffling truths of elementary mathematics is the following:

$$1 = 0.9999\ldots$$

Most people believe that $0.\overline{9}$ has to be less than 1, but this is not the case. The following argument shows why. Let $x = 0.9999\ldots$ Then

$$10x = 9.9999\ldots$$
$$\underline{x = 0.9999\ldots}$$
$$9x = 9 \qquad \text{Subtract.}$$
$$x = 1. \qquad \text{Divide.}$$

Therefore, $1 = 0.9999\ldots$ Similarly, it can be shown that any terminating decimal can be represented as a repeating decimal with an endless string of 9s. For example, $0.5 = 0.49999\ldots$ and $2.6 = 2.59999\ldots$ This is a way of justifying that any rational number may be represented as a repeating decimal.

Step 3 Subtract the expressions in Step 1 from the final expressions in Step 2.

$$100x = 85.858585\ldots \qquad \text{(Recall that } x = 1x \text{ and } 100x - x = 99x.\text{)}$$
$$\underline{x = 0.858585\ldots}$$
$$99x = 85 \qquad \text{Subtract.}$$

Step 4 Solve the equation $99x = 85$ by dividing both sides by 99.

$$99x = 85$$
$$\frac{99x}{99} = \frac{85}{99} \qquad \text{Divide by 99.}$$
$$x = \frac{85}{99} \qquad \tfrac{99x}{99} = x$$
$$0.\overline{85} = \frac{85}{99} \qquad x = 0.\overline{85}$$

When checking with a calculator, remember that the calculator will show only a finite number of decimal places and may round off in the final decimal place shown. ∎

FOR FURTHER THOUGHT

The Influence of Spanish Coinage on Stock Prices

Until August 28, 2000, when decimalization of the U.S. stock market began, market prices were reported with fractions having denominators that were powers of 2, such as $17\frac{3}{4}$ and $112\frac{5}{8}$. Did you ever wonder why this was done?

During the early years of the United States, prior to the minting of its own coinage, the Spanish eight-reales coin, also known as the Spanish milled dollar, circulated freely in the states. Its fractional parts, the four reales, two reales, and one real, were known as **pieces of eight** and were described as such in pirate and treasure lore. When the New York Stock Exchange was founded in 1792, it chose to use the Spanish milled dollar as its price basis, rather than the decimal base proposed by Thomas Jefferson that same year. All prices on the U.S. stock markets are now reported in decimals. (*Source:* "Stock price tables go to decimal listings," *The Times Picayune,* June 27, 2000.)

For Group or Individual Investigation

Consider this: Have you ever heard this old cheer? "Two bits, four bits, six bits, a dollar. All for the (home team), stand up and holler." The term **two bits** refers to 25 cents. Discuss how this cheer is based on the Spanish eight-reales coin.

6.3 EXERCISES

Basic Concepts of Fractions *Complete each of the following, based on the discussion in this section.*

1. A number that can be represented as a quotient of integers with denominator not 0 is called a(n) _____.

2. In the fraction $\frac{7}{13}$, _____ is the numerator and _____ is the denominator.

3. A number of the type in **Exercise 1** is in lowest terms if its numerator and denominator have _____ as their greatest common factor.

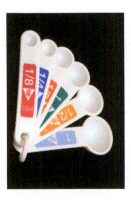

4. The fundamental property of rational numbers allows us to write $\frac{18}{27}$ as the equivalent $\frac{2}{3}$ because there is a common factor of _____ in the numerator and the denominator.

5. The fractions $\frac{13}{27}$ and $\frac{221}{459}$ are equivalent because the product of the extremes, 13 and _____, is equal to the product of the means, 27 and _____. Both products are _____.

6. In the mixed number $3\frac{1}{7}$ there is a(n) _____ sign understood between the whole number part and the fraction part.

7. To divide $\frac{7}{12}$ by $\frac{3}{4}$, we multiply _____ by the reciprocal of the divisor. That reciprocal is _____.

8. The rational numbers exhibit the density property, which says that between any two distinct rational numbers there exists _____.

9. The integers do not exhibit the density property. For example, between 5 and _____ there is no other integer.

10. A rational number will have a decimal representation that will show one of two patterns: It will _____ or it will _____.

Choose the expression(s) that is (are) equivalent to the given rational number.

11. $\frac{4}{8}$

    A. $\frac{1}{2}$ **B.** $\frac{8}{4}$ **C.** 0.5   **D.** $0.5\overline{0}$ **E.** $0.\overline{55}$

12. $\frac{2}{3}$

    A. 0.67 **B.** $0.\overline{6}$ **C.** $\frac{20}{30}$ **D.** 0.666... **E.** 0.6

13. $\frac{5}{9}$

    A. 0.56 **B.** 0.55 **C.** $0.\overline{5}$ **D.** $\frac{9}{5}$ **E.** $1\frac{4}{5}$

14. $\frac{1}{4}$

    A. 0.25 **B.** $0.24\overline{9}$ **C.** $\frac{25}{100}$ **D.** 4 **E.** $\frac{10}{400}$

Write each fraction in lowest terms.

15. $\frac{16}{48}$ 16. $\frac{21}{28}$ 17. $-\frac{15}{35}$ 18. $-\frac{8}{48}$

Write each fraction in three other ways.

19. $\frac{3}{8}$ 20. $\frac{9}{10}$

21. $-\frac{5}{7}$ 22. $-\frac{7}{12}$

23. **Fractional Parts** Write a fraction in lowest terms that represents the portion of each figure that is in color.

  (a) 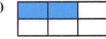   **(b)**

  (c)   **(d)**

24. **Fractional Parts** Write a fraction in lowest terms that represents the region described in parts (a)–(d).

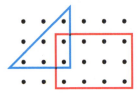

  (a) the dots in the rectangle as a part of the dots in the entire figure

  (b) the dots in the triangle as a part of the dots in the entire figure

  (c) the dots in the rectangle as a part of the dots in the union of the triangle and the rectangle

  (d) the dots in the intersection of the triangle and the rectangle as a part of the dots in the union of the triangle and the rectangle

25. Fractional Parts Refer to the figure for **Exercise 24,** and write a description of the region that is represented by the fraction $\frac{1}{12}$.

26. Batting Averages In a softball league, Paul got 8 hits in 20 at-bats, and Josh got 12 hits in 30 at-bats. Which player (if either) had the higher batting average?

27. Batting Averages After ten games, the following statistics were obtained.

Player	At-bats	Hits	Home Runs
Anne	40	9	2
Christine	36	12	3
Leah	11	5	1
Otis	16	8	0
Carol	20	10	2

Answer using estimation skills as necessary.

(a) Which player got a hit in exactly $\frac{1}{3}$ of his or her at-bats?

(b) Which player got a hit in just less than $\frac{1}{2}$ of his or her at-bats?

(c) Which player got a home run in just less than $\frac{1}{10}$ of his or her at-bats?

(d) Which player got a hit in just less than $\frac{1}{4}$ of his or her at-bats?

(e) Which two players got hits in exactly the same fractional parts of their at-bats? What was the fractional part, reduced to lowest terms?

28. Error Coin Refer to the margin note discussing the use of common fractions on early U.S. copper coinage. The photo here shows an error near the bottom that occurred on an 1802 large cent. Discuss the error and how it represents a mathematical impossibility.

Perform the indicated operations and express answers in lowest terms. Use the order of operations as necessary.

29. $\frac{3}{8}+\frac{1}{8}$ **30.** $\frac{7}{9}+\frac{1}{9}$ **31.** $\frac{5}{16}+\frac{7}{12}$

32. $\frac{1}{15}+\frac{7}{18}$ **33.** $\frac{2}{3}-\frac{7}{8}$ **34.** $\frac{13}{20}-\frac{5}{12}$

35. $\frac{5}{8}-\frac{3}{14}$ **36.** $\frac{19}{15}-\frac{7}{12}$ **37.** $\frac{3}{4}\cdot\frac{9}{5}$

38. $\frac{3}{8}\cdot\frac{2}{7}$ **39.** $-\frac{2}{3}\cdot\left(-\frac{5}{8}\right)$ **40.** $-\frac{2}{4}\cdot\frac{3}{9}$

41. $\frac{5}{12}\div\frac{15}{4}$ **42.** $\frac{15}{16}\div\frac{30}{8}$

43. $-\frac{9}{16}\div\left(-\frac{3}{8}\right)$ **44.** $-\frac{3}{8}\div\left(-\frac{5}{4}\right)$

45. $\left(\frac{1}{3}\div\frac{1}{2}\right)+\frac{5}{6}$ **46.** $\frac{2}{5}\div\left(-\frac{4}{5}\div\frac{3}{10}\right)$

Convert each mixed number to a fraction, and convert each fraction to a mixed number.

47. $4\frac{1}{3}$ **48.** $3\frac{7}{8}$ **49.** $2\frac{9}{10}$

50. $\frac{18}{5}$ **51.** $\frac{27}{4}$ **52.** $\frac{19}{3}$

Perform each operation and express your answer as a mixed number.

53. $3\frac{1}{4}+2\frac{7}{8}$ **54.** $6\frac{1}{5}-2\frac{7}{15}$

55. $-4\frac{7}{8}\cdot3\frac{2}{3}$ **56.** $-4\frac{1}{6}\div1\frac{2}{3}$

Carpenter Calculations *Carpentry applications often require calculations involving fractions and mixed numbers with denominators of* 2, 4, 8, *and* 16. *Perform each of the following, and express each answer as a mixed number.*

57. $\frac{9}{16}+2\left(\frac{5}{8}\right)-\frac{1}{2}$ **58.** $\frac{5}{16}+3\left(\frac{3}{8}\right)-\frac{1}{4}$

59. $9\left(2\frac{1}{2}\right)+3\left(5\frac{1}{8}\right)$ **60.** $8\left(4\frac{1}{4}\right)-3\left(7\frac{1}{2}\right)$

61. $5\left(\frac{1}{16}\right)+3\left(2\frac{1}{2}\right)-\frac{7}{16}$ **62.** $8\left(\frac{3}{16}\right)+2\left(6\frac{1}{4}\right)-\frac{13}{16}$

63. $4\left[3\left(\frac{1}{8}\right)+2\left(\frac{7}{16}\right)\right]-2\left(\frac{1}{2}-\frac{1}{16}\right)$

64. $6\left[2\left(\frac{5}{8}\right)+3\left(\frac{3}{16}\right)\right]-3\left(\frac{3}{8}-\frac{3}{16}\right)$

Solve each problem.

65. Socket Wrench Measurements A hardware store sells a 22-piece socket wrench set. The measure of the largest socket is $\frac{3}{4}$ in., while the measure of the smallest socket is $\frac{3}{16}$ in. What is the difference between these measures?

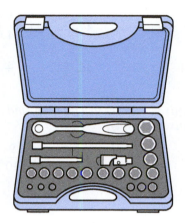

66. TV Guide First published in 1953, the digest-sized *TV Guide* has changed to a "full-sized" magazine, as shown in the figure. What is the difference in their heights? (*Source: TV Guide.*)

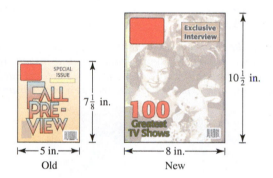

Old — 5 in. — $7\frac{1}{8}$ in.

New — 8 in. — $10\frac{1}{2}$ in.

67. Desk Height The diagram appears in a woodworker's book. Find the height of the desk to the top of the writing surface.

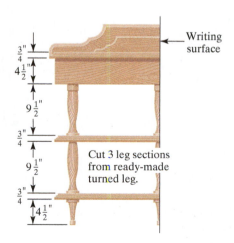

$\frac{3}{4}$" $4\frac{1}{2}$" $9\frac{1}{2}$" $\frac{3}{4}$" $9\frac{1}{2}$" $\frac{3}{4}$" $4\frac{1}{2}$"

Writing surface

Cut 3 leg sections from ready-made turned leg.

68. Legs of a Desk The desk in **Exercise 67** has four legs, each of which consists of three individual pieces. What is the total length of these twelve pieces?

69. Golf Tees The Pride Golf Tee Company, the only U.S. manufacturer of wooden golf tees, has created the Professional Tee system shown in the figure. (*Source: The Gazette.*)

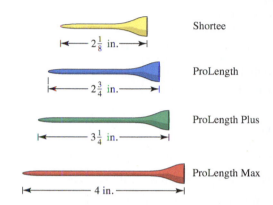

Shortee — $2\frac{1}{8}$ in.

ProLength — $2\frac{3}{4}$ in.

ProLength Plus — $3\frac{1}{4}$ in.

ProLength Max — 4 in.

(a) Find the difference between the lengths of the ProLength Plus and the once-standard Shortee.

(b) The ProLength Max tee is the longest tee allowed by the U.S. Golf Association's Rules of Golf. How much longer is the ProLength Max than the Shortee?

70. Recipe for Grits The following chart appears on a package of Quaker® Quick Grits.

	Microwave		Stove Top	
Servings	**1**	**1**	**4**	**6**
Water	$\frac{3}{4}$ cup	1 cup	3 cups	4 cups
Grits	3 Tbsp	3 Tbsp	$\frac{3}{4}$ cup	1 cup
Salt (optional)	dash	dash	$\frac{1}{4}$ tsp	$\frac{1}{2}$ tsp

(a) How many cups of water would be needed for 6 microwave servings?

(b) How many cups of grits would be needed for 5 stove-top servings? (*Hint:* 5 is halfway between 4 and 6.)

71. Cuts on a Board A board is 48 in. long. It must be divided into four pieces of equal length, and each cut will cause a waste of $\frac{3}{16}$ in. How long will each piece be after the cuts are made?

72. Cuts on a Board A board is 72 in. long. It must be divided into six pieces of equal length, and each cut will cause a waste of $\frac{1}{8}$ in. How long will each piece be after the cuts are made?

73. Cake Recipe A cake recipe calls for $1\frac{3}{4}$ cups of sugar. A caterer has $15\frac{1}{2}$ cups of sugar on hand. How many cakes can he make?

74. Fabric It takes $2\frac{1}{4}$ yd of fabric to cover a chair of a particular design. How many chairs can be covered with $23\frac{2}{3}$ yd of fabric?

75. Fabric It takes $2\frac{3}{8}$ yd of fabric to make a costume for a school play. How much fabric would be needed for seven costumes?

76. Cookie Recipe A cookie recipe calls for $2\frac{2}{3}$ cups of sugar. How much sugar would be needed to make four batches of cookies?

77. Public 4-Year Institution Costs The table shows the average amount of tuition and fees at public 4-year institutions for in-state students for five academic years. What is the average cost for this period (rounded to the nearest dollar)? (*Source:* National Center for Education Statistics.)

Academic Year	Amount of Tuition and Fees
2007–08	$5943
2008–09	6312
2009–10	6695
2010–11	7136
2011–12	7701

78. Public 2-Year Institution Costs The table shows the average amount of tuition and fees at public 2-year institutions for in-state students for five academic years. What is the average cost for this period (rounded to the nearest dollar)? (*Source:* National Center for Education Statistics.)

Academic Year	Amount of Tuition and Fees
2007–08	$2061
2008–09	2136
2009–10	2285
2010–11	2439
2011–12	2647

Find the rational number halfway between the two given rational numbers.

79. $\frac{1}{2}, \frac{3}{4}$

80. $\frac{1}{3}, \frac{5}{12}$

81. $\frac{3}{5}, \frac{2}{3}$

82. $\frac{7}{12}, \frac{5}{8}$

83. $-\frac{2}{3}, -\frac{5}{6}$

84. $-\frac{5}{16}, -\frac{5}{2}$

*Use the method of **Example 9** to decide whether each rational number would yield a repeating or a terminating decimal. (Hint: Write in lowest terms before trying to decide.)*

85. $\frac{8}{15}$

86. $\frac{8}{35}$

87. $\frac{13}{125}$

88. $\frac{3}{24}$

89. $\frac{22}{55}$

90. $\frac{24}{75}$

Convert each rational number into either a repeating or a terminating decimal. Use a calculator if your instructor so allows.

91. $\frac{3}{4}$

92. $\frac{7}{8}$

93. $\frac{3}{16}$

94. $\frac{9}{32}$

95. $\frac{3}{11}$

96. $\frac{9}{11}$

97. $\frac{2}{7}$

98. $\frac{11}{15}$

Convert each decimal into a quotient of integers, written in lowest terms.

99. 0.4

100. 0.9

101. 0.85

102. 0.105

103. 0.934

104. 0.7984

105. $0.\overline{67}$

106. $0.\overline{53}$

107. $0.0\overline{42}$

108. $0.0\overline{86}$

109. $0.0\overline{1}$

110. $0.1\overline{2}$

111. Hard to Believe? Follow through on all parts of this exercise in order.

(a) Find the decimal for $\frac{1}{3}$.

(b) Find the decimal for $\frac{2}{3}$.

(c) By adding the decimal expressions obtained in parts (a) and (b), obtain a decimal expression for $\frac{1}{3} + \frac{2}{3} = \frac{3}{3} = 1$.

(d) State your result. Read the margin note on terminating and repeating decimals in this section, which refers to this idea.

112. Hard to Believe? It is a fact that $\frac{1}{3} = 0.333\ldots$. Multiply both sides of this equation by 3. Does your answer bother you? See the margin note on terminating and repeating decimals in this section.

6.4 IRRATIONAL NUMBERS AND DECIMAL REPRESENTATION

OBJECTIVES

1 Understand how irrational numbers differ from rational numbers in their decimal representations.

2 Follow the proof that $\sqrt{2}$ is an irrational number.

3 Use a calculator to find square roots.

4 Apply the product and quotient rules for square roots.

5 Rationalize a denominator.

6 Understand the relevance of the irrational numbers π, ϕ, and e in mathematics.

Definition and Basic Concepts

Every rational number has a decimal form that terminates or repeats, and every repeating or terminating decimal represents a rational number. However,

$$0.102001000200001000002\ldots$$

does not terminate and does not repeat. (It is true that there is a pattern in this decimal, but no single block of digits repeats indefinitely.)*

IRRATIONAL NUMBERS

Irrational numbers $= \{x \mid x$ is a number represented by a nonrepeating, nonterminating decimal$\}$

As the name implies, an irrational number cannot be represented as a quotient of integers.

The decimal number mentioned at the top of this page is an irrational number. Other irrational numbers include $\sqrt{2}$, $\frac{1 + \sqrt{5}}{2}$ (ϕ, from **Section 5.5**), π (the ratio of the circumference of a circle to its diameter), and e (a constant *approximately equal to* 2.71828). There are infinitely many irrational numbers.

Irrationality of $\sqrt{2}$ and Proof by Contradiction

Figure 13 illustrates how a point with coordinate $\sqrt{2}$ can be located on a number line. (This was first mentioned in **Section 6.1**.)

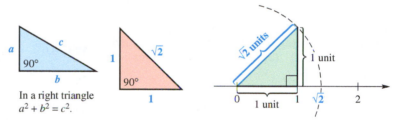

In a right triangle $a^2 + b^2 = c^2$.

Figure 13

The irrational number $\sqrt{2}$ was discovered by the Pythagoreans in about 500 B.C. This discovery was a great setback to their philosophy that everything is based on the whole numbers. The Pythagoreans kept their findings secret, and legend has it that members of the group who divulged this discovery were sent out to sea, and, according to Proclus (410–485), "perished in a shipwreck, to a man."

The proof that $\sqrt{2}$ is irrational is a classic example of a **proof by contradiction.** We begin by assuming that $\sqrt{2}$ is rational, which leads to a contradiction, or absurdity. The method is also called **reductio ad absurdum** (Latin for "reduce to the absurd"). In order to understand the proof, we consider three preliminary facts:

1. When a rational number is written in lowest terms, the greatest common factor of the numerator and denominator is 1.

2. If an integer is even, then it has 2 as a factor and may be written in the form $2k$, where k is an integer.

3. If a perfect square is even, then its square root is even.

---

*In this section, we will assume that the digits of a number such as this continue indefinitely in the pattern established. The next few digits would be 000000100000002, and so on.

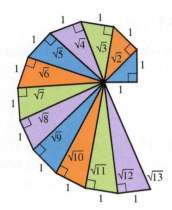

An interesting way to represent the lengths corresponding to $\sqrt{2}$, $\sqrt{3}$, $\sqrt{4}$, $\sqrt{5}$, and so on, is shown in the figure. Use the **Pythagorean theorem** to verify the lengths in the figure.

THEOREM

Statement: $\sqrt{2}$ is an irrational number.

Proof: Assume that $\sqrt{2}$ is a rational number. Then by definition,

$$\sqrt{2} = \frac{p}{q}, \quad \text{for some integers } p \text{ and } q.$$

Furthermore, assume that $\frac{p}{q}$ is the form of $\sqrt{2}$ that is written in lowest terms, so the greatest common factor of p and q is 1.

$$2 = \frac{p^2}{q^2} \qquad \text{\color{blue}Square both sides of the equation.}$$

$$2q^2 = p^2 \qquad \text{\color{blue}Multiply by } q^2.$$

The last equation, $2q^2 = p^2$, indicates that 2 is a factor of p^2. So p^2 is even, and thus p is even. Since p is even, it may be written in the form $2k$, where k is an integer.

Now, substitute $2k$ for p in the last equation and simplify.

$$2q^2 = (2k)^2 \qquad \text{\color{blue}Let } p = 2k.$$

$$2q^2 = 4k^2 \qquad \text{\color{blue}} (2k)^2 = 2k \cdot 2k = 4k^2$$

$$q^2 = 2k^2 \qquad \text{\color{blue}Divide by 2.}$$

Since q^2 has 2 as a factor, q^2 must be even, and thus q must be even. This leads to a contradiction: p and q cannot both be even because they would then have a common factor of 2. It was assumed that their greatest common factor is 1. The original assumption that $\sqrt{2}$ is rational has led to a contradiction, so $\sqrt{2}$ is irrational. ∎

Operations with Square Roots

In everyday mathematical work, nearly all of our calculations deal with rational numbers, usually in decimal form. However, we must sometimes perform operations with irrational numbers. ***Recall that $\sqrt{a}$, for $a \geq 0$, is the nonnegative number whose square is a. That is, $\left(\sqrt{a}\right)^2 = a$.***

$$\sqrt{2}, \quad \sqrt{3}, \quad \text{and} \quad \sqrt{13} \qquad \text{\color{blue}Examples of square roots that are irrational}$$

$$\sqrt{4} = 2, \quad \sqrt{36} = 6, \quad \text{and} \quad \sqrt{100} = 10 \qquad \text{\color{blue}Examples of square roots that are rational}$$

If n is a positive integer that is not the square of an integer, then $\sqrt{n}$ is an irrational number.

A calculator with a square root key can give approximations of square roots of numbers that are not perfect squares. We use the $\approx$ symbol to indicate "is approximately equal to." Sometimes, for convenience, the = symbol is used even if the statement is actually one of approximation, such as $\pi = 3.14$.

EXAMPLE 1 Using a Calculator to Approximate Square Roots

Use a calculator to verify the following approximations.

(a) $\sqrt{2} \approx 1.414213562$ **(b)** $\sqrt{6} \approx 2.449489743$ **(c)** $\sqrt{1949} \approx 44.14748011$

Solution

Use a calculator, such as the ones found on today's smartphones, to verify these approximations. Depending on the model, fewer or more digits may be displayed, and because of different rounding procedures, final digits may differ slightly. ∎

EXAMPLE 2 Applying a Formula for Pipe Flow

The following formula gives the number N of smaller pipes of diameter d that are necessary to supply the same total flow as one larger pipe of diameter D.

$$N = \sqrt{\left(\frac{D}{d}\right)^5}$$

This formula takes into account the extra friction caused by the smaller pipes. Use this formula to determine the number of $\frac{1}{2}$-inch pipes that will provide the same flow as one $1\frac{1}{2}$-inch pipe. (*Source:* Saunders, Hal M., and Robert A. Carman. *Mathematics for the Trades—A Guided Approach.* Pearson, 2015.)

Solution

Let $d = \frac{1}{2}$ and $D = 1\frac{1}{2} = \frac{3}{2}$ in the formula.

$$N = \sqrt{\left(\frac{\frac{3}{2}}{\frac{1}{2}}\right)^5} \qquad d = \tfrac{1}{2},\, D = \tfrac{3}{2}$$

$$= \sqrt{3^5} \qquad \tfrac{3}{2} \div \tfrac{1}{2} = \tfrac{3}{2} \cdot \tfrac{2}{1} = 3$$

$$= \sqrt{243} \qquad 3^5 = 243$$

$$= 15.59 \qquad \text{Approximate with a calculator.}$$

In this case the answer must be a whole number, so we round 15.59 up to 16. A total of 16 pipes measuring $\frac{1}{2}$ inch will be needed. ∎

We will now look at some simple operations with square roots. Notice that

$$\sqrt{4} \cdot \sqrt{9} = 2 \cdot 3 = 6$$

and

$$\sqrt{4 \cdot 9} = \sqrt{36} = 6.$$

Thus, $\sqrt{4} \cdot \sqrt{9} = \sqrt{4 \cdot 9}$. This is a particular case of the following rule.

PRODUCT RULE FOR SQUARE ROOTS

For nonnegative real numbers a and b, the following is true.

$$\sqrt{a} \cdot \sqrt{b} = \sqrt{a \cdot b}$$

Just as every rational number $\frac{a}{b}$ can be written in *lowest terms* (by using the fundamental property of rational numbers), every square root radical has a *simplified form*.

CONDITIONS FOR A SIMPLIFIED SQUARE ROOT RADICAL

A square root radical is in **simplified form** if the following three conditions are met.

1. The number under the radical (**radicand**) has no factor (except 1) that is a perfect square.

2. The radicand has no fractions.

3. No denominator contains a radical.

Tsu Ch'ung-chih (about A.D. 500), the Chinese mathematician honored on the above stamp, investigated the digits of π. **Aryabhata,** his Indian contemporary, gave 3.1416 as the value.

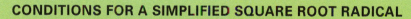

EXAMPLE 3 **Simplifying a Square Root Radical (Product Rule)**

Simplify $\sqrt{27}$.

Solution

Since 9 is a factor of 27, and 9 is a perfect square, $\sqrt{27}$ is not in simplified form. The first condition for simplified form is not met. We simplify as follows.

$$\sqrt{27} = \sqrt{9 \cdot 3}$$
$$= \sqrt{9} \cdot \sqrt{3} \qquad \text{Use the product rule.}$$
$$= 3\sqrt{3} \qquad \sqrt{9} = 3 \text{ because } 3^2 = 9. \qquad \blacksquare$$

Simplified form

Expressions such as $\sqrt{27}$ and $3\sqrt{3}$ represent the *exact value* of the square root of 27. If we use the square root key of a calculator, we find

$$\sqrt{27} \approx 5.196152423.$$

If we find $\sqrt{3}$ and then multiply the result by 3, we obtain

$$3\sqrt{3} \approx 3(1.732050808) \approx 5.196152423.$$

These approximations are the same, as we would expect. The work in **Example 3** provides the mathematical justification that they are indeed equal.

QUOTIENT RULE FOR SQUARE ROOTS

For nonnegative real numbers a and positive real numbers b, the following is true.

$$\frac{\sqrt{a}}{\sqrt{b}} = \sqrt{\frac{a}{b}}$$

EXAMPLE 4 **Simplifying Square Root Radicals (Quotient Rule)**

Simplify each radical.

(a) $\sqrt{\dfrac{25}{9}}$ **(b)** $\sqrt{\dfrac{3}{4}}$ **(c)** $\sqrt{\dfrac{1}{2}}$

Solution

(a) The radicand contains a fraction, so the radical expression is not simplified.

$$\sqrt{\frac{25}{9}} = \frac{\sqrt{25}}{\sqrt{9}} = \frac{5}{3} \qquad \text{Use the quotient rule.}$$

(b) $\sqrt{\dfrac{3}{4}} = \dfrac{\sqrt{3}}{\sqrt{4}} = \dfrac{\sqrt{3}}{2}$ Use the quotient rule.

(c) $\sqrt{\dfrac{1}{2}} = \dfrac{\sqrt{1}}{\sqrt{2}} = \dfrac{1}{\sqrt{2}}$ This is not yet simplified.

Condition 3 for simplified form is not met. To find an equivalent expression with no radical in the denominator, we **rationalize the denominator.**

Carl Gauss Square roots of negative numbers were called **imaginary numbers** by early mathematicians. Eventually the symbol *i* came to represent the **imaginary unit** $\sqrt{-1}$, and numbers of the form *a* + *bi*, where *a* and *b* are real numbers, were named **complex numbers.**

In about 1831, **Carl Gauss** was able to show that numbers of the form *a* + *bi* can be represented as points on the plane just as real numbers are. He shared this contribution with **Robert Argand**, a bookkeeper in Paris, who wrote an essay on the geometry of the complex numbers in 1806. This went unnoticed at the time.

The complex number *i* has the property that its whole number powers repeat in a cycle, with four values.

$$i^0 = 1 \quad i^1 = i \quad i^2 = -1 \quad i^3 = -i$$

The pattern continues on and on this way.

To rationalize the denominator here, we multiply $\frac{1}{\sqrt{2}}$ by $\frac{\sqrt{2}}{\sqrt{2}}$, which is a form of 1, the identity element for multiplication.

$$\frac{1}{\sqrt{2}} = \frac{1}{\sqrt{2}} \cdot \frac{\sqrt{2}}{\sqrt{2}} = \frac{\sqrt{2}}{2} \qquad \sqrt{2} \cdot \sqrt{2} = 2$$

> This is the simplified form of $\sqrt{\frac{1}{2}}$.

Is $\sqrt{4} + \sqrt{9} = \sqrt{4 + 9}$ a true statement? The answer is *no*, because

$$\sqrt{4} + \sqrt{9} = 2 + 3 = 5, \quad \text{while} \quad \sqrt{4 + 9} = \sqrt{13}, \quad \text{and} \quad 5 \neq \sqrt{13}.$$

Square root radicals may be combined, however, if they have the same radicand. Such radicals are called **like radicals.** We add (and subtract) like radicals using the distributive property.

EXAMPLE 5 **Adding and Subtracting Square Root Radicals**

Add or subtract as indicated.

(a) $3\sqrt{6} + 4\sqrt{6}$ **(b)** $\sqrt{18} - \sqrt{32}$

Solution

(a) Since both terms contain $\sqrt{6}$, they are like radicals, and may be combined.

$$3\sqrt{6} + 4\sqrt{6} = (3 + 4)\sqrt{6} \qquad \text{Distributive property}$$

$$= 7\sqrt{6} \qquad \text{Add.}$$

(b) If we simplify $\sqrt{18}$ and $\sqrt{32}$, then this operation can be performed.

$$\sqrt{18} - \sqrt{32} = \sqrt{9 \cdot 2} - \sqrt{16 \cdot 2} \qquad \text{Factor so that perfect squares are in the radicands.}$$

$$= \sqrt{9} \cdot \sqrt{2} - \sqrt{16} \cdot \sqrt{2} \qquad \text{Product rule}$$

$$= 3\sqrt{2} - 4\sqrt{2} \qquad \text{Take square roots.}$$

$$= (3 - 4)\sqrt{2} \qquad \text{Distributive property}$$

$$= -1\sqrt{2} \qquad \text{Subtract.}$$

$$= -\sqrt{2} \qquad -1 \cdot a = -a$$

Like radicals may be added or subtracted by adding or subtracting their coefficients (the numbers by which they are multiplied) and keeping the same radical.

Examples: $9\sqrt{7} + 8\sqrt{7} = 17\sqrt{7}$ (because $9 + 8 = 17$)

$4\sqrt{3} - 12\sqrt{3} = -8\sqrt{3}$ (because $4 - 12 = -8$)

The Irrational Numbers π, ϕ, and *e*

Figure 14 shows approximations for three important irrational numbers. The first, π, represents the ratio of the circumference of a circle to its diameter. The second, ϕ, is the Golden Ratio. Its exact value is $\frac{1 + \sqrt{5}}{2}$. The third is *e*, a fundamental number in our universe. It is the base of the *natural exponential* and *natural logarithmic* functions. The letter *e* was chosen to honor Leonhard Euler, who published extensive research on the number in 1746.

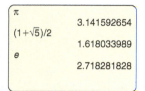

π	3.141592654
$(1+\sqrt{5})/2$	1.618033989
e	2.718281828

Figure 14

This poem, dedicated to **Archimedes** ("the immortal Syracusan"), allows us to learn the first 31 digits of the decimal representation of π. By replacing each word with the number of letters it contains, with a decimal point following the initial 3, the decimal is found. The poem was written by A. C. Orr, and it appeared in the *Literary Digest* in 1906.

> *Now I, even I, would celebrate*
> *In rhymes unapt, the great*
> *Immortal Syracusan, rivaled nevermore,*
> *Who in his wondrous lore*
> *Passed on before,*
> *Left men his guidance*
> *How to circles mensurate.*

Pi (π)

$$\pi \approx 3.14159265358979323846264338327\,9$$

The computation of the digits of π has fascinated mathematicians since ancient times. Archimedes was the first to explore it extensively. As of August 5, 2014, its value had been computed to 12 trillion decimal places. Some of today's foremost researchers of the digits of π are Yasumasa Kanada, Gregory and David Chudnovsky, and the team of Alexander J. Yee and Shigeru Kondo.

A History of π, by Petr Beckmann, is a classic that is now in its third edition. Numerous Web sites, including the following, are devoted to the history and methods of computation of pi.

www.joyofpi.com/

www.math.utah.edu/~alfeld/Archimedes/Archimedes.html

www.super-computing.org/

www.pbs.org/wgbh/nova/sciencenow/3210/04.html

www.numberworld.org/misc_runs/pi-5t/details.html

One of the methods of computing pi involves the topic of *infinite series*.

▌EXAMPLE 6   Computing the Digits of Pi Using an Infinite Series

It is shown in higher mathematics that the *infinite series*

$$1 - \frac{1}{3} + \frac{1}{5} - \frac{1}{7} + \frac{1}{9} + \ldots \text{ "converges" to } \frac{\pi}{4}.$$

That is, as more and more terms are considered, its value becomes closer and closer to $\frac{\pi}{4}$. With a calculator, approximate the value of pi using twenty-one terms of this series.

Solution

Figure 15 shows the necessary calculation on the TI-83/84 Plus calculator. The sum of the first twenty-one terms is multiplied by 4, to obtain the approximation

$$3.189184782.$$

This series converges slowly. Although this is correct only to the first decimal place, better approximations are obtained using more terms of the series. ∎

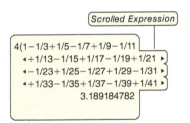

Scrolled Expression

4(1−1/3+1/5−1/7+1/9−1/11
◄+1/13−1/15+1/17−1/19+1/21 ►
◄−1/23+1/25−1/27+1/29−1/31 ►
◄+1/33−1/35+1/37−1/39+1/41 ►
3.189184782

Figure 15

A rectangle that satisfies the condition that the ratio of its length to its width is equal to the ratio of the sum of its length and width to its length is called a **Golden Rectangle**. See **Figure 16**. This ratio is called the **Golden Ratio**. Its exact value is the irrational number $\frac{1 + \sqrt{5}}{2}$. It is represented by the Greek letter ϕ (phi).

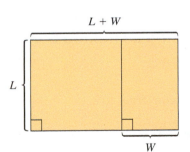

$L + W$

L

W

In a golden rectangle,

$$\frac{L + W}{L} = \frac{L}{W}.$$

Figure 16

Phi (ϕ)

$$\phi = \frac{1 + \sqrt{5}}{2} \approx 1.6180339887498948482045868343\,65$$

In 1767 **J. H. Lambert** proved that π is irrational (and thus its decimal will never terminate and never repeat). Nevertheless, the 1897 Indiana state legislature considered a bill that would have *legislated* the value of π. In one part of the bill, the value was stated to be 4, and in another part, 3.2. Amazingly, the bill passed the House, but the Senate postponed action on the bill indefinitely.

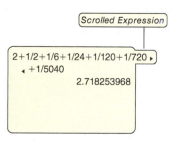

233/144	
	1.618055556
377/233	
	1.618025751
610/377	
	1.618037135

Figure 17

Two books on phi are *The Divine Proportion, A Study in Mathematical Beauty,* by H. E. Huntley, and the more recent *The Golden Ratio,* by Mario Livio. Web sites devoted to this irrational number include the following.

www.mcs.surrey.ac.uk/Personal/R.Knott/Fibonacci/

www.goldennumber.net/

www.mathforum.org/dr.math/faq/faq.golden.ratio.html

www.geom.uiuc.edu/~demo5337/s97b/art.htm

EXAMPLE 7 Computing the Digits of Phi Using the Fibonacci Sequence

The first twelve terms of the Fibonacci sequence are

$$1, 1, 2, 3, 5, 8, 13, 21, 34, 55, 89, 144.$$

Each term after the first two terms is obtained by adding the two previous terms. Thus, the thirteenth term is $89 + 144 = 233$. As one goes farther out in the sequence, the ratio of a term to its predecessor gets closer to ϕ. How far out must one go in order to approximate ϕ so that the first five decimal places agree?

Solution

After 144, the next three Fibonacci numbers are 233, 377, and 610. **Figure 17** shows that $\frac{610}{377} \approx 1.618037135$, which agrees with ϕ to the fifth decimal place. ∎

The irrational number e is a fundamental constant in mathematics.

e
$e \approx 2.718281828459045235360287471353$

Because of its nature, e is less well understood by the layperson than π (or even ϕ, for that matter). The 1994 book *e: The Story of a Number,* by Eli Maor, attempted to rectify this situation. The following Web sites also give information on e.

www.mathforum.org/dr.math/faq/faq.e.html

www-groups.dcs.st-and.ac.uk/~history/HistTopics/e.html

http://antwrp.gsfc.nasa.gov/htmltest/gifcity/e.1mil

www.math.toronto.edu/mathnet/answers/ereal.html

Example 8 illustrates another infinite series, this one converging to e.

EXAMPLE 8 Computing the Digits of e Using an Infinite Series

The infinite series

$$2 + \frac{1}{1 \cdot 2} + \frac{1}{1 \cdot 2 \cdot 3} + \frac{1}{1 \cdot 2 \cdot 3 \cdot 4} + \ldots \text{ converges to } e.$$

Use a calculator to approximate e using the first seven terms of this series.

Solution

Figure 18 shows the sum of the first seven terms. (The denominators have all been multiplied out.) The sum is 2.718253968, which agrees with e to four decimal places. This series converges more rapidly than the one for π in **Example 6.** ∎

Scrolled Expression

$2+1/2+1/6+1/24+1/120+1/720$ ►

◄ $+1/5040$

2.718253968

Figure 18

6.4 EXERCISES

Identify each number as rational *or* irrational.

1. $\dfrac{4}{9}$ **2.** $\dfrac{7}{8}$ **3.** $\sqrt{10}$ **4.** $\sqrt{14}$

5. 1.618   **6.** 2.718   **7.** $0.\overline{41}$ **8.** $0.\overline{32}$

9. π **10.** e **11.** 3.14159   **12.** $\dfrac{22}{7}$

13. 0.8787787778877778 . . .   **14.** $\dfrac{1 + \sqrt{5}}{2}$

In Exercises 15 and 16, work parts (a) and (b) in order.

15. (a) Find the sum.   0.272772777277772 . . .

$\qquad\qquad\qquad\qquad$ +0.616116111611116 . . .

(b) Based on the result of part (a), we can conclude that the sum of two _____ numbers may be a(n) _____ number.

16. (a) Find the sum.   0.010110111011110 . . .

$\qquad\qquad\qquad\qquad$ +0.252552555255552 . . .

(b) Based on the result of part (a), we can conclude that the sum of two _____ numbers may be a(n) _____ number.

Use a calculator to find a rational decimal approximation for each irrational number.

17. $\sqrt{39}$ **18.** $\sqrt{44}$ **19.** $\sqrt{15.1}$ **20.** $\sqrt{33.6}$

21. $\sqrt{884}$ **22.** $\sqrt{643}$ **23.** $\sqrt{\dfrac{9}{8}}$ **24.** $\sqrt{\dfrac{6}{5}}$

25. Plumbing Use the formula from **Example 2,**

$$N = \sqrt{\left(\dfrac{D}{d}\right)^5},$$

to find the number of $\tfrac{3}{4}$-inch pipes that would be necessary to provide the same total flow as a single $1\tfrac{1}{2}$-inch pipe.

26. Plumbing Use the formula from **Example 2** to find the number of $\tfrac{5}{8}$-inch pipes that would be necessary to provide the same total flow as a single 2-inch pipe.

Solve each problem. Use a calculator as necessary, and give approximations to the nearest tenth unless specified otherwise.

27. Allied Health Body surface area (BSA) is used in the allied health field to calculate proper dosages of drugs based on a patient's height and weight. The formula

$$\text{BSA} = \sqrt{\dfrac{\text{weight in kg} \times \text{height in cm}}{3600}}$$

will give the patient's BSA in square meters. (*Source:* Saunders, Hal M., and Robert A. Carman. *Mathematics for the Trades—A Guided Approach, Tenth Edition.* Pearson, 2015.) What is the BSA for a person who weighs 100 kilograms and is 200 centimeters tall?

28. Allied Health Use the formula from **Exercise 27** to find the BSA of a person who weighs 94 kilograms and is 190 centimeters tall.

29. Diagonal of a Box The length of the diagonal of a box is given by

$$D = \sqrt{L^2 + W^2 + H^2},$$

where L, W, and H are the length, width, and height of the box. Find the length of the diagonal, D, of a box that is 4 feet long, 3 feet wide, and 2 feet high.

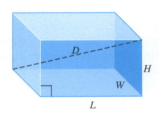

30. Distance to the Horizon A meteorologist provided this answer to one of the readers of his newspaper column, a 6-foot man who lives 150 feet above the ground.

To find the distance to the horizon in miles, take the square root of the height of your view and multiply that result by 1.224. That will give you the number of miles to the horizon.

Assuming the viewer's eyes are 6 feet above his floor, the total height from the ground is $150 + 6 = 156$ feet. How far can he see to the horizon?

31. Period of a Pendulum The period of a pendulum, in seconds, depends on its length, L, in feet, and is given by the formula

$$P = 2\pi\sqrt{\dfrac{L}{32}}.$$

If a pendulum is 5.1 feet long, what is its period? Use 3.14 for π.

32. Radius of an Aluminum Can The radius of the circular top or bottom of an aluminum can with surface area S in square cm and height h in cm is given by

$$r = \dfrac{-h + \sqrt{h^2 + 0.64S}}{2}.$$

What radius should be used to make a can with height 12 cm and surface area 400 square cm?

33. Electronics Formula The formula

$$I = \sqrt{\frac{2P}{L}}$$

relates the coefficient of self-induction L (in henrys), the energy P stored in an electronic circuit (in joules), and the current I (in amps). Find the value of I if $P = 120$ joules and $L = 80$ henrys.

34. Law of Tensions In the study of sound, one version of the law of tensions is

$$f_1 = f_2\sqrt{\frac{F_1}{F_2}}.$$

Find f_1 to the nearest unit if $F_1 = 300$, $F_2 = 60$, and $f_2 = 260$.

Accident Reconstruction *Police sometimes use the following procedure to estimate the speed at which a car was traveling at the time of an accident. A police officer drives the car involved in the accident under conditions similar to those during which the accident took place and then skids to a stop. If the car is driven at 30 miles per hour, then the speed at the time of the accident is given by*

$$s = 30\sqrt{\frac{a}{p}}.$$

Here, a is the length of the skid marks left at the time of the accident, and p is the length of the skid marks in the police test. Find s, to the nearest unit, for the following values of a and p.

35. $a = 862$ feet; $p = 156$ feet

36. $a = 382$ feet; $p = 96$ feet

37. $a = 84$ feet; $p = 26$ feet

38. $a = 90$ feet; $p = 35$ feet

39. Area of the Bermuda Triangle **Heron's formula** gives a method of finding the area of a triangle if the lengths of its sides are known. Suppose that a, b, and c are the lengths of the sides. Let s denote one-half of the perimeter of the triangle (called the **semiperimeter**); that is,

$$s = \frac{1}{2}(a + b + c).$$

Then the area $\mathcal{A}$ of the triangle is given by

$$\mathcal{A} = \sqrt{s(s - a)(s - b)(s - c)}.$$

Find the area of the Bermuda Triangle, if the "sides" of this triangle measure approximately 850 miles, 925 miles, and 1300 miles. Give your answer to the nearest thousand square miles.

40. Area Enclosed by the Vietnam Veterans' Memorial The Vietnam Veterans' Memorial in Washington, D.C., is in the shape of an unenclosed isosceles triangle with equal sides of length 246.75 feet. If the triangle were enclosed, the third side would have length 438.14 feet. Use Heron's formula from the previous exercise to find the area of this enclosure to the nearest hundred square feet. (*Source:* Information pamphlet obtained at the Vietnam Veterans' Memorial.)

41. Perfect Triangles A **perfect triangle** is a triangle whose sides have whole number lengths and whose area is numerically equal to its perimeter. Use Heron's formula to show that the triangle with sides of length 9, 10, and 17 is perfect.

42. Heron Triangles A **Heron triangle** is a triangle having integer sides and area. Use Heron's formula to show that each of the following is a Heron triangle.

(a) $a = 11, b = 13, c = 20$

(b) $a = 13, b = 14, c = 15$

(c) $a = 7, b = 15, c = 20$

*Use the methods of **Examples 3 and 4** to simplify each expression. Then, use a calculator to approximate both the given expression and the simplified expression. (Both should be the same.)*

43. $\sqrt{50}$ **44.** $\sqrt{32}$ **45.** $\sqrt{75}$

46. $\sqrt{150}$ **47.** $\sqrt{288}$ **48.** $\sqrt{200}$

49. $\dfrac{5}{\sqrt{6}}$ **50.** $\dfrac{3}{\sqrt{2}}$

51. $\sqrt{\dfrac{7}{4}}$ **52.** $\sqrt{\dfrac{8}{9}}$

53. $\sqrt{\dfrac{7}{3}}$ **54.** $\sqrt{\dfrac{14}{5}}$

*Use the method of **Example 5** to perform the indicated operations.*

55. $\sqrt{17} + 2\sqrt{17}$ **56.** $3\sqrt{19} + \sqrt{19}$

57. $5\sqrt{7} - \sqrt{7}$ **58.** $3\sqrt{27} - \sqrt{27}$

59. $3\sqrt{18} + \sqrt{2}$ **60.** $2\sqrt{48} - \sqrt{3}$

61. $-\sqrt{12} + \sqrt{75}$ **62.** $2\sqrt{27} - \sqrt{300}$

Irrational Investigations *Exercises 63–78 deal with the irrational numbers* π, ϕ, e, *and* $\sqrt{3}$. *Use a calculator or computer as necessary.*

63. Move one matchstick to make the equation approximately true. (*Source:* www.joyofpi.com)

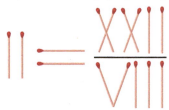

64. Find the square root of $\frac{2143}{22}$ using a calculator. Then find the square root of that result. Compare your result to the decimal given for π in the margin note. What do you notice?

65. Use a calculator to find the first eight digits in the decimal for $\frac{355}{113}$. Compare the result to the decimal for π given in the text. What do you notice?

66. You may have seen the statements

"Use $\frac{22}{7}$ for π"   and   "Use 3.14 for π."

Since $\frac{22}{7}$ is the quotient of two integers, and 3.14 is a terminating decimal, do these statements suggest that π is rational?

67. In the Bible (I Kings 7:23), a verse describes a circular pool at King Solomon's temple, about 1000 B.C. The pool is said to be ten cubits across, "and a line of 30 cubits did compass it round about." What value of π does this imply?

68. The ancient Egyptians used a method for finding the area of a circle that is equivalent to a value of 3.1605 for π. Write this decimal as a mixed number.

69. The computation of π has fascinated mathematicians and others for centuries. In the nineteenth century, the British mathematician William Shanks spent many years of his life calculating π to 707 decimal places. It turned out that only the first 527 were correct. Use an Internet search to find the 528th decimal digit of π (following the whole number part 3).

70. A **mnemonic device** is a scheme whereby one is able to recall facts by memorizing something completely unrelated to the facts. One way of learning the first few digits of the decimal for π is to memorize a sentence (or several sentences) and count the letters in each word of the sentence. For example,

"See, I know a digit,"

will give the first 5 digits of π: "See" has 3 letters, "I" has 1 letter, "know" has 4 letters, "a" has 1 letter, and "digit" has 5 letters. So the first five digits are 3.1415.

Verify that the following mnemonic devices work.

(a) "May I have a large container of coffee?"

(b) "See, I have a rhyme assisting my feeble brain, its tasks ofttimes resisting."

(c) "How I want a drink, alcoholic of course, after the heavy lectures involving quantum mechanics."

71. In the second season of the original *Star Trek* series, the episode "Wolf in the Fold" told the story of an alien entity that had taken over the computer of the starship *Enterprise*. To drive the entity out of the computer, Mr. Spock gave the alien the compulsory directive to compute π to the last digit. Explain why this strategy proved successful.

72. *Northern Exposure* ran between 1990 and 1995 on the CBS network. In the episode "Nothing's Perfect," the local disc jockey Chris Stevens meets and develops a relationship with a mathematician after accidentally running over her dog. Her area of research is computation of the decimal digits of pi. She mentions that a string of eight 8s appears in the decimal relatively early in the expansion. Use an Internet search to determine the position at which this string appears.

73. Use a calculator to find the decimal approximations for

$$\phi = \frac{1 + \sqrt{5}}{2} \quad \text{and its } \textbf{conjugate,} \quad \frac{1 - \sqrt{5}}{2}.$$

Comment on the similarities and differences in the two decimals.

74. In some literature, the Golden Ratio is defined to be the reciprocal of

$$\frac{1 + \sqrt{5}}{2}, \quad \text{which is} \quad \frac{2}{1 + \sqrt{5}}.$$

Use a calculator to find a decimal approximation for $\frac{2}{1 + \sqrt{5}}$, and compare it to ϕ as defined in this text. What do you observe?

75. Near the end of the 2008 movie *Harold & Kumar Escape from Guantanamo Bay*, Kumar (Kal Penn), recites a poem dealing with the square root of 3, written by the late David Feinberg. The text of the poem can be found on the Internet by searching

"David Feinberg Square Root of 3."

There are references to irrational numbers, integers, an approximation for the square root of 3, and the product rule for square root radicals. Why would Kumar prefer the 3 to be a 9?

76. See **Exercise 75.** What is the decimal approximation for $\sqrt{3}$ (to 4 decimal places) given in Feinberg's poem?

77. An approximation for e is

$$2.718281828.$$

A student noticed that there seems to be a repetition of four digits in this number (1, 8, 2, 8) and concluded that e is rational, because repeating decimals represent rational numbers. Was the student correct? Why or why not?

78. Use a calculator with an exponential key to find values for the following:

$$(1.1)^{10}, \quad (1.01)^{100}, \quad (1.001)^{1000},$$

$$(1.0001)^{10,000}, \quad \text{and} \quad (1.00001)^{100,000}.$$

Compare your results to the approximation given for e in this section. What do you find?

Roots Other Than Square Roots   *The concept of square (second) root can be extended to* **cube (third) root, fourth root,** *and so on. For example,*

$$\sqrt[3]{8} = 2 \quad because \quad 2^3 = 8,$$

$$\sqrt[3]{1000} = 10 \quad because \quad 10^3 = 1000,$$

$$\sqrt[4]{81} = 3 \quad because \quad 3^4 = 81, \quad and\ so\ on.$$

If $n \geq 2$ and a is a nonnegative number, then $\sqrt[n]{a}$ represents the nonnegative number whose nth power is a.

Find each root.

79. $\sqrt[3]{64}$ **80.** $\sqrt[3]{125}$

81. $\sqrt[3]{343}$ **82.** $\sqrt[3]{729}$

83. $\sqrt[3]{216}$ **84.** $\sqrt[3]{512}$

85. $\sqrt[4]{1}$ **86.** $\sqrt[4]{16}$

87. $\sqrt[4]{256}$ **88.** $\sqrt[4]{625}$

89. $\sqrt[4]{4096}$ **90.** $\sqrt[4]{2401}$

Use a calculator to approximate each root. (Hint: To find the fourth root, find the square root of the square root.)

91. $\sqrt[3]{43}$ **92.** $\sqrt[3]{87}$

93. $\sqrt[3]{198}$ **94.** $\sqrt[4]{2107}$

95. $\sqrt[4]{10,265.2}$ **96.** $\sqrt[4]{863.5}$

97. $\sqrt[4]{968.1}$ **98.** $\sqrt[4]{12,966.4}$

6.5   APPLICATIONS OF DECIMALS AND PERCENTS

OBJECTIVES

1 Perform operations of arithmetic with decimal numbers.

2 Round whole numbers and decimals to a given place value.

3 Perform computations using percent.

4 Convert among forms of fractions, decimals, and percents.

5 Find percent increase and percent decrease.

6 Apply formulas involving fractions and decimals from the allied health industry.

Operations with Decimals

Because calculators have, for the most part, replaced paper-and-pencil methods for operations with decimals and percent, we will only briefly mention these latter methods. *We strongly suggest that the work in this section be done with a calculator at hand.*

ADDITION AND SUBTRACTION OF DECIMALS

To add or subtract decimal numbers, line up the decimal points in a column and perform the operation.

EXAMPLE 1   **Adding and Subtracting Decimal Numbers**

Find each of the following.

(a) $0.46 + 3.9 + 12.58$ **(b)** $12.1 - 8.723$

Solution

(a)
$$\begin{array}{r} 0.46 \\ 3.90 \\ +12.58 \\ \hline 16.94 \end{array}$$
Line up decimal points.
Attach a zero as a placeholder.
← Sum

(b)
$$\begin{array}{r} 12.100 \\ -8.723 \\ \hline 3.377 \end{array}$$
Attach zeros.
← Difference

Recall that when two numbers are multiplied, the numbers are **factors** and the answer is the **product.** When two numbers are divided, the number being divided is the **dividend,** the number doing the dividing is the **divisor,** and the answer is the **quotient.**

MULTIPLICATION AND DIVISION OF DECIMALS

Multiplication   To multiply decimals, multiply in the same manner as integers are multiplied. The number of decimal places to the right of the decimal point in the product is the *sum* of the numbers of places to the right of the decimal points in the factors.

Division   To divide decimals, move the decimal point to the right the same number of places in the divisor and the dividend so as to obtain a whole number in the divisor. Divide in the same manner as integers are divided. The number of decimal places to the right of the decimal point in the quotient is the same as the number of places to the right of the decimal in the dividend.

EXAMPLE 2 Multiplying and Dividing Decimal Numbers

Find each of the following.

(a) 4.613×2.52 **(b)** $65.175 \div 8.25$

Solution

(a)

$$
\begin{array}{r}
4.613 \quad \leftarrow \text{3 decimal places} \\
\times \quad 2.52 \quad \leftarrow \text{2 decimal places} \\
\hline
9226 \\
23065 \\
9226 \\
\hline
11.62476 \quad \leftarrow 3 + 2 = \text{5 decimal places}
\end{array}
$$

(b)

$8.25\overline{)65.175} \rightarrow$

$$
\begin{array}{r}
7.9 \\
825\overline{)6517.5} \\
5775 \\
\hline
7425 \\
7425 \\
\hline
0
\end{array}
$$

Bring the decimal point straight up in the answer.

Rounding Methods

To round, or approximate, a whole number or a decimal number to a given place value, use the following procedure. (You may wish to refer to the diagram in **Section 6.3** that describes place values determined by powers of 10.)

RULES FOR ROUNDING

Whole Numbers

Step 1   Locate the **place** to which the whole number is being rounded.

Step 2   Look at the next **digit to the right** of the place to which the number is being rounded.

Step 3A   If the digit to the right is **less than 5,** replace it and all digits following it with zeros. Do not change the digit in the place to which the number is being rounded.

Step 3B   If the digit to the right is **5 or greater,** replace it and all digits following it with zeros, but also add 1 to the digit in the place to which the number is being rounded. (If adding 1 to a 9, replace the 9 with 0 and add 1 to the next digit to the left. If that next digit is also 9, repeat this procedure; and so on, to the left.)

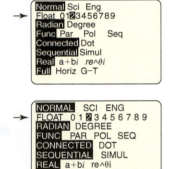

"**Technology** pervades the world outside school. There is no question that students will be expected to use calculators in other settings; this technology is now part of our culture . . . students no longer have the same need to perform these [paper-and-pencil] procedures with large numbers of lengthy expressions that they might have had in the past without ready access to technology."

From *Computation, Calculators, and Common Sense (A Position Paper of the National Council of Teachers of Mathematics).*

RULES FOR ROUNDING

Decimal Numbers

Steps 1 and 2 are the same as for whole numbers.

Step 3A   If the digit to the right is **less than 5,** drop all digits to the right of the place to which the number is being rounded. Do not change the digit in the place to which it is being rounded.

Step 3B   If the digit to the right is **5 or greater,** drop it and all digits following it, but also add 1 to the digit in the place to which the number is being rounded. (When adding 1 to a 9, use the same procedure as for whole numbers.)

Note: Some disciplines have guidelines specifying that in certain cases, a *downward* roundoff is made. We will not investigate these.

EXAMPLE 3 Rounding Numbers

Round to the place values indicated.

(a) 0.9175 to the nearest tenth, hundredth, and thousandth

(b) 1358 to the nearest thousand, hundred, and ten

Solution

(a) Round 0.9175 as follows: 0.9 to the nearest tenth,
0.92 to the nearest hundredth,
0.918 to the nearest thousandth.

(b) Round 1358 as follows: 1000 to the nearest thousand,
1400 to the nearest hundred,
1360 to the nearest ten.   ■

The calculator rounds 0.9175 to the nearest hundredth.

The calculator rounds 0.9175 *up* to 0.918.

Percent

The word **percent** means **"per hundred."** The symbol % represents "percent."

PERCENT

$$1\% = \frac{1}{100} = 0.01$$

EXAMPLE 4 Converting Percents to Decimals

Convert each percent to a decimal.

(a) 98%   **(b)** 3.4%   **(c)** 0.2%   **(d)** 150%

Solution

(a) $98\% = 98(1\%) = 98(0.01) = 0.98$

(b) $3.4\% = 3.4(1\%) = 3.4(0.01) = 0.034$

(c) $0.2\% = 0.2(1\%) = 0.2(0.01) = 0.002$

(d) $150\% = 150(1\%) = 150(0.01) = 1.5$ ■

EXAMPLE 5 **Converting Decimals to Percents**

Convert each decimal to a percent.

(a) 0.13   **(b)** 0.532   **(c)** 2.3   **(d)** 0.07

Solution

Reverse the procedure used in **Example 4.**

(a) $0.13 = 13(\mathbf{0.01}) = 13(\mathbf{1\%}) = 13\%$

(b) $0.532 = 53.2(0.01) = 53.2(1\%) = 53.2\%$

(c) $2.3 = 230(0.01) = 230(1\%) = 230\%$

(d) $0.07 = 7(0.01) = 7(1\%) = 7\%$ ∎

Examples 4 and 5 suggest the following shortcut methods for converting.

CONVERTING BETWEEN DECIMALS AND PERCENTS

To convert a percent to a decimal, drop the percent symbol (%) and move the decimal point two places to the left, inserting zeros as placeholders if necessary.

To convert a decimal to a percent, move the decimal point two places to the right, inserting zeros as placeholders if necessary, and attach the percent symbol (%).

EXAMPLE 6 **Converting Fractions to Percents**

Convert each fraction to a percent.

(a) $\dfrac{3}{5}$ **(b)** $\dfrac{14}{25}$ **(c)** $2\dfrac{7}{10}$

Solution

(a) First write $\dfrac{3}{5}$ as a decimal. Dividing 3 by 5 gives $\dfrac{3}{5} = 0.6 = 60\%$.

(b) $\dfrac{14}{25} = \dfrac{14 \cdot \mathbf{4}}{25 \cdot \mathbf{4}} = \dfrac{56}{100} = 0.56 = 56\%$

(c) $2\dfrac{7}{10} = 2.7 = 270\%$ ∎

CONVERTING A FRACTION TO A PERCENT

To convert a fraction to a percent, convert the fraction to a decimal, and then convert the decimal to a percent.

The percent symbol, %, probably evolved from a symbol introduced in an Italian manuscript of 1425. Instead of "per 100," "P 100," or "P cento," which were common at that time, the author used "Pℝ." By about 1650 the ℝ had become $\frac{0}{0}$, so "per $\frac{0}{0}$" was often used. Finally the "per" was dropped, leaving $\frac{0}{0}$ or %.

(*Source: Historical Topics for the Mathematics Classroom,* the Thirty-first Yearbook of the National Council of Teachers of Mathematics, 1969.)

In the following examples involving percents, three methods are shown. They illustrate some basic ideas of solving equations. The second method in each case involves using cross-products. The third method involves the percent key of a basic calculator. (Keystrokes may vary among models.)

EXAMPLE 7 Finding a Percent of a Number

A publisher requires that an author reduce the page count in her next edition by 18%. The current edition has 250 pages. How many pages must the author cut from her book?

Solution

This problem can be reworded as follows: Find 18% of 250.

Method 1   The key word "of" translates as "times."

$$18\%(250) = 0.18(250) = 45$$

Method 2   Think "18 is to 100 as what (x) is to 250?" This translates as follows.

$$\frac{18}{100} = \frac{x}{250}$$

$$100x = 18 \cdot 250 \qquad \tfrac{a}{b} = \tfrac{c}{d} \text{ leads to } ad = bc.$$

$$x = \frac{18 \cdot 250}{100} \qquad \text{Divide by 100.}$$

$$x = 45 \qquad \text{Simplify.}$$

Method 3   Use the percent key on a calculator with the following keystrokes.

Thus, 18% of 250 is 45. The author must reduce her page count by 45 pages.   ■

EXAMPLE 8 Finding What Percent One Number Is of Another

A kindergarten teacher has submitted 6 of the required 50 class preparations to the principal of her school. What percent of her requirement has she completed?

Solution

This problem can be reworded as follows: What percent of 50 is 6?

Method 1   Let the phrase "what percent" be represented by $x \cdot 1\%$ or $0.01x$. Again the word "of" translates as "times," while "is" translates as "equals."

$$0.01x \cdot 50 = 6$$

$$0.50x = 6 \qquad \text{Multiply on the left side.}$$

$$50x = 600 \qquad \text{Multiply by 100 to clear decimals.}$$

$$x = 12 \qquad \text{Divide by 50.}$$

Method 2   Think "What (x) is to 100 as 6 is to 50?"

$$\frac{x}{100} = \frac{6}{50}$$

$$50x = 600 \qquad \text{Cross-products}$$

$$x = 12 \qquad \text{Divide by 50.}$$

Method 3   Use the following keystrokes on a calculator.

Thus, 6 is 12% of 50. She has completed 12% of her requirement.   ■

WHEN   Will I   Ever USE   This   ?

Cynthia (not her real name) had graduated in fashion merchandising and worked for eight years in an upscale women's clothing store in New Orleans. She decided to leave her position to go back to school with the goal of becoming a veterinarian.

Her curriculum required a course in college algebra, and Cynthia finished with the highest average in her class. Along the way she realized something that would have saved her a lot of time in her previous job. One day her instructor (one of the authors) made an offhand comment about percent increase. He said something like this:

If you want to mark up an item by a given percent, such as 15%, all you need to do is multiply the original price by 1.15. That's your final amount.

After class, she approached him and admitted that when she marked up items for retail sale, as her first step she would find the amount of markup on her calculator, and then in a separate step add it to the wholesale cost of the item.

For example, if the cost was $25.80, she would calculate **15% of $25.80** on her calculator in the first step, to obtain **$3.87**. Then in her second step, she would add

$3.87 + $25.80   to obtain $29.67. ← Selling price

She only now realized that the selling price could be obtained in a single step by multiplying

$25.80 · 1.15   to obtain $29.67. ← Selling price

She realized that she could have saved time had she known this procedure.

The reason that this method works can be explained by the identity property of multiplication and the distributive property. The explanation below uses 15%, but any percent could be used with the same result.

Suppose that C represents the wholesale cost of an item. Then $0.15C$ represents 15% of C, the amount of the markup. The selling price S is found by adding wholesale cost to the amount of markup.

$S = C + 0.15C$	Selling price = wholesale cost + markup
$S = 1C + 0.15C$	Identity property of multiplication
$S = (1 + 0.15)C$	Distributive property
$S = 1.15C$	$1 + 0.15 = 1.15$

A similar situation holds if a discount is to be applied. For example, to mark an item down 25%, just subtract from 100% to get 75%, and then multiply the original amount by 0.75. The result is the discounted price.

Cynthia went on to get her doctor's degree in veterinary medicine and open her own practice in the New Orleans area.

EXAMPLE 9   Finding a Number That Is a Given Percent of a Given Number

A government employee working for the county judicial system chooses 5% of a jury pool to question for possible service. This amounts to 38 people. What is the size of the entire jury pool?

Solution

This problem can be reworded as follows: 38 is 5% of what number?

Method 1

$$38 = 0.05x \quad \text{Let } x \text{ represent the number.}$$

$$x = \frac{38}{0.05} \quad \text{Divide by 0.05.}$$

$$x = 760 \quad \text{Simplify.}$$

Method 2   Think "38 is to what number (x) as 5 is to 100?"

$$\frac{38}{x} = \frac{5}{100}$$

$$5x = 3800 \quad \text{Cross-products}$$

$$x = 760 \quad \text{Divide by 5.}$$

Method 3   Use the following keystrokes on a calculator.

$$\boxed{3}\ \boxed{8}\ \boxed{\div}\ \boxed{5}\ \boxed{\%} \qquad 760 \quad \text{◁ Final display}$$

Each method shows us that 38 is 5% of 760. There are 760 in the jury pool.   ■

Consumers often encounter figures in the media that involve **percent change,** which includes **percent increase** and **percent decrease.** The following guidelines are helpful in understanding these concepts.

FINDING PERCENT INCREASE OR DECREASE

1. To find the **percent increase from *a* to *b*,** where $b > a$, subtract *a* from *b*, and divide this result by *a*. Convert to a percent.

   Example: The percent increase from 4 to 7 is $\frac{7-4}{4} = \frac{3}{4} = 75\%$.

2. To find the **percent decrease from *a* to *b*,** where $b < a$, subtract *b* from *a*, and divide this result by *a*. Convert to a percent.

   Example: The percent decrease from 8 to 6 is $\frac{8-6}{8} = \frac{2}{8} = \frac{1}{4} = 25\%$.

EXAMPLE 10   Solving Problems about Percent Change

Solve each problem involving percent change.

(a) An electronics store marked up a laptop computer from the store's cost of $1200 to a selling price of $1464. What was the percent markup?

(b) The enrollment in a community college declined from 12,750 during one school year to 11,350 the following year. Find the percent decrease to the nearest tenth of a percent.

It's Time to End Decimal Point Abuse
The use of an erroneous decimal point with a ¢ symbol is seen almost on a daily basis. Think about it: $0.99 represents $\frac{99}{100}$ of a dollar, or 99 cents, while 99¢ also represents 99 cents (because ¢ is the symbol for *cent*). So what does .99¢ represent? That's right: $\frac{99}{100}$ of one cent!

Look at these photos and see if you can find the obvious errors.

Solution

(a) "Markup" is a name for an increase. Let x = the percent increase (as a decimal).

$$\text{percent increase} = \frac{\text{amount of increase}}{\text{original amount}}$$

Subtract to find the amount of increase.

$$x = \frac{1464 - 1200}{1200} \quad \text{Substitute the given values.}$$

Use the original cost.

$$x = \frac{264}{1200}$$

$$x = 0.22 \quad \text{Use a calculator.}$$

The computer was marked up 22%.

(b) Let x = the percent decrease (as a decimal).

$$\text{percent decrease} = \frac{\text{amount of decrease}}{\text{original amount}}$$

Subtract to find the amount of decrease.

$$x = \frac{12{,}750 - 11{,}350}{12{,}750} \quad \text{Substitute the given values.}$$

Use the original enrollment.

$$x = \frac{1400}{12{,}750}$$

$$x \approx 0.110 \quad \text{Use a calculator.}$$

The college enrollment decreased by about 11.0%. ■

When calculating a percent increase or a percent decrease, use the original number (**before** *the increase or decrease) as the base.* A common error is to use the final number (*after* the increase or decrease) in the denominator of the fraction. Applications of decimal numbers sometimes involve formulas.

EXAMPLE 11 Determining a Child's Body Surface Area

If a child weighs k kilograms, then the child's body surface area S in square meters (m²) is determined by the following formula.

$$S = \frac{4k + 7}{k + 90}$$

What is the body surface area of a child who weighs 40 pounds? (*Source:* Hegstad, Lorrie N., and Wilma Hayek. *Essential Drug Dosage Calculations,* 4th ed. Prentice Hall, 2001.)

Solution

Tables and Web sites for conversion between the metric and English systems are readily available. Using www.metric-conversions.org, we find that 40 pounds is approximately 18.144 kilograms. Use this value for k in the formula.

$$S = \frac{4(\mathbf{18.144}) + 7}{\mathbf{18.144} + 90} \quad \text{Let } k = 18.144 \text{ in the formula.}$$

$$S = 0.74 \quad \text{Use a calculator. Round to the nearest hundredth.}$$

The body surface area is approximately 0.74 m². ■

EXAMPLE 12 **Determining a Child's Dose of a Drug**

If D represents the usual adult dose of a drug, the corresponding child's dose C is calculated by the following formula.

$$C = \frac{\text{body surface area in square meters}}{1.7} \times D$$

Determine the appropriate dose for a child weighing 40 pounds if the usual adult dose is 50 milligrams. (*Source*: Hegstad, Lorrie N., and Wilma Hayek. *Essential Drug Dosage Calculations*, 4th ed. Prentice Hall, 2001.)

Solution

From **Example 11,** the body surface area of a child weighing 40 pounds is 0.74 m². Apply the formula for the child's dose.

$$C = \frac{0.74}{1.7} \times 50 \qquad \text{Body surface area = 0.74, } D = 50.$$

$$C = 22 \qquad \text{Use a calculator. Round to the nearest unit.}$$

The child's dose is 22 milligrams. ∎

6.5   EXERCISES

Concepts of Percent *Decide whether each statement is* true *or* false.

1. 300% of 12 is 36.

2. 25% of a quantity is the same as $\frac{1}{4}$ of that quantity.

3. To find 50% of a quantity, we may simply divide the quantity by 2.

4. A soccer team that has won 12 games and lost 8 games has a winning percentage of 60%.

5. If 70% is the lowest passing grade on a quiz that has 50 items of equal value, then answering at least 35 items correctly will assure you of a passing grade.

6. 30 is more than 40% of 120.

7. .99¢ = 99 cents

8. If an item usually costs $70.00 and it is discounted 10%, then the discount price is $7.00.

Calculate each of the following using either a calculator or paper-and-pencil methods, as directed by your instructor.

9. 8.53 + 2.785

10. 9.358 + 7.2137

11. 8.74 − 12.955

12. 2.41 − 3.997

13. 25.7 × 0.032

14. 45.1 × 8.344

15. 1019.825 ÷ 21.47

16. −262.563 ÷ 125.03

17. $\dfrac{118.5}{1.45 + 2.3}$

18. 2.45(1.2 + 3.4 − 5.6)

Personal Finance *Solve each problem.*

19. Andrew has $48.35 in his checking account. He uses his debit card to make purchases of $35.99 and $20.00, which overdraws his account. His bank charges his account an overdraft fee of $28.50. He then deposits his paycheck for $66.27 from his part-time job at Arby's. What is the balance in his account?

20. Kayla has $37.60 in her checking account. She uses her debit card to make purchases of $25.99 and $19.34, which overdraws her account. Her bank charges her account an overdraft fee of $25.00. She then deposits her paycheck for $58.66 from her part-time job at Subway. What is the balance in her account?

21. Ahmad owes $382.45 on his Visa account. He returns two items costing $25.10 and $34.50 for credit. Then he makes purchases of $45.00 and $98.17.

    (a) How much should his payment be if he wants to pay off the balance on the account?

    (b) Instead of paying off the balance, he makes a payment of $300 and then incurs a finance charge of $24.66. What is the balance on his account?

22. Sabrina owes $237.59 on her MasterCard account. She returns one item costing $47.25 for credit and then makes two purchases of $12.39 and $20.00.

    (a) How much should her payment be if she wants to pay off the balance on the account?

    (b) Instead of paying off the balance, she makes a payment of $75.00 and incurs a finance charge of $32.06. What is the balance on her account?

23. Bank Account Balance In August, Kimberly began with a bank account balance of $904.89. Her withdrawals and deposits for August are given below:

Withdrawals	Deposits
$35.84	$85.00
$26.14	$120.76
$3.12	$205.00
$21.46	

Assuming no other transactions, what was her account balance at the end of August?

24. Bank Account Balance In September, David began with a bank account balance of $904.89. His withdrawals and deposits for September are given below:

Withdrawals	Deposits
$41.29	$80.59
$13.66	$276.13
$84.40	$550.00
$93.00	

Assuming no other transactions, what was his account balance at the end of September?

Rounding *Round each number to the place value indicated. For example, in part (a) of Exercise 25, round 54,793 to the nearest ten thousand. (Hint: Always round from the original number.)*

25. 54,793

    (a) ten thousand   **(b)** thousand

    (c) hundred   **(d)** ten

26. 453,258

    (a) hundred thousand   **(b)** thousand

    (c) hundred   **(d)** ten

27. 0.892451

    (a) hundred-thousandth

    (b) ten-thousandth

    (c) thousandth

    (d) hundredth

    (e) tenth

    (f) one or unit

28. 22.483956

    (a) hundred-thousandth

    (b) ten-thousandth

    (c) thousandth

    (d) hundredth

    (e) tenth

    (f) ten

Use the concept of percent in Exercises 29–32.

29. Match the fractions in Group II with their equivalent percents in Group I.

I		II	
(a) 25%	**(b)** 10%	**A.** $\frac{1}{3}$	**B.** $\frac{1}{50}$
(c) 2%	**(d)** 20%	**C.** $\frac{3}{4}$	**D.** $\frac{1}{10}$
(e) 75%	**(f)** $33\frac{1}{3}$%	**E.** $\frac{1}{4}$	**F.** $\frac{1}{5}$

30. Fill in each blank with the correct numerical response.

    (a) 5% means _____ in every 100.

    (b) 25% means 6 in every _____ .

    (c) 200% means _____ for every 4.

    (d) 0.5% means _____ in every 100.

    (e) _____ % means 12 for every 2.

31. The figures in **Exercise 23** of **Section 6.3** are reproduced here. Express the fractional parts represented by the shaded areas as percents.

(a)

(b)

(c)

(d)

32. The Venn diagram shows the numbers of elements in the four regions formed.

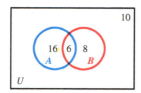

(a) What percent of the elements in the universe are in $A \cap B$?

(b) What percent of the elements in the universe are in A but not in B?

(c) What percent of the elements in $A \cup B$ are in $A \cap B$?

(d) What percent of the elements in the universe are in neither A nor B?

Convert each decimal to a percent.

33. 0.42   **34.** 0.87   **35.** 0.365

36. 0.792   **37.** 0.008   **38.** 0.0093

39. 2.1   **40.** 8.9

Convert each percent to a decimal.

41. 96%   **42.** 23%   **43.** 5.46%

44. 2.99%   **45.** 0.3%   **46.** 0.6%

47. 400%   **48.** 260%   **49.** $\frac{1}{2}\%$

50. $\frac{2}{5}\%$ **51.** $3\frac{1}{2}\%$ **52.** $8\frac{3}{4}\%$

Convert each fraction to a percent.

53. $\frac{1}{5}$ **54.** $\frac{2}{5}$ **55.** $\frac{1}{100}$ **56.** $\frac{1}{50}$

57. $\frac{3}{8}$ **58.** $\frac{5}{6}$ **59.** $\frac{3}{2}$ **60.** $\frac{7}{4}$

61. Explain the difference between $\frac{1}{2}$ of a quantity and $\frac{1}{2}\%$ of the quantity.

62. Explain the difference between 9% and .9%.

Work each problem involving percent.

63. What is 26% of 480?

64. What is 38% of 12?

65. What is 10.5% of 28?

66. What is 48.6% of 19?

67. What percent of 30 is 45?

68. What percent of 48 is 20?

69. 25% of what number is 150?

70. 12% of what number is 3600?

71. 0.392 is what percent of 28?

72. 78.84 is what percent of 292?

Solve each problem involving percent increase or decrease.

73. *Percent Increase* After one year on the job, Grady got a raise from $10.50 per hour to $11.34 per hour. What was the percent increase in his hourly wage?

74. *Percent Discount* Clayton bought a ticket to a rock concert at a discount. The regular price of the ticket was $70.00, but he paid only $59.50. What was the percent discount?

75. *Percent Decrease* Between 2000 and 2012, the estimated population of Toledo, Ohio, declined from 313,619 to 284,012. What was the percent decrease to the nearest tenth? (*Source:* U.S. Census Bureau.)

76. *Percent Increase* Between 2000 and 2012, the estimated population of Oklahoma City, Oklahoma, grew from 506,132 to 599,199. What was the percent increase to the nearest tenth? (*Source:* U.S. Census Bureau.)

77. *Percent Discount* The DVD of the Original Broadway production of the musical *Memphis* was available at www.amazon.com for $16.22. The list price of this DVD was $17.98. To the nearest tenth, what was the percent discount? (*Source:* www.amazon.com)

78. *Percent Discount* The Blu-ray of the movie *Django Unchained* had a list price of $39.99 and was for sale at www.amazon.com at $14.99. To the nearest tenth, what was the percent discount? (*Source:* www.amazon.com)

79. *Value of 1916-D Mercury Dime* The 1916 Mercury dime minted in Denver is quite rare. In 1979 its value in Extremely Fine condition was $625. The 2014 value had increased to $6200. What was the percent increase in the value of this coin from 1979 to 2014? (*Sources: A Guide Book of United States Coins; Coin World Coin Values.*)

80. *Value of 1903-O Morgan Dollar* In 1963, the value of a 1903 Morgan dollar minted in New Orleans in typical Uncirculated condition was $1500. Due to a discovery of a large hoard of these dollars late that year, the value plummeted. Its value in 2014 was $650. What was the percent decrease in its value from 1963 to 2014? (*Sources: A Guide Book of United States Coins; Coin World Coin Values.*)

Gasoline Prices *The line graph shows the average price, adjusted for inflation, that Americans paid for a gallon of gasoline for selected years between 1958 and 2008. Use this information in Exercises 81 and 82.*

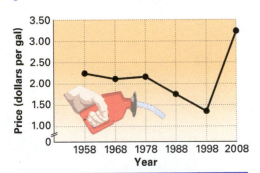

Average Gasoline Prices

Source: www.inflationdata.com

81. By what percent did prices increase from 1998 to 2008?

82. By what percent did prices decrease from 1978 to 1988?

Business Earnings Report *An article in the March 1, 2014, business section of* The Gazette *covered various facets of the earnings report of a major corporation. Answer each of the following questions about a particular phase of the report. Round to the nearest tenth if applicable.*

83. In 2013 profit fell from $1.56 billion to $1.23 billion. What was the percent decrease in profit?

84. Sales rose 2.3% to $8.7 billion. What were the sales prior to this rise?

85. Profit in one division dropped 24% to $680 million. What was the profit prior to this drop?

86. Shares of the company closed on one recent day at $17.98 per share, down $1.19. What percent decrease did this represent?

Use mental techniques to answer the questions in Exercises 87–92. Try to avoid using paper and pencil or a calculator.

87. *Allowance Increase* Carly's allowance was raised from $4.00 per week to $5.00 per week. What was the percent of the increase?

   A. 25%   **B.** 20%   **C.** 50%   **D.** 30%

88. *Boat Purchase and Sale* Susan bought a boat five years ago for $5000 and sold it this year for $2000. What percent of her original purchase did she lose on the sale?

   A. 40%   **B.** 50%   **C.** 20%   **D.** 60%

89. *Population of Alabama* A 2012 report indicated that the population of Alabama was 4,822,023, with 26.0% represented by African Americans. What is the best estimate of the African American population in Alabama? (*Source:* U.S. Census Bureau.)

   A. 500,000   **B.** 1,500,000

   C. 1,250,000   **D.** 750,000

90. *Population of Hawaii* A 2012 report indicated that the population of Hawaii was 1,392,313, with 19.4% of the population being of two or more races. What is the best estimate of this racial demographic population of Hawaii? (*Source:* U.S. Census Bureau.)

   A. 280,000 **B.** 300,000 **C.** 21,400 **D.** 24,000

91. *Discount and Markup* Suppose that an item regularly costs $100.00 and is discounted 20%. If it is then marked up 20%, is the resulting price $100.00? If not, what is it?

92. *Computing a Tip* Suppose that you have decided that you will always tip 20% when dining in restaurants. By what whole number should you divide the bill to find the amount of the tip?

Body Surface Area *Use the formula for determining a child's body surface area*

$$S = \frac{4k + 7}{k + 90}$$

in **Example 11** *to work Exercises 93–96. Round to the nearest hundredth.*

93. Find the body surface area S in square meters of a child who weighs 20 kilograms.

94. Find the body surface area S in square meters of a child who weighs 26 kilograms.

95. Find the body surface area S in square meters of a child who weighs 20 pounds. (Use $k = 9.07$.)

96. Find the body surface area S in square meters of a child who weighs 26 pounds. (Use $k = 11.79$.)

Child's Medication Dosage   *Use the formula for determining a child's dose of a drug*

$$C = \frac{\text{body surface area in square meters}}{1.7} \times D$$

in **Example 12** *to work Exercises 97–100. Round to the nearest unit.*

97. If the usual adult dose D of a drug is 250 mg, what is the child's dose for a child weighing 20 kilograms? (*Hint:* Use the result of **Exercise 93.**)

98. If the usual adult dose D of a drug is 250 mg, what is the child's dose for a child weighing 26 kilograms? (*Hint:* Use the result of **Exercise 94.**)

99. If the usual adult dose D of a drug is 500 mg, what is the child's dose for a child weighing 20 pounds? (*Hint:* Use the result of **Exercise 95.**)

100. If the usual adult dose D of a drug is 500 mg, what is the child's dose for a child weighing 26 pounds? (*Hint:* Use the result of **Exercise 96.**)

Metabolic Units   *One way to measure a person's cardio fitness is to calculate how many METs, or metabolic units, he or she can reach at peak exertion. One MET is the amount of energy used when sitting quietly. To calculate ideal METs, we can use one of the following formulas.*

$$\text{MET} = 14.7 - \text{age} \cdot 0.13$$

$$\text{MET} = 14.7 - \text{age} \cdot 0.11$$

(*Source: New England Journal of Medicine.*)

101. A 40-year-old woman wishes to find her ideal MET.

    (a) Write the expression using her age.

    (b) Calculate her ideal MET.

    (c) Researchers recommend that people reach about 85% of their MET when exercising. Calculate 85% of the ideal MET from part (b). Then refer to the following table. What activity can the woman do that is approximately this value?

Activity	METs	Activity	METs
Golf (with cart)	2.5	Skiing (water or downhill)	6.8
Walking (3 mph)	3.3	Swimming	7.0
Mowing lawn (power mower)	4.5	Walking (5 mph)	8.0
Ballroom or square dancing	5.5	Jogging	10.2

Source: Harvard School of Public Health.

102. Repeat **Exercise 101** for a 55-year-old man.

Exercises 103–106 are based on formulas found in Auto Math Handbook: Mathematical Calculations, Theory, and Formulas for Automotive Enthusiasts, *by John Lawlor (1991, HP Books).*

103. ***Blood Alcohol Concentration***   The blood alcohol concentration (BAC) of a person who has been drinking is given by the formula

$$\text{BAC} = \frac{(\text{ounces} \times \text{percent alcohol} \times 0.075)}{\text{body weight in lb}}$$
$$- \,(\text{hours of drinking} \times 0.015).$$

Suppose a policeman stops a 190-pound man who, in two hours, has ingested four 12-ounce beers, each having a 3.2% alcohol content. The formula would then read

$$\text{BAC} = \frac{[(4 \times 12) \times 3.2 \times 0.075]}{190} - (2 \times 0.015).$$

    (a) Find this BAC.

    (b) Find the BAC for a 135-pound woman who, in three hours, has drunk three 12-ounce beers, each having a 4.0% alcohol content.

104. ***Approximate Automobile Speed***   The approximate speed of an automobile in miles per hour (mph) can be found in terms of the engine's revolutions per minute (rpm), the tire diameter in inches, and the overall gear ratio by the following formula.

$$\text{mph} = \frac{\text{rpm} \times \text{tire diameter}}{\text{gear ratio} \times 336}$$

If a certain automobile has an rpm of 5600, a tire diameter of 26 inches, and a gear ratio of 3.12, what is its approximate speed (mph)?

105. ***Engine Horsepower***   Horsepower can be found from mean effective pressure (mep) in pounds per square inch, engine displacement in cubic inches, and revolutions per minute (rpm) using the following formula.

$$\text{Horsepower} = \frac{\text{mep} \times \text{displacement} \times \text{rpm}}{792,000}$$

An engine has displacement of 302 cubic inches and indicated mep of 195 pounds per square inch at 4000 rpm. What is its approximate horsepower?

106. *Torque Approximation* To determine the torque at a given value of rpm, the following formula applies.

$$\text{Torque} = \frac{5252 \times \text{horsepower}}{\text{rpm}}$$

If the horsepower of a certain vehicle is 400 at 4500 rpm, what is the approximate torque?

Win-Loss Record *Exercises 107 and 108 deal with winning percentage in the standings of sports teams.*

107. At the end of the regular 2013 Major League Baseball season, the standings of the East Division of the American League were as shown. Winning percentage is commonly expressed as a decimal rounded to the nearest thousandth. To find the winning percentage of a team, divide the number of wins (W) by the total number of games played (W + L). Find the winning percentage of each team.

(a) Boston **(b)** Tampa Bay
(c) New York Yankees **(d)** Toronto

Team	W	L	Pct.
Boston	97	65	
Tampa Bay	92	71	
Baltimore	85	77	.525
New York Yankees	85	77	
Toronto	74	88	

Source: World Almanac and Book of Facts.

108. Repeat **Exercise 107** for the following standings for the East Division of the National League.

(a) Washington **(b)** New York Mets
(c) Philadelphia **(d)** Miami

Team	W	L	Pct.
Atlanta	96	66	.593
Washington	86	76	
New York Mets	74	88	
Philadelphia	73	89	
Miami	62	100	

Source: World Almanac and Book of Facts.

Tipping Procedure *It is customary in our society to "tip" the wait staff when dining in restaurants. One common rate for tipping is 15%. A quick way of figuring a tip that will give a close approximation of 15% follows.*

Step 1 Round off the bill to the nearest dollar.

Step 2 Find 10% of this amount by moving the decimal point one place to the left.

Step 3 Take half of the amount obtained in Step 2 and add it to the result of Step 2.

These steps will give approximately 15% of the bill. The amount obtained in Step 3 is 5%, and

$$10\% + 5\% = 15\%.$$

Use the method above to find an approximation of 15% for each restaurant bill.

109. $29.57 **110.** $38.32

111. $5.15 **112.** $7.89

For Excellent Service *Suppose that you get extremely good service and decide to tip 20%. You can use the first two steps listed earlier and then, in Step 3, double the amount you obtained in Step 2. Use this method to find an approximation of 20% for each restaurant bill.*

113. $59.96 **114.** $40.24

115. $180.43 **116.** $199.86

Postage Stamp Pricing *Refer to the margin note on decimal point abuse. At one time, the United States Postal Service sold rolls of 33-cent stamps that featured fruit berries. One such stamp is suggested on the left. On the right is a photo of the pricing information found on the cellophane wrapper of such a roll.*

100 STAMPS PSA
.33¢ ea. TOTAL $33.00
FRUIT BERRIES
ITEM 7757
BCA

117. Look at the second line of the pricing information. According to the price listed *per stamp*, how many stamps should you be able to purchase for one cent?

118. The total price listed is the amount the Postal Service actually charges. If you were to multiply the listed price *per stamp* by the number of stamps, what should the total price be?

Pricing of Pie and Coffee *The photos here were taken at a flea market near Natchez, Mississippi. The handwritten signs indicate that a piece of pie costs .10¢ and a cup of coffee ("ffee") costs .5¢. Assuming these are the actual prices, answer the questions in Exercises 119–122.*

119. How much will 10 pieces of pie and 10 cups of coffee cost?

120. How much will 20 pieces of pie and 10 cups of coffee cost?

121. How many pieces of pie can you get for $1.00?

122. How many cups of coffee can you get for $1.00?

123. ***Producer Percent*** In the 1967 movie *The Producers*, Leo Bloom (Gene Wilder) and Max Bialystock (Zero Mostel) scheme to make a fortune by overfinancing what they think will be a Broadway flop. After enumerating the percent of profits all of Max's little old ladies have been offered in the production, reality sets in. Watch the movie to see what percent of the profits was sold.

124. ***Willy Wonka and Percent*** There are several appearances of percent in the 1971 movie *Willy Wonka and the Chocolate Factory*. In one of them, upon preparing a mixture in his laboratory, Willy Wonka (Gene Wilder) states an impossible percent analysis as he drinks his latest concoction. Watch the movie to see what percent the ingredients total up to be.

CHAPTER 6 SUMMARY

KEY TERMS

6.1

natural numbers
whole numbers
number line
origin
negative numbers
positive numbers
signed numbers
integers
graph
rational numbers
irrational numbers
real numbers
additive inverses
   (negatives, opposites)
absolute value

6.2

sum
addends (terms)
difference
minuend
subtrahend
product
factors
quotient
dividend (numerator)
divisor (denominator)
positive change
negative change

6.3

extremes
means
cross-product test
   (for equality of fractions)
least common denominator
   (LCD)
mixed number
arithmetic mean (average)
terminating decimal
repeating decimal

6.4

irrational number
radicand
like radicals
Golden Rectangle
Golden Ratio

6.5

factors
product
dividend
divisor
quotient
percent

NEW SYMBOLS

$\{x \mid x \text{ has a certain property}\}$	set-builder notation	$a > b$	a is greater than b
%	percent	$a \geq b$	a is greater than or equal to b
¢	cents	$\lvert x \rvert$	absolute value of x
$a = b$	a is equal to b	π	irrational number $\quad \pi \approx 3.14159$
$a < b$	a is less than b	ϕ	irrational number $\quad \phi \approx 1.618$
$a \leq b$	a is less than or equal to b	e	irrational number $\quad e \approx 2.71828$

TEST YOUR WORD POWER

See how well you have learned the vocabulary in this chapter.

1. The **absolute value** of a real number is
  A. the same as the opposite of the number in all cases.
  B. the same as the number itself in all cases.
  C. never zero.
  D. its undirected distance from zero on the number line.

2. The **sum** of two numbers is the result obtained by
  A. addition.   **B.** subtraction.
  C. multiplication.   **D.** division.

3. The **identity element for multiplication** is
  A. 0.   **B.** 1.   **C.** –1.   **D.** –*a*, for the real number *a*.

4. An example of an **irrational number** is
  A. ϕ.   **B.** *e*.
  C. π.   **D.** all of these.

5. An example of a rational number that has a **terminating decimal** representation is
  A. $\frac{1}{3}$.   **B.** $\frac{1}{7}$.   **C.** $\frac{1}{2}$.   **D.** $\frac{1}{9}$.

6. The irrational number π represents
  A. the circumference of a circle divided by its radius.
  B. the radius of a circle divided by its diameter.
  C. the diameter of a circle divided by its radius.
  D. the circumference of a circle divided by its diameter.

ANSWERS
  1. D   **2.** A   **3.** B   **4.** D   **5.** C   **6.** D

QUICK REVIEW

Concepts	*Examples*

6.1 Real Numbers, Order, and Absolute Value

Sets of Numbers

Natural Numbers   {1, 2, 3, 4, . . . }

10, 25, 143   Natural numbers

Whole Numbers   {0, 1, 2, 3, 4, . . . }

0, 8, 47   Whole numbers

Integers   { . . . , −2, −1, 0, 1, 2, . . . }

−22, −7, 0, 4, 9   Integers

Rational Numbers
$\{x \mid x$ is a quotient of two integers, with denominator not equal to $0\}$

$-\frac{2}{3}, -0.14, 0, \frac{15}{8},$

$6, 0.33333 \ldots, \sqrt{4}$ Rational numbers

Irrational Numbers
$\{x \mid x$ is a number on the number line that is not rational$\}$

$-\sqrt{22}, \sqrt{3}, \pi$ Irrational numbers

Real Numbers
$\{x \mid x$ is a number that can be represented by a point on the number line$\}$

$-3, -\frac{2}{7}, 0.7, \pi, \sqrt{11}$ Real numbers

Order in the Real Numbers
Suppose that *a* and *b* are real numbers.

$a = b$ if they are represented by the same point on the number line.

$a < b$ if the graph of *a* lies to the left of the graph of *b*.

$a > b$ if the graph of *a* lies to the right of the graph of *b*.

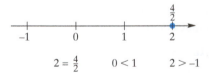

$2 = \frac{4}{2}$ $0 < 1$ $2 > -1$

Additive Inverse
The additive inverse of *a* is −*a*. For all *a*,
$$a + (-a) = (-a) + a = 0.$$

$-(-5) = 5 \qquad -\left(-\frac{2}{3}\right) = \frac{2}{3}$

$5 + (-5) = 0$

Absolute Value
$$|x| = \begin{cases} x & \text{if } x \geq 0 \\ -x & \text{if } x < 0 \end{cases}$$

$|-1| = 1 \qquad |0| = 0 \qquad |2| = 2$

Concepts	Examples

6.2 Operations, Properties, and Applications of Real Numbers

Addition
Like Signs: Add the absolute values. The sum has the same sign as the given numbers.

Unlike Signs: Find the absolute values of the numbers, and subtract the lesser absolute value from the greater. The sum has the same sign as the number with the greater absolute value.

Add.
$$9 + 4 = 13$$
$$-8 + (-5) = -13$$
$$7 + (-12) = -5$$
$$-5 + 13 = 8$$

Subtraction
For all real numbers a and b,
$$a - b = a + (-b).$$

Subtract.
$$-3 - 4 = -3 + (-4) = -7$$
$$-2 - (-6) = -2 + 6 = 4$$
$$13 - (-8) = 13 + 8 = 21$$

Multiplication and Division
Like Signs: The product or quotient of two numbers with like signs is positive.

Unlike Signs: The product or quotient of two numbers with different signs is negative.

Multiply or divide.
$$6 \cdot 5 = 30 \quad -7(-8) = 56 \quad \frac{20}{4} = 5 \quad \frac{-24}{-6} = 4$$
$$-6(5) = -30 \quad 6(-5) = -30$$
$$\frac{-18}{9} = -2 \quad \frac{49}{-7} = -7$$

Division Involving 0
$\frac{a}{0}$ is undefined for all a. $\quad \frac{0}{a} = 0$ for all nonzero a.

$\frac{5}{0}$ is undefined. $\quad \frac{0}{5} = 0$

Order of Operations
If parentheses or square brackets are present:

Step 1 Work separately above and below any fraction bar.

Step 2 Use the rules below within each set of parentheses or square brackets. Start with the innermost set and work outward.

If no parentheses or brackets are present:

Step 1 Apply any exponents.

Step 2 Do any multiplications or divisions in the order in which they occur, working from left to right.

Step 3 Do any additions or subtractions in the order in which they occur, working from left to right.

Simplify.
$$(-6)[2^2 - (3 + 4)] + 3 \qquad \text{Work inside the innermost parentheses.}$$
$$= (-6)[2^2 - 7] + 3$$
$$= (-6)[4 - 7] + 3 \qquad \text{Work inside the brackets.}$$
$$= (-6)[-3] + 3$$
$$= 18 + 3 \qquad \text{Multiply.}$$
$$= 21 \qquad \text{Add.}$$

Properties of Real Numbers

If a and b are real numbers, then

Closure Properties

$a + b$ and ab are real numbers.

$3 + 4$ and $3 \cdot 4$ are real numbers.

Commutative Properties
$$a + b = b + a$$
$$ab = ba$$

$$7 + (-1) = -1 + 7$$
$$5(-3) = (-3)5$$

Associative Properties
$$(a + b) + c = a + (b + c)$$
$$(ab)c = a(bc)$$

$$(3 + 4) + 8 = 3 + (4 + 8)$$
$$[-2(6)]4 = -2[(6)4]$$

Concepts	Examples

Identity Properties
$$a + 0 = a \qquad 0 + a = a$$
$$a \cdot 1 = a \qquad 1 \cdot a = a$$

$-7 + 0 = -7 \quad 0 + (-7) = -7$

$9 \cdot 1 = 9 \qquad 1 \cdot 9 = 9$

Inverse Properties
$$a + (-a) = 0 \qquad -a + a = 0$$
$$a \cdot \frac{1}{a} = 1 \qquad \frac{1}{a} \cdot a = 1 \quad (a \neq 0)$$

$7 + (-7) = 0 \qquad -7 + 7 = 0$

$-2\left(-\dfrac{1}{2}\right) = 1 \quad -\dfrac{1}{2}(-2) = 1$

Distributive Property
$$a(b + c) = ab + ac$$
$$(b + c)a = ba + ca$$

$5(4 + 2) = 5(4) + 5(2)$

$(4 + 2)5 = 4(5) + 2(5)$

6.3 Rational Numbers and Decimal Representation

Fundamental Property of Rational Numbers
$$\frac{a \cdot k}{b \cdot k} = \frac{a}{b} \quad (b \neq 0, k \neq 0)$$

Write $\frac{8}{12}$ in lowest terms.
$$\frac{8}{12} = \frac{2 \cdot 4}{3 \cdot 4} = \frac{2}{3}$$

Cross-Product Test for Equality of Rational Numbers
$$\frac{a}{b} = \frac{c}{d} \text{ if and only if } a \cdot d = b \cdot c \quad (b \neq 0, d \neq 0)$$

Is $\frac{25}{36} = \frac{5}{6}$ a true statement?
$$25 \cdot 6 = 150 \quad \text{and} \quad 36 \cdot 5 = 180$$
Because $150 \neq 180$, the statement is false.

Adding and Subtracting Rational Numbers
$$\frac{a}{b} + \frac{c}{d} = \frac{ad + bc}{bd} \quad \text{and} \quad \frac{a}{b} - \frac{c}{d} = \frac{ad - bc}{bd}$$

In practice, we usually find the least common denominator to add and subtract fractions.

Add or subtract.
$$\frac{2}{5} + \frac{7}{5} = \frac{2 + 7}{5} = \frac{9}{5}, \quad \text{or} \quad 1\frac{4}{5}$$
$$\frac{2}{3} - \frac{1}{2} = \frac{4}{6} - \frac{3}{6} \qquad \text{6 is the LCD.}$$
$$= \frac{1}{6}$$

Multiplying and Dividing Rational Numbers
$$\frac{a}{b} \cdot \frac{c}{d} = \frac{ac}{bd} \quad (b \neq 0, d \neq 0)$$
$$\frac{a}{b} \div \frac{c}{d} = \frac{a}{b} \cdot \frac{d}{c} = \frac{ad}{bc} \quad (b \neq 0, c \neq 0, d \neq 0)$$

Multiply or divide.
$$\frac{4}{3} \cdot \frac{5}{6} = \frac{20}{18} = \frac{10}{9}, \quad \text{or} \quad 1\frac{1}{9}$$
$$\frac{6}{5} \div \frac{1}{4} = \frac{6}{5} \cdot \frac{4}{1} = \frac{24}{5}, \quad \text{or} \quad 4\frac{4}{5}$$

Density Property of the Rational Numbers
If r and t are distinct rational numbers, with $r < t$, then there exists a rational number s such that
$$r < s < t.$$

Find the average, or mean, of $\frac{2}{3}$ and $\frac{3}{4}$.
$$\frac{2}{3} + \frac{3}{4} = \frac{8 + 9}{12} = \frac{17}{12} \qquad \text{Add.}$$
$$\frac{1}{2} \cdot \frac{17}{12} = \frac{17}{24} \qquad \text{Find half of the sum.}$$
$$\leftarrow \text{Average of } \tfrac{2}{3} \text{ and } \tfrac{3}{4}$$

The decimal representation of a rational number will either terminate or will repeat indefinitely in a "block" of digits.

$$\frac{2}{3} = 0.\overline{6}, \text{ or } 0.666\ldots \qquad \frac{3}{16} = 0.1875$$

Concepts	Examples

6.4 Irrational Numbers and Decimal Representation

Decimal Representation
The decimal for an irrational number neither terminates nor repeats.

$$\sqrt{2}, \quad \sqrt{10}, \quad 0.10110111011110\ldots$$

Approximations
Approximations for irrational numbers (for example, square roots of whole numbers that are not perfect squares) can be found with a calculator.

$$\sqrt{2} \approx 1.414213562 \qquad \sqrt{10} \approx 3.16227766$$

Product and Quotient Rules for Square Roots

$$\sqrt{a} \cdot \sqrt{b} = \sqrt{a \cdot b} \quad (a \geq 0, b \geq 0)$$

$$\frac{\sqrt{a}}{\sqrt{b}} = \sqrt{\frac{a}{b}} \quad (a \geq 0, b > 0)$$

Simplify.

$$\sqrt{54} = \sqrt{9 \cdot 6} = \sqrt{9} \cdot \sqrt{6} = 3\sqrt{6}$$

$$\frac{\sqrt{36}}{\sqrt{4}} = \sqrt{\frac{36}{4}} = \sqrt{9} = 3$$

Conditions Necessary for the Simplified Form of a Square Root Radical

1. The number under the radical (radicand) has no factor (except 1) that is a perfect square.

2. The radicand has no fractions.

3. No denominator contains a radical.

Simplify.

$$\sqrt{\frac{5}{6}} = \frac{\sqrt{5}}{\sqrt{6}} = \frac{\sqrt{5} \cdot \sqrt{6}}{\sqrt{6} \cdot \sqrt{6}} = \frac{\sqrt{30}}{6}$$

Three Important Irrational Numbers

$$\pi \approx 3.14159265358979323846264338327 9$$

$$\phi = \frac{1 + \sqrt{5}}{2} \approx 1.61803398874989484820458683436 5$$

$$e \approx 2.71828182845904523536028747135 3$$

The rational number $\frac{355}{113}$ gives an excellent approximation for the irrational number π.

The Golden Ratio ϕ appears throughout nature.

The irrational number e is important in higher mathematics.

6.5 Applications of Decimals and Percents

Addition and Subtraction of Decimals
To add or subtract decimal numbers, line up the decimal points in a column and perform the operation.

Add or subtract.

$$1.2 + 36.158 + 9.26 \qquad 93.86 - 42.9142$$

$$
\begin{array}{r}
1.200 \\
36.158 \\
9.260 \\
\hline
46.618
\end{array}
\qquad
\begin{array}{r}
93.8600 \\
-\,42.9142 \\
\hline
50.9458
\end{array}
$$

Concepts	**Examples**

Multiplication and Division of Decimals

Multiplication   To multiply decimals, multiply in the same manner as integers are multiplied. The number of decimal places to the right of the decimal point in the product is the *sum* of the numbers of places to the right of the decimal points in the factors.

Division   To divide decimals, move the decimal point to the right the same number of places in the divisor and the dividend so as to obtain a whole number in the divisor. Divide in the same manner as integers are divided. The number of decimal places to the right of the decimal point in the quotient is the same as the number of places to the right of the decimal point in the dividend.

Multiply or divide.

51.6×2.3

```
    51.6
     2.3
    1548
   1032
  118.68
```

$35.38 \div 6.1$

```
         5.8
    61)353.8
        305
        488
        488
          0
```

Rules for Rounding
See pages 269–270.

Round 745.2935 to the given place value.

hundreds: 700

ones or units: 745

tenths: 745.3

thousandths: 745.294

Percent
The word **percent** means "per hundred." The symbol % represents "percent."

$$1\% = \frac{1}{100} = 0.01$$

Converting between Decimals and Percents
To convert a percent to a decimal, drop the percent symbol and move the decimal point two places to the left, inserting zeros as placeholders if necessary.

To convert a decimal to a percent, move the decimal point two places to the right, inserting zeros as placeholders if necessary, and attach the percent symbol.

Convert 0.8% to a decimal.
$$0.8\% = 0.008$$

Convert 0.8 to a percent.
$$0.8 = 0.80 = 80\%$$

Converting a Fraction to a Percent
To convert a fraction to a percent, convert the fraction to a decimal, and then convert the decimal to a percent.

Convert $\frac{11}{4}$ to a percent.
$$\frac{11}{4} = 2\frac{3}{4} = 2.75 = 275\%$$

Finding Percent Increase or Decrease
To find the percent increase from a to b, where $b > a$, subtract a from b, and divide this result by a. Convert to a percent.

To find the percent decrease from a to b, where $b < a$, subtract b from a, and divide this result by a. Convert to a percent.

The sales of a textbook decreased from $3,500,000 in its third edition to $2,975,000 in its fourth edition. What was the percent decrease?

$$\text{Percent decrease} = \frac{\$3,500,000 - \$2,975,000}{\$3,500,000}$$
$$= \frac{\$525,000}{\$3,500,000}$$
$$= 0.15$$

The percent decrease was 15%.

1. Consider $\{-4, -\sqrt{5}, -\frac{3}{2}, -0.5, 0, \sqrt{3}, 4.1, 12\}$. List the elements of the set that belong to each of the following.

   (a) natural numbers

   (b) whole numbers

   (c) integers

   (d) rational numbers

   (e) irrational numbers

   (f) real numbers

2. Match each set in (a)–(d) with the correct set-builder notation description in A–D.

   (a) $\{\ldots, -4, -3, -2, -1\}$

   (b) $\{3, 4, 5, 6, \ldots\}$

   (c) $\{1, 2, 3, 4, \ldots\}$

   (d) $\{-12, \ldots, -2, -1, 0, 1, 2, \ldots, 12\}$

   A. $\{x \mid x$ is an integer with absolute value less than or equal to $12\}$

   B. $\{x \mid x$ is an integer greater than $2.5\}$

   C. $\{x \mid x$ is a negative integer$\}$

   D. $\{x \mid x$ is a positive integer$\}$

3. Decide whether each statement is *true* or *false*.

   (a) The absolute value of a number must be positive.

   (b) $|-7| = -(-7)$

   (c) $\frac{2}{5}$ is an example of a real number that is not an integer.

   (d) Every real number is either positive or negative.

Perform the indicated operations. Use the order of operations as necessary.

4. $6^2 - 4(9 - 1)$

5. $\dfrac{(-8 + 3) - (5 + 10)}{7 - 9}$

6. $(-3)(-2) - [5 + (8 - 10)]$

7. $5(-6) - (3 - 8)^2 + 12$

8. **Temperature Extremes** The record high temperature in the United States was 134° Fahrenheit, recorded at Death Valley, California, in 1913. The record low was −80°F, at Prospect Creek, Alaska, in 1971. How much greater was the highest temperature than the lowest temperature? (*Source: The World Almanac and Book of Facts.*)

9. **Altitude of a Plane** The surface of the Dead Sea has altitude 1299 ft below sea level. A pilot is flying 80 ft above that surface. How much altitude must she gain to clear a 3852-ft pass by 225 ft? (*Source: The World Almanac and Book of Facts.*)

10. Match each statement in (a)–(f) with the property that justifies it in A–F.

   (a) $7 \cdot (8 \cdot 5) = (7 \cdot 8) \cdot 5$

   (b) $3x + 3y = 3(x + y)$

   (c) $8 \cdot 1 = 1 \cdot 8 = 8$

   (d) $7 + (6 + 9) = (6 + 9) + 7$

   (e) $9 + (-9) = -9 + 9 = 0$

   (f) $5 \cdot 8$ is a real number.

   A. Distributive property

   B. Identity property

   C. Closure property

   D. Commutative property

   E. Associative property

   F. Inverse property

11. **Basketball Shot Statistics** Six players on the local high school basketball team had shooting statistics as shown in the table below. Answer each question, using estimation skills as necessary.

Player	Field Goal Attempts	Field Goals Made
Ed	40	13
Jack	10	4
Chuck	20	8
Ben	6	4
Charlie	7	2
Jason	7	6

   (a) Which players made more than half of their attempts?

   (b) Which players made just less than $\frac{1}{3}$ of their attempts?

   (c) Which player made exactly $\frac{2}{3}$ of his attempts?

   (d) Which two players made the same fractional parts of their attempts? What was the fractional part, reduced to lowest terms?

   (e) Which player made the greatest fractional part of his attempts?

Perform each operation. Write your answer in lowest terms.

12. $\dfrac{3}{16} + \dfrac{1}{2}$

13. $\dfrac{9}{20} - \dfrac{3}{32}$

14. $\dfrac{3}{8} \cdot \left(-\dfrac{16}{15}\right)$

15. $\dfrac{7}{9} \div 1\dfrac{3}{4}$

16. *Foreign-born Population* Approximately 40 million people living in the United States in 2010 were born in other countries. The graph gives the fractional number for each region of birth for these people.

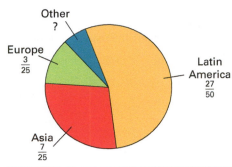

U.S. Foreign-Born Population by Region of Birth

Other ?

Europe $\dfrac{3}{25}$

Latin America $\dfrac{27}{50}$

Asia $\dfrac{7}{25}$

Source: U.S. Census Bureau.

(a) What fractional part of the U.S. foreign-born population was from a region other than the ones specified?

(b) What fractional part of the foreign-born population was from Latin America or Asia?

(c) About how many people (in millions) were born in Europe?

17. Convert each rational number into a repeating or terminating decimal. Use a calculator if your instructor so allows.

(a) $\dfrac{9}{20}$

(b) $\dfrac{5}{12}$

18. Convert each decimal into a quotient of integers, reduced to lowest terms.

(a) 0.72

(b) $0.\overline{58}$

19. Identify each number as rational or irrational.

(a) $\sqrt{10}$

(b) $\sqrt{16}$

(c) 0.01

(d) $0.\overline{01}$

(e) $0.0101101110\ldots$

(f) π

For each of the following, **(a)** *use a calculator to find a decimal approximation and* **(b)** *simplify the radical according to the guidelines in this chapter.*

20. $\sqrt{150}$

21. $\dfrac{13}{\sqrt{7}}$

22. $2\sqrt{32} - 5\sqrt{128}$

23. *Rate of Return on an Investment* If an investment of P dollars grows to A dollars in two years, the annual rate of return on the investment is given by

$$r = \dfrac{\sqrt{A} - \sqrt{P}}{\sqrt{P}}.$$

What is the rate if an investment of $50,000 grows to $58,320?

24. Work each of the following using either a calculator or paper-and-pencil methods, as directed by your instructor.

(a) $4.6 + 9.21$

(b) $12 - 3.725 - 8.59$

(c) $86(0.45)$

(d) $236.439 \div (-9.73)$

25. Round 346.0449 to the given place values.

(a) tens   **(b)** hundredths   **(c)** thousandths

26. *Sale Price* A dress that originally sold for $100.00 is discounted 40% for a sale. Then the owner decides to offer an additional 10% off of the sale price. What is the final price of the dress?

27. Consider the figure.

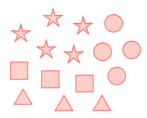

(a) What percent of the total number of shapes are circles?

(b) What percent of the total number of shapes are not stars?

28. *Sales of Books* Use estimation techniques to answer the following: In 2013, Carol sold $300,000 worth of books. In 2014, she sold $900,000. Her 2014 sales were _____ of her 2013 sales.

A. 30%   **B.** $33\dfrac{1}{3}$%   **C.** 3%   **D.** 300%

29. *Creature Comforts*   From a list of "everyday items" often taken for granted, adults were asked to indicate those items they wouldn't want to live without. Complete the results shown in the table if 2400 adults were surveyed.

Item	Percent That Wouldn't Want to Live Without	Number That Wouldn't Want to Live Without
Toilet paper	69%	
Zipper	42%	
Frozen food		384
Self-stick note pads		144

(Other items included tape, hairspray, pantyhose, paper clips, and Velcro.)

Source: Market Facts for Kleenex Cottonelle.

30. *Child's Drug Dosage*   If D represents the usual adult dose of a drug, the corresponding child's dose C is calculated by the following formula.

$$C = \frac{\text{body surface area in square meters}}{1.7} \times D$$

Determine the appropriate dose for a child with body surface area 0.85 m^2 if the usual adult dose is 150 mg.